Lecture Notes in Physics

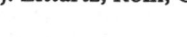

W0055431

Springer-Verlag
Berlin Heidelberg GmbH

The Editorial Policy for Proceedings

The series Lecture Notes in Physics reports new developments in physical research and teaching – quickly, informally, and at a high level. The proceedings to be considered for publication in this series should be limited to only a few areas of research, and these should be closely related to each other. The contributions should be of a high standard and should avoid lengthy redraftings of papers already published or about to be published elsewhere. As a whole, the proceedings should aim for a balanced presentation of the theme of the conference including a description of the techniques used and enough motivation for a broad readership. It should not be assumed that the published proceedings must reflect the conference in its entirety. (A listing or abstracts of papers presented at the meeting but not included in the proceedings could be added as an appendix.)

When applying for publication in the series Lecture Notes in Physics the volume's editor(s) should submit sufficient material to enable the series editors and their referees to make a fairly accurate evaluation (e.g. a complete list of speakers and titles of papers to be presented and abstracts). If, based on this information, the proceedings are (tentatively) accepted, the volume's editor(s), whose name(s) will appear on the title pages, should select the papers suitable for publication and have them refereed (as for a journal) when appropriate. As a rule discussions will not be accepted. The series editors and Springer-Verlag will normally not interfere with the detailed editing except in fairly obvious cases or on technical matters.

Final acceptance is expressed by the series editor in charge, in consultation with Springer-Verlag only after receiving the complete manuscript. It might help to send a copy of the authors' manuscripts in advance to the editor in charge to discuss possible revisions with him. As a general rule, the series editor will confirm his tentative acceptance if the final manuscript corresponds to the original concept discussed, if the quality of the contribution meets the requirements of the series, and if the final size of the manuscript does not greatly exceed the number of pages originally agreed upon. The manuscript should be forwarded to Springer-Verlag shortly after the meeting. In cases of extreme delay (more than six months after the conference) the series editors will check once more the timeliness of the papers. Therefore, the volume's editor(s) should establish strict deadlines, or collect the articles during the conference and have them revised on the spot. If a delay is unavoidable, one should encourage the authors to update their contributions if appropriate. The editors of proceedings are strongly advised to inform contributors about these points at an early stage.

The final manuscript should contain a table of contents and an informative introduction accessible also to readers not particularly familiar with the topic of the conference. The contributions should be in English. The volume's editor(s) should check the contributions for the correct use of language. At Springer-Verlag only the prefaces will be checked by a copy-editor for language and style. Grave linguistic or technical shortcomings may lead to the rejection of contributions by the series editors. A conference report should not exceed a total of 500 pages. Keeping the size within this bound should be achieved by a stricter selection of articles and not by imposing an upper limit to the length of the individual papers. Editors receive jointly 30 complimentary copies of their book. They are entitled to purchase further copies of their book at a reduced rate. As a rule no reprints of individual contributions can be supplied. No royalty is paid on Lecture Notes in Physics volumes. Commitment to publish is made by letter of interest rather than by signing a formal contract. Springer-Verlag secures the copyright for each volume.

The Production Process

The books are hardbound, and the publisher will select quality paper appropriate to the needs of the author(s). Publication time is about ten weeks. More than twenty years of experience guarantee authors the best possible service. To reach the goal of rapid publication at a low price the technique of photographic reproduction from a camera-ready manuscript was chosen. This process shifts the main responsibility for the technical quality considerably from the publisher to the authors. We therefore urge all authors and editors of proceedings to observe very carefully the essentials for the preparation of camera-ready manuscripts, which we will supply on request. This applies especially to the quality of figures and halftones submitted for publication. In addition, it might be useful to look at some of the volumes already published. As a special service, we offer free of charge L^AT_EX and T_EX macro packages to format the text according to Springer-Verlag's quality requirements. We strongly recommend that you make use of this offer, since the result will be a book of considerably improved technical quality. To avoid mistakes and time-consuming correspondence during the production period the conference editors should request special instructions from the publisher well before the beginning of the conference. Manuscripts not meeting the technical standard of the series will have to be returned for improvement.

For further information please contact Springer-Verlag, Physics Editorial Department II, Tiergartenstrasse 17, D-69121 Heidelberg, Germany

Zalán Horváth   László Palla (Eds.)

Conformal Field Theories and Integrable Models

Lectures Held at the Eötvös Graduate Course,
Budapest, Hungary, 13–18 August 1996

 Springer

Editors

Zalán Horváth
László Palla
Institute for Theoretical Physics
Eötvös University
Puskin u. 5–7
H-1088 Budapest, Hungary

Cataloging-in-Publication Data applied for.

Die Deutsche Bibliothek - CIP-Einheitsaufnahme

Conformal field theories and integrable models : lectures held at the Eötvös graduate course, Budapest, Hungary 13 - 18 August 1996 / Zalán Horváth ; László Palla (ed.).
(Lecture notes in physics ; Vol. 498)
ISBN 978-3-662-14111-3   ISBN 978-3-540-69613-1 (eBook)
DOI 10.1007/978-3-540-69613-1

ISSN 0075-8450
ISBN 978-3-662-14111-3

Typesetting: Camera-ready by the authors/editors
Cover design: *design & production* GmbH, Heidelberg
SPIN: 10643842   55/3144-543210 - Printed on acid-free paper

Preface

In 1992 the Institute of Physics of Eötvös University started the series **International Summer Schools** for graduate students on various subjects in theoretical and experimental physics. These schools last approximately one week. The organizers – who are professors of the Institute – invite the best international experts on the subject in order to guarantee the high standards of the courses. Since 1994 the courses have taken place at the Bolyai College of the University.

This book contains the notes of almost all lectures given at the 1996 summer school of this series held 13–18 August. The title of this school was **Conformal Field Theories and Integrable Models**.

In the last few years we witnessed an upsurge of interest in exactly solvable quantum field theoretical models in many branches of theoretical physics ranging from mathematical physics through high energy physics to solid states. The aim of the school was to bring together and serve the interests of graduate students and young research workers interested in these disciplines by providing them with high quality lectures that can lead them to the front line of research. The topics included conformal field theory and W algebras, the special features of 2d scattering theory as embodied in the exact S matrices and the form factor studies built on them, the Yang–Baxter equations, and the various aspects of the Bethe Ansatz systems.

Collected here are the written versions of five of the seven 3–4-hour-long lectures given at the school. The sixth one, Prof. H. Grosse's lecture, was not given there, as his lumbago made it impossible for him to come, but he kindly agreed to include it to make the book more complete. In addition to the lectures presented here Prof. G. Mussardo spoke about "Form factor studies in integrable models" and Prof. Al.B. Zamolodchikov on "The sine-Gordon model on and off the mass shell".

It is a pleasant obligation to thank both the lecturers and the participants for their presence and their contributions to makeing the school a success. We also thank Prof. I. Kondor, the director of Bolyai College for providing the good atmosphere and the excellent facilities in the college.

Budapest, August 1997   Zalán Horváth and László Palla

Contents

List of Participants

Carlo Acerbi
  International School for Advanced Studies (SISSA) Italy,
Francois Barbarin
  Laboratoire de Physique Theorique ENSLAPP France,
Lorenzo Belardinelli
  Dipartimento di Fisica, Università degli studi di Milano Italy,
Marco Bertola
  International School for Advanced Studies (SISSA) Italy,
Omduth Coceal
  Theory Group, Physics Department, Queen Mary and Westfield College
  United Kingdom,
Mats Granath
  Institute of Theoretical Physics Chalmers University of Technology Sweden,
David Gustafsson
  Institute of Theoretical Physics Chalmers University of Technology Sweden,
Laurent Gallot
  Laboratoire de Physique Theorique ENSLAPP France,
Antonio Liguori
  Dipartimento di Fisica, Università di Pisa Italy,
Andrew Pocklington
  Dept. Mathematical Sciences University of Durham United Kingdom,
Predrag Prester
  Theoretical Physics Dept., PMF, University of Zagreb Croatia,
Ciré Sow
  Laboratoire de Modèles de Physique Mathématique faculté des sciences et
  techniques Université de Tours France,
Michael Stiller
  II. Institute for Theoretical Physics, University Hamburg Germany,
Robert Talbot
  Dept. Mathematical Sciences University of Durham United Kingdom,
Kevin Thompson
  Dept. Mathematical Sciences University of Durham United Kingdom,
Fabrizio Tinebra
  ROMA Italy,

Angelo Valleriani
    SISSA Italy,
Zoltán Bajnok
    Institute for Theoretical Physics, Eötvös University, ELTE Hungary
Gábor Etesi
    ELTE,
József Gyűrűsi
    ELTE,
Tamás Hauer
    ELTE-MIT,
Róbert Karp
    ELTE,
Zoltán Németh
    ELTE,
Gábor Takács
    ELTE,
János Varró
    ELTE

Lectures on Conformal Field Theory and Kac–Moody Algebras

Jürgen Fuchs

DESY, Notkestraße 85, D – 22603 Hamburg, Germany

Abstract. This is an introduction to the basic ideas and to a few further selected topics in conformal quantum field theory and in the theory of Kac-Moody algebras.

1 Conformal Field Theory

1.1 Conformal Quantum Field Theory

Over the years, quantum field theory has enjoyed a great number of successes. Nevertheless, from a conceptual point of view the situation is far from satisfactory, and in some respects it even resembles a disaster. Indeed, beyond perturbation theory many of the methods used in phenomenologically successful models cannot be justified rigorously. For instance, a thorough understanding of path integrals is essentially only available for free fields, or for topological 'field' theories which have a finite-dimensional configuration space. Also, the relation between the elementary 'point-like' quantum *fields* and the actual *particle* contents of a theory is often obscure. (The elementary fields may be regarded as a coordinatization of the field space, while the particles have a meaning independent of a choice of coordinates; typically a distinguished and / or manageable choice of field coordinates which directly correspond to particles is not available.) On the other hand, frameworks whose foundations are mathematically sound, like e.g. Wightman field theory ([Streater and Wightman 1989]) or C^*-algebraic approaches ([Haag 1992]), are difficult to relate to concrete models, and indeed have traditionally been plagued with a scarcity of models to which they apply. For example, the proper treatment of $U(1)$-charges (which are coulombic rather than localized) in the framework of algebras of local observables is still problematic.

It is worth stressing that 'quantum field theory' is not a protected term, but is used for a variety of rather different concepts. [1] In particular there are two basically different points of view of the relation between 'quantum' versus 'classical' field theories. One may either try to quantize a classical theory (say by canonical quantization or using path integrals), or directly start at the quantum level, e.g. by describing the field and observable algebras, in which case a classical

---

[1] While a lot of physicists believe to know exactly what quantum field theory is, you will probably get two different answers if you ask any two of them. And not only the methods and results are under debate, but even the basic questions to be asked.

theory may be obtained by performing a suitable limit. It is far from obvious that the resulting quantum, respectively classical, theory is uniquely determined. Another important difference concerns the description of observables. In some approaches these are characterized quite concretely (e.g. as gauge respectively BRS invariant combinations of the elementary field variables of the classical theory), in others they appear more abstractly as operators on a Hilbert space which at an initial stage are only known via their general algebraic properties.

Conformal field theory admits formulations which directly approach the exact theory at the quantum level, but nevertheless describe observables quite explicitly and handle large classes of models simultaneously. In particular many quantities of interest can be *calculated exactly* rather than only in a (typically at best asymptotic) perturbation series. In addition, for a large number of models there exists (or is at least conjectured) a Lagrangian / path integral realization, so that in principle one can study such models in several guises and e.g. compare with conventional perturbation theory. Conformal field theory can therefore help to bridge the gap between various approaches to quantum field theory.

Here and below, by the term conformal field theory I refer to models in one or two space-time dimensions which are relativistic quantum field theories and possess conformal symmetry. What precisely this characterization means will be studied shortly. Again one must note that the term 'conformal field theory' is not protected and can stand for a variety of different formulations. Here I approach this subject from the perspective of quantum field theory. However, in this introductory exposition I will not attempt to build exclusively on general field theoretic considerations, but rather I concentrate on certain basic algebraic structures and assume that the reader has already some faint acquaintance with fundamental field theoretic notions that will play a rôle, such as the operator product algebra, correlation functions and fusion rules. On the other hand, in a sense I start from scratch, in that at least I try to *indicate* each logical step explicitly, even though I do not provide sufficient explanation for all of them.

The motivations to study conformal field theories are manifold, and I am content to just drop a few keywords: [2] systems at the critical point of a second order phase transition in statistical mechanics; vacuum configurations of strings and superstrings; braid group statistics; topological field theory; invariants of knots and links, and of three-manifolds; integrable systems as perturbed conformal models; the fractional quantum Hall effect; high temperature superconductivity.

In the study of conformal field theories, the following two issues prove to be most fundamental. First, the classification programme; and second, the complete solution, by reconstruction from certain basic data, of specific models. We will encounter various manifestations of these two aspects of conformal field theory in the sequel. They indicate that these theories are very special indeed: while (contrary to the impression that is occasionally given) a classification of *all* conformal field theories or a complete solution of any *arbitrary* given conformal field

---

[2] For more details I refer to the references listed in Sect. 5.1, to other lectures at this school, as well as to the huge number of works where such motivations are advertised.

theory model is totally out of reach, one *can* classify large classes of conformal field theories, and one *can* solve very many models to a large extent.

1.2   Observables: The Chiral Symmetry Algebra

One of the ultimate goals of a physical theory is to predict the outcome of all possible measurements. Accordingly, a basic step in the investigation of a conformal field theory consists of characterizing its *observables*. In quantum physics, one thinks of observables as operators acting on a Hilbert space of physical states. Moreover, in relativistic quantum field theory it is usually necessary to consider operators which are bounded (only these can be continuous) and are localized in some compact space-time region, i.e. act as the identity outside ([Haag 1992]). Observables localized in space-like separated regions commute (*locality* principle). In contrast, below I will work with non-localized unbounded operators; that this does not lead to any severe problems has its origin in the fact that these observables enjoy an underlying *Lie algebra* structure. To obtain localized bounded operators one would, roughly speaking, have to smear these operators with test functions of compact support and form bounded functions thereof.

The most important subset of these observables of a conformal field theory is provided by the conformal symmetry itself. Conformal transformations of space-time – i.e. of a (pseudo-)Riemannian manifold – are by definition those general coordinate transformations which preserve the angles between any two vectors, or what is the same, which scale the metric locally by an over-all factor (Weyl transformations). In a flat space-time of $D > 2$ dimensions the infinitesimal conformal transformations consist of translations, rotations (respectively Lorentz boosts), a dilation and so-called special conformal transformations; these generate a $(D+1)(D+2)/2$-dimensional Lie algebra, which is a real form of $\mathfrak{so}(D+2, \mathbb{C})$ (e.g. $\mathfrak{so}(D,2)$ in the case of Minkowski space). In contrast, for $D = 2$ any arbitrary holomorphic mapping of the (compactified) complex plane is angle-preserving. Thus in complex coordinates z, $\bar{z}$, infinitesimal conformal transformations are generated by mappings which transform z as $z \mapsto z + \eta(z)$ by functions which do not depend on $\bar{z}$ – a basis of which is given by $\eta(z) = -z^{n+1}\epsilon$ for $n \in \mathbb{Z}$ – and by analogous mappings of $\bar{z}$ with functions which do not depend on z. On functions of z, $\bar{z}$ these mappings are generated by differential operators

$$l_n = -z^{n+1}\partial_z \qquad \text{and} \qquad \bar{l}_n = -\bar{z}^{n+1}\partial_{\bar{z}}, \qquad (1)$$

respectively. The operators l_n satisfy the commutator relations

$$[l_m, l_n] = (m - n)\, l_{m+n}, \qquad (2)$$

and an analogous formula holds for the $\bar{l}_n$. In mathematical terms, this means that both the l_n and the $\bar{l}_n$ span an infinite-dimensional Lie algebra; moreover, these two algebras are combined as a direct sum, i.e. one has $[l_m, \bar{l}_n] = 0$. The Lie algebra defined by (2) is known as the *Witt algebra*.

The formula (2) was obtained by purely classical considerations. In quantum theory, there are analogous operators L_n with $n \in \mathbb{Z}$, but now they satisfy

$$[L_m, L_n] = (m - n) L_{m+n} + \tfrac{1}{12} (m^3 - m) \delta_{m+n,0} C. \qquad (3)$$

That is, they again give rise to a Lie algebra, but the symmetry is only realized projectively: there appears an additional generator C which is *central*, i.e. obeys

$$[C, L_n] = 0 \qquad (4)$$

and hence has zero Lie bracket with the whole algebra. The Lie algebra [3] defined by (3) and (4) is called the *Virasoro algebra* and denoted by $\mathcal{V}ir$. In short, the Lie brackets of $\mathcal{V}ir$ differ from their classical counterpart only in the term with C; the generator C, which in mathematics is known as the canonical central element or central charge, is therefore in physics also referred to as the conformal *anomaly*. As in the classical situation one deals in fact with the direct sum of two Virasoro algebras, the second one being generated by operators $\bar{L}_n$, $n \in \mathbb{Z}$, and C.

In a purely mathematical context, the relations (3) arise – by solving a cohomology problem ([Fuks 1986]) – as the unique non-trivial central extension of (2). Field-theoretically, (3) is obtained by making a general ansatz for the equal-time commutator of the 'energy-momentum tensor', whose Fourier–Laurent components are the operators L_n and $\bar{L}_n$ (see (25) below). When one imposes the requirements that the Wightman axioms are satisfied, that the system is dilation invariant, and that the energy-momentum tensor is a conserved (Noether) current, this equal-time commutator becomes a finite sum of products of (derivatives of) the δ-function multiplied with local operators, and this result – known as the Lüscher–Mack theorem ([Furlan et al. 1989]) – is equivalent to (3).

Let me point out once more that in a quantum field theoretic setting the physical meaning of the generators of conformal symmetry is as providing (non-bounded, non-localized) observables, in the same spirit as e.g. momentum and angular momentum are observables in ordinary one-particle quantum mechanics. A brief characterization of a conformal field theory is therefore that the observable algebra contains (the direct sum of two copies of) the Virasoro algebra. In general, the full observable algebra of a conformal field theory may be larger, though. But by considerations based on properties of conserved currents, similarly as in the proof of the Lüscher–Mack theorem, one can argue that the observables (regarded as Fourier–Laurent components of conserved currents) still generate a Lie algebra, and the direct sum structure persists as well. In short, the observables of a conformal field theory form an infinite-dimensional Lie algebra $\mathcal{W}_{tot}$ over $\mathbb{C}$ with countable basis, which can be written as

$$\mathcal{W}_{tot} = \mathcal{W} \oplus \overline{\mathcal{W}'} \quad \text{with} \quad \mathcal{W} \supset \mathcal{V}ir \quad \text{and} \quad \overline{\mathcal{W}'} \supset \overline{\mathcal{V}ir}. \qquad (5)$$

---

[3] Let me remind you at this point that, by definition, a *Lie algebra* over $\mathbb{C}$ (or similarly over $\mathbb{R}$, or more generally over any field of scalars $\{\xi\}$), is a vector space with a bilinear product – called the Lie bracket and denoted by $[\cdot, \cdot]$ – which is antisymmetric and satisfies the Jacobi identity. In physics, the most common realization of a Lie bracket is as the commutator with respect to an underlying associative product.

There is a variety of names for the two direct summands W and $\overline{W}'$: the holomorphic / anti-holomorphic, or chiral / antichiral, or left / right subalgebras. When one considers only one of these subalgebras, one refers to the observables as the *chiral symmetry algebra*. As a consequence of the direct sum structure, for many purposes one can restrict ones attention to one 'chiral half' of the two-dimensional theory, and I will do so for the time being. (For aspects of the *two*-dimensional theory see e.g. Sects. 1.7, 2.2, 2.7. Also note that one needs not necessarily require that the left and right chiral halves have isomorphic chiral symmetry algebras; when $\overline{W}' \not\cong W$, then one speaks of a 'heterotic' theory.)

Unless stated otherwise, from now on it will be assumed that W is *maximal*, i.e. that it already includes *all* (chiral) observables. W has a basis of the form $\{W_n^i\} \cup \{C_\ell\}$, where the index n of W_n^i takes values in $\mathbb{Z}$, while the index i takes values in a set $I \equiv \{0, 1, 2, ...\}$ which may be possibly infinite, and where the C_ℓ are central elements satisfying $[C_\ell, \cdot] = 0$. The Virasoro generators can be identified as $L_n \equiv W_n^0$. At least for $m \in \{0, \pm 1\}$ any generator W_n^i then satisfies

$$[W_m^0, W_n^i] \equiv [L_m, W_n^i] = ((\Delta_i - 1)m - n) W_{m+n}^i, \tag{6}$$

with certain numbers Δ_i which must be positive integers. In particular, the subscript n already determines the Lie bracket with the Virasoro generator L_0,

$$[L_0, W_n^i] = -n W_n^i. \tag{7}$$

In fact, this subscript even supplies us with a $\mathbb{Z}$-grading of W, i.e. the structure constants of W, defined by [4] $[W_m^i, W_n^j] = \sum_{p \in \mathbb{Z}} \sum_{k \in I} f_{mn,k}^{ij,p} W_p^k$ are subject to

$$f_{mn,k}^{ij,p} = 0 \quad \text{for} \quad p \neq m+n. \tag{8}$$

Also, central elements have grade 0.

For the time being I will not be more specific about the algebra W. Later on, we will encounter important examples, the so-called current algebras. Other possibilities are supersymmetric extensions [5] of the Virasoro algebra, and many more chiral algebras are described in detail in the lectures by Gerard Watts.

Before proceeding, let me point out that the restriction to one chiral half requires in particular to regard z and $\bar{z}$ as two *independent* complex variables; $\bar{z}$ is to be identified with the complex conjugate of z only once the two chiral halves of a two-dimensional theory are combined. That this is a sensible way to proceed is far from obvious, and actually the interpretation of this prescription

---

[4] In principle, on the right hand side infinitely many terms may appear. In this case strictly speaking one does not deal with a proper basis of a Lie algebra, and a completion with respect to a suitable topology is needed. However, as it turns out, when applied to any vector in the relevant space $\mathcal{H}$ of physical states, only a finite number of terms is non-zero, owing to the fact that the generators W_m^i act 'locally nilpotently' in $\mathcal{H}$. This is also one of the reasons why considering formally infinite series as in (24) below does not lead to any serious problems.

[5] The extension is by normal ordered products of supercurrents; the supercurrents themselves have half integral Δ, i.e. are not observables and are not contained in W.

depends in part on the context in which the conformal symmetry is considered:
- When arising in minkowskian quantum field theory, the variables z and $\bar{z}$ take their values on circles that are obtained as the two compactified light-cones $x^0 = \pm x^1$ of the two-dimensional theory, from which they are extended to the punctured plane $\mathbb{C}\backslash\{0\}$. Identification of $\bar{z}$ as the conjugate of z thus amounts to considering an analytical continuation of the original minkowskian theory to euclidean signature which does not coincide with the usual Wick rotation.
- In applications to string theory, the punctured plane arises as the image of the cylindrical world sheet swept out by a closed relativistic string. In the description of phase transitions, z is the complex coordinate of a euclidean two-dimensional field theory. The meaning of separating z and $\bar{z}$ is in these cases less clear.
- At the level of the chiral algebra, taking z and $\bar{z}$ as independent amounts to regarding $\mathcal{W}$ and $\overline{\mathcal{W}}$ as algebras over the complex numbers. It is tempting to interpret the identification of $\bar{z}$ with the complex conjugate of z as considering a suitable real form of the complex Lie algebra, but this is not quite correct.

1.3 Physical States: Highest Weight Modules

Once we know the observables, the next logical step is to investigate how they 'act' on a *space $\mathcal{H}$ of physical states*. Ideally, this way in the end we describe all properties of a model in terms of its chiral symmetry algebra $\mathcal{W}$. Now in the conventional operator approaches to quantum field theory, $\mathcal{H}$ is a separable Hilbert space and the observables are contained in the algebra of bounded linear operators on this Hilbert space. As already pointed out, here we are in a somewhat different situation. Nevertheless, the following basic requirements are still implicit in the concept of a space of physical states:

(H1) $\mathcal{H}$ is a representation space of the observable algebra, i.e. of $\mathcal{W}$.

In the following I will employ the shorter term *module* common in mathematics as a synonym for 'representation space'; elements of $\mathcal{H}$ are denoted by $v, w, \ldots$.

(H2) The representation of $\mathcal{W}$ on $\mathcal{H}$ is unitary. Correspondingly, $\mathcal{H}$ is endowed with a positive hermitian product $(\cdot | \cdot)$, so that it becomes a pre-Hilbert space.

(H3) The *spectrum condition*: The energy is bounded from below.

In order that the qualification 'unitary' in (H2) makes sense, $\mathcal{W}$ must be endowed with a generalized complex conjugation, i.e. a mapping $x \mapsto x^*$ satisfying

$$(x^*)^* = x, \qquad (\xi x)^* = \bar{\xi}\, x, \qquad (xy)^* = y^* \, x^* \tag{9}$$

(in mathematical terms: $\mathcal{W}$ must be a *-algebra), such that

$$(v \,|\, xw) = (x^* v \,|\, w) \tag{10}$$

holds for the hermitian product on $\mathcal{H}$. In the last of the relations (9) (and, likewise, often implicitly below), one works in fact within the universal enveloping algebra of $\mathcal{W}$, i.e. with the associative algebra $\mathfrak{U}(\mathcal{W})$ that (roughly) can be described as consisting of formal products of elements of $\mathcal{W}$, with the Lie bracket of $\mathcal{W}$ equal to the commutator in $\mathfrak{U}(\mathcal{W})$.

In formulating the spectrum condition (H3), it is in particular assumed that among the observables there is an energy operator. In chiral conformal field theory this operator is provided by the Virasoro generator L_0. Briefly, L_0 acts as $-z\frac{\partial}{\partial z}$ and hence measures the mass (or scaling) dimension, i.e. $L_0 \mathcal{O} = \Delta \cdot \mathcal{O}$ for operators $\mathcal{O} \sim (\text{length})^{-\Delta} \sim (\text{mass})^{\Delta} \sim (\text{energy})^{\Delta}$. Also, it is implicit in formulating (H3) that the action of L_0 on $\mathcal{H}$ can be diagonalized, so that $(L_0)^* = L_0$ by (H2). The eigenvalue Δ of L_0 on an eigenvector in $\mathcal{H}$ is called the *conformal dimension* or the conformal *weight* of the vector. Furthermore, together with the relation $[L_0, W_n^i] = -n W_n^i$ (7), (H3) implies that every vector in $\mathcal{H}$ is annihilated by all W_n^i with sufficiently large n, and (because L_1 augments the L_0-eigenvalue by 1) that any vector $v \in \mathcal{H}$ satisfies $(L_1)^N v = 0$ for large enough N.

Because of its graded structure (see (8)), as a vector space W is the direct sum $W = W^- \oplus W^0 \oplus W^+$ of subalgebras $W^\pm$ and W^0 which are given by

$$W^\pm := \text{span}\{W_n^i \mid i \in I, \pm n > 0\}, \qquad W^0 := \text{span}\{W_0^i \mid i \in I\}. \qquad (11)$$

Further, again because of (7) (namely, $[L_0, W_n^i] \neq 0$ for $n \neq 0$), the *zero mode subalgebra* W^0 contains a maximal abelian subalgebra $W_0 \subset W$ with $L_0 \in W_0$, and there are subalgebras $W_\pm \subseteq W^0 \oplus W^\pm$ such that

$$W = W_- \oplus W_0 \oplus W_+ \qquad (12)$$

(direct sum of vector spaces) is a *triangular decomposition* of W, which means that

$$[W_\pm, W_0 \oplus W_\pm] \subseteq W_\pm. \qquad (13)$$

Lie algebras which enjoy a triangular decomposition possess a distinguished class of modules, the *highest weight modules*. Such modules V are by definition generated by a *highest weight vector* $v_{\text{h.w.}}$. This means the following. First,

$$V \subseteq \mathfrak{U}(W_-) v_{\text{h.w.}}, \qquad (14)$$

i.e. all vectors in a highest weight module V can be obtained from the highest weight vector $v_{\text{h.w.}} \in V$ by acting with the universal enveloping algebra of W_-; thus the elements of W_- play the rôle of creation operators. Second, all elements of W_+ act on $v_{\text{h.w.}}$ as annihilation operators:

$$x_+ v_{\text{h.w.}} = 0 \quad \text{for all } x_+ \in W_+. \qquad (15)$$

And third, $v_{\text{h.w.}}$ is an eigenvector of the abelian algebra W_0. In other words, for any highest weight vector v_A there is a linear function $\lambda_A : W_0 \to \mathbb{C}$ such that

$$x_0 v_A = \lambda_A(x_0) \cdot v_A \qquad (16)$$

for all $x_0 \in W_0$; I will call λ_A the *weight* of v_A with respect to W_0.

Owing to $L_0 \in W_0$ and relation (7), every highest weight module of W satisfies the spectrum condition (H3). However, not any arbitrary highest weight module of W qualifies as a subspace of $\mathcal{H}$: the unitarity property (H2) must be satisfied as well.[6] Now every highest weight module V is already endowed with a natural hermitian product; this is uniquely specified by the value on the highest weight vector, say $(v_{h.w.} | v_{h.w.}) = 1$, namely by choosing a basis of $\mathcal{W}$ in such a way that

$$(W_n^i)^* = W_{-n}^{\bar{i}}, \tag{17}$$

where $i \mapsto \bar{i}$ is some involutive permutation of the index set, and then implementing the property (14). If the so obtained product is degenerate, then the subspace of V consisting of all *null vectors*, i.e. vectors v_{null} which are orthogonal with respect to the form $(\cdot | \cdot)$ to all of V (i.e. $(v_{null} | w) = 0$ for all $w \in V$), is a submodule of V, and hence V is reducible (but not fully reducible). On the other hand, every unitary highest weight module V is in fact also irreducible.

Now for any vector v_A, a simple way to construct a highest weight module is to act *freely* with the enveloping algebra $\mathfrak{U}(\mathcal{W}_-)$; the module

$$\mathcal{V}_A = \mathfrak{U}(\mathcal{W}) \otimes_{\mathcal{W}_0 \oplus \mathcal{W}_+} v_A \cong \mathfrak{U}(\mathcal{W}_-) \otimes_{\mathbb{C}} v_A. \tag{18}$$

obtained by this construction is known as the *Verma module* generated by v_A. Verma modules are typically neither irreducible nor unitary; however, in the cases of interest in conformal field theory the submodule of $\mathcal{V}_A$ consisting of null vectors is a maximal submodule, and by 'setting the null vectors to zero' (or, in more mathematical terms, by taking the quotient of $\mathcal{V}_A$ by this submodule) one arrives at a module $\mathcal{H}_A$ which is unitary and irreducible. Also, any highest weight module with highest weight vector v_A can be understood as some quotient of $\mathcal{V}_A$. It follows that each irreducible sub-$\mathcal{W}$-module of the space $\mathcal{H}$ can be obtained in this fashion, with A some index labelling the module.

The conformal weight of the highest weight state v_A of $\mathcal{H}_A$ will be denoted by Δ_A. When $v \in \mathcal{H}_A$ is an eigenvector of $R_A(L_0)$[7] of eigenvalue Δ, then $m := \Delta_A - \Delta$ is a non-negative integer, which for $v = W_{-m_1}^{i_1} \cdots W_{-m_l}^{i_l} v_A$ ($m_p \geq 0$ for $p = 1, 2, ..., l$) can be computed as $m = \sum_{p=1}^{l} m_p$; m is called the *grade* of v.

1.4   Sectors: The Spectrum

One can summarize some of the results reviewed above by stating that (H1) to (H3) imply that $\mathcal{H}$ has the structure of a (possibly infinite) direct sum

$$\mathcal{H} = \bigoplus_A \mathcal{H}_A \tag{19}$$

of unitary irreducible highest weight modules $\mathcal{H}_A$ of $\mathcal{W}$. However, these requirements do *not* specify which of the various unitary irreducible highest weight modules of $\mathcal{W}$ appear in the sum (19). The set of irreducible modules which do

---

[6] While unitarity is an obvious requirement in quantum field theory, in statistical mechanics systems it is not always necessary. Nevertheless the $\mathcal{W}$-modules arising in applications to statistical mechanics are highest weight modules (except in so-called 'logarithmic' conformal field theories), for a reason which is still mysterious to me.

[7] Henceforth, the symbol R will almost always be suppressed.

appear constitutes the *spectrum* of the conformal field theory; these spaces are also called the superselection sectors of the theory, or briefly the *sectors*.

Any observable which appears as a central element of $\mathcal{W}$ not only acts as a constant in each sector $\mathcal{H}_A$ (this is just a corollary of Schur's lemma), but in fact it must act by one and the *same* constant on all sectors. This condition constitutes a strong restriction on the spectrum. Its origin is that central charges cannot be localized; in the context of local observables which are required to form a simple associative algebra, any central charge must therefore be a scalar multiple of the unit operator. Frequently this requirement also arises, at a more practical level, from the fact that one must work with expressions which only make sense when these charges are regarded as numbers; cf. e.g. the Sugawara formula (112) below.

It is an important result, known as *naturality*, that once the eigenvalues of all central elements of $\mathcal{W}$ are fixed (and provided that $\mathcal{W}$ is maximal), in the spectrum of a chiral half of a conformal field theory each (isomorphism class of) unitary irreducible highest weight module appears precisely once. (The derivation of this assertion ([Moore and Seiberg 1989]) is beyond the scope of these lectures.) When the number of sectors is finite, then the conformal field theory is called a *rational* theory. This name originates from the observation that in a rational conformal field theory both the eigenvalues of (canonically normalized) central elements and all conformal dimensions Δ_A are rational numbers.

Among the spectrum of a conformal field theory there is a distinguished sector $\mathcal{H}_0$ with $\Delta_0 = 0$. Its presence is implied by another basic property of quantum field theory, namely the existence of a vacuum state. This is a vector v_0 in $\mathcal{H}_0$ (unique up to scalar multiplication) which is invariant under all unbroken symmetries of the theory. In the present context this means that v_0 is in particular $\mathcal{W}_0$-neutral, i.e. $x_0 v_0 = 0$ for all $x_0 \in \mathcal{W}_0$, or in short, $\lambda_0 \equiv 0$. Naively one might expect that v_0 also has all other symmetries of the theory, in particular that it is conformally invariant, i.e. is annihilated by *all* L_n. But for $c \neq 0$ this would be incompatible with the relations (3) of $\mathcal{V}\text{ir}$. Rather, one can only require that v_0 be annihilated by L_n with $n \geq -1$. Similar remarks apply to the action of other $\mathcal{W}_n^i \in \mathcal{W}$ on the vacuum; in short, v_0 respects the maximally possible number of symmetries of the theory. As a consequence, v_0 is a highest weight vector of $\mathcal{W}$.

That v_0 is annihilated by L_n with $n \geq -1$ means that it is both a highest weight vector with respect to $\mathcal{V}\text{ir}$ and invariant under the subalgebra

$$\mathcal{P} := \mathcal{V}\text{ir}_{0,\pm 1} \equiv \text{span}\{L_1, L_0, L_{-1}\} \cong \mathfrak{sl}(2) \tag{20}$$

of $\mathcal{V}\text{ir}$ (the *projective*, or *Möbius*, transformations); this subalgebra is isomorphic to $\mathfrak{sl}(2)$.[8] that the projective transformations are precisely those whose classical counterpart is well-defined on the whole Riemann sphere so that they can be

----

[8] When the field of scalars is restricted from $\mathbb{C}$ to $\mathbb{R}$, one obtains the real form $\mathfrak{su}(1,1) \cong \mathfrak{sl}(2,\mathbb{R})$ of $\mathfrak{sl}(2)$. Together with the anti-holomorphic counterparts $\bar{L}_0$, $\bar{L}_{\pm 1}$ one arrives at a real form of $\mathfrak{so}(4)$; this is precisely the conformal algebra that one would get when 'naively' extrapolating from arbitrary $D > 2$ to $D = 2$.Note

exponentiated so as to provide finite conformal transformations. Namely, the action of $-z^{n+1}\partial_z$ is non-singular at $z = 0$ only for $n \geq -1$, and non-singular at $z = \infty$ (where it acts as $w^{1-n}\partial_w$ with $w = 1/z$) only for $n \leq 1$.

Let me also remark that in some applications one may well need the full conformal invariance, so that one needs $c = 0$. This happens e.g. in string theory. In that case there is an extra contribution to c from the ghost fields which are employed to gauge-fix two-dimensional gravity, so that the requirement is that $c_{matter} = -c_{ghost}$, e.g. $c_{matter} = 26$ for the bosonic string. [9]

For unitarity, the numbers Δ_A, and hence the conformal weights of all L_0-eigenvectors in $\mathcal{H}$, must be non-negative. Namely, consider first vectors v which are *quasi-primary*, which means that they are annihilated by L_1. Then due to $[L_1, L_{-1}] = 2L_0$ and $(L_{-1})^* = L_1$ one has $\|L_{-1}v\|^2 = 2\Delta_v\|v\|^2$, which by unitarity implies $\Delta_v \geq 0$. Further, for any arbitrary $v \in \mathcal{H}$ let N be the smallest natural number such that $(L_1)^N v = 0$ (which exists, see before (11)); then the vector $\tilde{v} := (L_1)^{N-1}v$ is non-zero and also quasi-primary. Since $\Delta_{\tilde{v}} = \Delta_v - (N-1)$, $\Delta_v < 0$ would imply that $\Delta_{\tilde{v}} < 0$, which as already discussed cannot happen.

A similar argument shows that $\|L_{-n}v_0\|^2 = (n^3 - n)\cdot c/12$ for all $n \in \mathbb{N}$, so that (taking $n > 1$) for unitarity it is required that $c > 0$. Further constraints on the spectrum are obtained by studying the positivity of $\|W^{i_1}_{-n_1}\cdots W^{i_p}_{-n_p}v_A\|$ systematically. That is, at any fixed grade one considers the matrix of inner products of all basis states; a necessary condition for unitarity is that each such matrix has non-negative determinant (one needs not impose strict positivity, because one can 'decouple' null vectors, which corresponds to quotienting the Verma module $\mathcal{V}_A$). To analyze this condition one writes the inner products as expectation values of $\mathcal{W}$-generators with respect to the highest weight vector, which can be calculated by using the bracket relations of $\mathcal{W}$. In practice this can become very hard; for $\mathcal{W} = \mathcal{V}ir$ the result is that for $c \geq 1$ there is no restriction on the conformal dimensions, while below $c = 1$ only the discrete set

$$c = c_m := 1 - \frac{6}{m(m+1)} \qquad \text{with } m \geq 3 \tag{21}$$

of c-values is allowed, and for each of these specific values only a finite set

$$\Delta = \Delta_{m;p,q} := \frac{((m+1)p-mq)^2 - 1}{4m(m+1)} \qquad \text{with } 1 \leq p,q \leq m-1 \tag{22}$$

of conformal weights is possible; the theories with these values of c and Δ are known as the minimal unitary models of the Virasoro algebra. Thus all unitary

---

[9] In this context it is less obvious that (the matter part of) the conformal field theory should be unitary. (What one has to achieve is unitarity on a certain cohomology of $\mathcal{V}ir$; for this it is certainly helpful to start from a unitary conformal field theory, but I do not know whether this is really necessary.) Also, typically one 'compactifies' the string theory by stipulating that c_{matter} is the sum of a contribution from space-time and one from some 'internal' theory. When the space-time is flat, each space-time dimension corresponds to one free boson and hence (see Sect. 2.5) contributes 1 to c; but it is not clear to me why in our world where space-time is only asymptotically flat its contribution to c should have precisely the integral value $c = 4$.

theories with $c < 1$ are rational. But even when the unitarity requirement is dropped, one still needs $c < 1$ for rationality. In other words, for rational theories with central charge $c \geq 1$, the chiral algebra $\mathcal{W}$ is always larger than $\mathcal{V}ir$.

1.5   Conformal Fields

Even though one is ultimately interested in observable quantities only, in quantum field theory it is often convenient to employ non-observable objects in intermediate steps of calculations. When such objects are operators on $\mathcal{H}$, they are referred to as *fields*. Generic fields are distinguished from the observables by the fact that in terms of the decomposition (19) of $\mathcal{H}$, they act as $\mathcal{H}_A \to \mathcal{H}_B$ with $B \neq A$, while observables act within each individual subspace $\mathcal{H}_A$. [10] In conformal field theory, one realizes field operators φ via the so-called *state-field correspondence*; the basic idea is that one generates all vectors in $\mathcal{H}$ from the vacuum v_0 by applying suitable point-like *conformal fields* $\varphi(z, \bar{z})$, according to

$$v_{\Delta, \bar{\Delta}} = \lim_{z, \bar{z} \to 0} \varphi_{\Delta, \bar{\Delta}}(z, \bar{z}) \, v_0 . \tag{23}$$

Here the labels indicate the conformal weights of the fields, respectively of the vectors $v_{\Delta, \bar{\Delta}}$ (usually more labels are needed to distinguish the fields, especially when $\mathcal{W} \supset \mathcal{V}ir$). The formula (23) is often read as the definition of the vector $v_{\Delta, \bar{\Delta}} \in \mathcal{H}$; in the present spirit, it rather specifies the rôle played by the fields φ.

The precise meaning of the prescription (23) and of the formal objects φ introduced there is not at all obvious, and depending on the framework adopted to interpret these quantities it can be a delicate issue to verify that the limit in (23) makes sense. To investigate this one best starts with the case where $v_{\Delta, \bar{\Delta}}$ lies in the vacuum sector $\mathcal{H}_0$, so that the fields are observables. Now the maximality requirement imposed on the chiral symmetry algebra $\mathcal{W}$ means in particular that any operator on $\mathcal{H}$ which does not belong to $\mathcal{W}$ changes the sector; thus in order for $v_{\Delta, \bar{\Delta}}$ to lie in $\mathcal{H}_0$, the fields must be suitable combinations of the generators W_n^i. The correct interpretation of the fields is then as generating functions

$$W^i(z) = \sum_{n \in \mathbb{Z}} z^{-n-\Delta_i} W_n^i , \tag{24}$$

or in other words, the generators of $\mathcal{W}$ are regarded as moments or *modes* of $W^i(z)$. In the case of the Virasoro generators, this generating function

$$T(z) = \sum_{n \in \mathbb{Z}} z^{-n-2} L_n , \tag{25}$$

plays the rôle of the *energy-momentum tensor*; that is, in models which can be formulated in a Lagrangian setting, it describes the response of the Lagrangian density to a variation of the space-time metric. [11] One must note that in (24)

---

[10] Such a distinction only makes sense if one demands that the superposition principle is not universally valid. In other words, only those vectors in $\mathcal{H}$ which lie in one of the subspaces $\mathcal{H}_A$ are regarded as corresponding to proper physical states.

[11] This makes sense even for theories on flat space-times, because the operations of variation and of taking the limit that the metric becomes flat do not commute.

the variable z is introduced just as a formal indeterminate. When z is interpreted in terms of a world sheet, and hence is regarded as a complex variable, then the W_n^i are just the Fourier–Laurent modes of $W^i(z)$, and one must worry about the convergence of expressions like (24). One way to attack this problem is to think of fields as operator valued distributions, and to work with localized bounded operators (by smearing with test functions and exponentiating), which can e.g. be analyzed in the framework of Wightman field theory ([Streater and Wightman 1989]) or of the algebraic theory of superselection sectors ([Haag 1992]). Here I rather describe the (much less rigorous) formulation that is based on the work of Belavin, Polyakov and Zamolodchikov ([1984]) and is sometimes called the *bootstrap approach*. Another alternative, at least as far as the mathematical aspects of the theory, rather than its physical interpretation, are concerned, is to think of z as a *formal* (in the well-defined mathematical sense) variable; this leads to the concept of a *vertex operator algebra* ([Frenkel et al. 1988]).

Interpreting z as a complex variable, one can invert (24), so as to write the Laurent modes as contour integrals of fields multiplied with powers of z,

$$W_m^i = \tfrac{1}{2\pi i} \oint_0 \mathrm{d}z \, z^{m+\Delta_i-1} \, W^i(z) \qquad (26)$$

(recall that $\Delta_i \in \mathbb{Z}$), where integration is over some curve encircling zero.

When it comes to genuine fields $\varphi(z,\bar{z})$ rather than observables, it is worth to return to the issue of which properties characterize a 'conformal field theory'. What one has to demand is that the 'space of all fields' carries a representation of the conformal group, respectively of the conformal algebra – analogously to what one is used to from Lorentz covariant theories in which the fields carry an action of the Lorentz group and in which, as a consequence, the excitations can be classified by representations of that group. But not all fields one can think of are on the same footing. In particular, the following ones are distinguished:

■ The ($\mathcal{V}i\mathfrak{r}$-) *primary fields*. These are defined by the requirement that the state $v_{\Delta,\bar{\Delta}}$ (23) is a $\mathcal{V}i\mathfrak{r}$-highest weight state, i.e. is annihilated by all L_n with $n > 0$.

■ The ($\mathcal{V}i\mathfrak{r}$-) *quasi-primary fields*. For these, $v_{\Delta,\bar{\Delta}}$ is annihilated by L_1, and hence is a highest weight state of the Möbius subalgebra $\mathcal{P}$ (20) of $\mathcal{V}i\mathfrak{r}$.

In the relations of the $\mathfrak{sl}(2)$-algebra $\mathcal{P}$ the central term is absent, so that one can identify the generators L_0, $L_{\pm 1}$ with the corresponding generators of the Witt algebra (2). Quasi-primary fields are therefore also called 'non-derivative' fields. In terms of the Laurent modes, the distinguished behavior of these types of fields is demonstrated by (6): when (6) holds for $m = 0$ and ± 1, then $W^i(z)$ is quasi-primary, while if it holds for all $m \in \mathbb{Z}$, then $W^i(z)$ is primary. Analogous relations hold for every (quasi-) primary field ϕ: $[L_m, \phi(z)] = z^{m+1}\partial\phi(z) + (m+1)\Delta_\phi z^m \phi(z)$; in particular, in the limit $z \to 0$ one obtains

$$[L_n, \phi(0)] = \begin{cases} 0 & \text{for } n > 0\,, \\ \Delta \cdot \phi(0) & \text{for } n = 0\,, \\ \partial\phi(0) & \text{for } n = -1\,. \end{cases} \qquad (27)$$

Non-primary fields are called *secondary* fields or *descendants*. Among the fields that correspond to the vectors in an irreducible highest weight module $\mathcal{H}_A$ of $\mathcal{W}$ there is precisely one primary field ϕ_A; it corresponds to the highest weight vector. The collection of these fields, consisting of the primary ϕ_A and all its descendants, is called the *family* of ϕ_A and is denoted by $[\phi_A]$. By the *grade* of a descendant of ϕ_A one means the grade of the corresponding vector in $\mathcal{H}_A$.

The simplest example of a primary field, present in every conformal field theory, is the *identity field* $\mathbf{1}$. It satisfies $\partial \mathbf{1} = 0$, and hence corresponds to v_0; its first non-trivial descendant is the energy-momentum tensor $T(z)$:

$$\lim_{z \to 0} T(z)\, v_0 = \sum_{n \in \mathbb{Z}} z^{-n-2} L_n v_0 = L_{-2} v_0 \,. \tag{28}$$

$T(z)$ is the prime example of a quasi-primary field. The Lie brackets of the Laurent components of all quasi-primary fields in the chiral algebra $\mathcal{W}$ read $[W_m^i, W_n^j] = d_{ij} p^i \, \delta_{m+n,0} + \sum_k C_k^{ij} q_{mn}^{ijk} W_{m+n}^k$, where the structure constants p^i and q_{mn}^{ijk} are rational numbers for which explicit expressions in terms of Δ_i, respectively, m, n and Δ_i, Δ_j, Δ_k are known ([Blumenhagen et al. 1991]). Here $d^{ij} := \langle W_{\Delta_i}^i, W_{-\Delta_j}^j \rangle$ and $C^{ijk} := \langle W_{\Delta_i}^i, W_{-\Delta_j}^j, W_{\Delta_j - \Delta_i}^k \rangle$ as well as $\sum_j d^{ij} d_{jk} = \delta_k^i$ and $C_k^{ij} := \sum_l C^{ijl} d_{lk}$, where $\langle \cdots \rangle$ denotes the *vacuum expectation value*, defined as

$$\langle X \rangle := (v_0 \,|\, X \, v_0)\,. \tag{29}$$

Bounded operators on a Hilbert space can be multiplied. Analogously one would like to multiply in a suitable way the fields $\varphi(z, \bar{z})$ that occur in the present setting. Just like in other approaches to quantum field theory there arises the problem that such products tend to be singular, at least in the limit of 'coinciding points'. A way out is to consider *normal ordered* expressions, i.e. set

$$:W_m^i W_n^j: \; = \begin{cases} W_m^i W_n^j & \text{for } n > 0, \\ W_n^j W_m^i & \text{for } n \leq 0 \end{cases} \tag{30}$$

(say) for bilinears in the modes W_n^i. (The normal ordering prescription adopted here is not at all the only possible one. By modifying (30) quite arbitrarily for only a finite number of terms, one obtains another sensible normal ordering.) The normal ordered products (30) share the property of the W_n^i to act locally nilpotently on $\mathcal{H}$. The normal ordering of fields $W^i(z)$ follows from the normal ordering of their modes W_n^i via the series expansion (24). Normal ordered products of quasi-primary fields are then generically not quasi-primary; but all commutators of normal ordered products of fields in $\mathcal{W}$ are fully determined by those of the energy-momentum tensor and of the Virasoro-primary fields in $\mathcal{W}$.

1.6 The Operator Product Algebra

In a chiral conformal field theory – a chiral half of a two-dimensional theory – the fields depend on a single complex variable z. As long as one deals with a single field, say $W^i(z)$, it suffices to describe it for values of z which lie on a closed curve encircling zero, say the unit circle, since then one recovers all its modes W_n^i by (26). However, as soon as one intends to analyze products of fields, one must at least 'smear' the circle. Namely, when expressed in terms of fields, the abstract Lie bracket of modes should correspond to the commutator with respect to a suitable associative product; now inserting (26) into a Lie bracket yields (with a suitable labelling of integration variables) the difference

$$[W_m^i, W_n^j] = [\tfrac{1}{2\pi i} \oint_0 dz \, \tfrac{1}{2\pi i} \oint_0 dw \, W^i(z) \, W^j(w) \\ - \tfrac{1}{2\pi i} \oint_0 dw \, \tfrac{1}{2\pi i} \oint_0 dz \, W^j(w) \, W^i(z)] \, z^{m+\Delta_i-1} w^{n+\Delta_j-1} \tag{31}$$

of two double contour integrals, where in each term the integrations must be performed in the indicated order. One would like to rewrite this expression in a form where there is a single contour integration over fields $W^k(z)$ multiplied with suitable powers of the position, so as to recover the result for the Lie bracket $[W_m^i, W_n^j]$ as a linear combination of modes W_p^k. To achieve this, one interprets the first pair of integrations as to be performed for $|z| > |w|$ and the second for $|w| > |z|$. Thus the formal product of fields appearing in (31) is to be understood as the *radially ordered product* $\Re$, defined for arbitrary fields φ_1 and φ_2 by

$$\Re(\varphi_1(z) \, \varphi_2(w)) := \begin{cases} \varphi_1(z) \, \varphi_2(w) & \text{for } |z| > |w|, \\ \varphi_2(w) \, \varphi_1(z) & \text{for } |w| > |z|. \end{cases} \tag{32}$$

For fixed value of w (say) one can now deform the two z-contours in (31) and merge them to a single one encircling the point w. One then arrives at the desired result for the commutator by postulating that the radially ordered product of two fields whose positions are sufficiently close can be written as a linear combination of fields, multiplied with suitable powers of $z-w$. The so obtained decomposition is called the *operator product expansion* of $W^i(z)$ and $W^j(w)$.

When one knows the structure constants of $\mathcal{W}$, one can determine the singular part of operator product expansions by requiring that the contour integrations $\oint_0 dw \oint_w dz$ yield the correct result. E.g. the Virasoro algebra amounts to

$$\Re(T(z) \, T(w)) = \tfrac{1/2}{(z-w)^4} \, C + \tfrac{2}{(z-w)^2} \, T(w) + \tfrac{1}{z-w} \, \partial_w T(w) + \mathcal{O}_{\text{reg}}. \tag{33}$$

Here $\mathcal{O}_{\text{reg}}$ stands for (an infinite power series of) terms which in the limit $z \to w$ are regular and hence do not affect the result for $[L_m, L_n]$. (More generally, only singular terms in expansions like (33) are relevant to the observables; the regular terms in fact do not have any meaning independent of the sector, e.g. their precise form depends on the choice of normal ordering prescription in the different sectors.) It is common to refer to radially ordered products (32) just as *operator products* and to omit the symbol $\Re$; I will usually follow this habit.

In the two-dimensional theory, we have both the chiral and the antichiral algebras $\mathcal{W}$ and $\overline{\mathcal{W}}$; they commute, and correspondingly the operator product of fields $W^i(z)$ and $\overline{W}^j(\bar{w})$ just coincides with their normal ordered product. But we may also consider operator products of the $W^i(z)$ with arbitrary fields φ. For instance, for the operator product of a primary field with T one finds

$$T(z)\,\phi(w,\bar{w}) = \tfrac{\Delta}{(z-w)^2}\,\phi(w,\bar{w}) + \tfrac{1}{z-w}\,\partial_w\phi(w,\bar{w}) + \mathcal{O}_{\text{reg}}\,, \tag{34}$$

which upon Laurent expansion correctly reproduces the formula (27) for the commutator $[L_n,\phi]$. The singular terms in an operator product $\varphi_1(z)\varphi_2(w)$ of the type above are called the contraction of φ_1 and φ_2 and are denoted by $\underline{\varphi_1(z)\varphi_2}(w)$. Also, the term of order $(z-w)^0$ is essentially the normal ordered product $:\varphi_1\,\varphi_2:$ – when the particular normal ordering prescription

$$:\varphi_1(z)\,\varphi_2(z): \; := \; \tfrac{1}{2\pi i}\oint_w \mathrm{d}z\,(w-z)^{-1}\,\Re(\varphi_1(z)\,\varphi_2(w)) \tag{35}$$

(which is slightly different from the one presented in (30)) is adopted, then it is indeed precisely the normal ordered product, i.e. one has

$$\varphi_1(z)\varphi_2(w) = \underline{\varphi_1(z)\varphi_2}(w) + :\varphi_1(z)\varphi_2(z): + O(z-w)\,. \tag{36}$$

Besides the conformal invariance, a second basic property of two-dimensional conformal field theories is that operator products cannot only be defined when one of the fields belongs to the chiral algebra, but for arbitrary pairs of fields. That is, there is a *closed operator product algebra* among *all* fields, and the operation involved is indeed a product in the sense that it is associative. It is expected that this structure is a direct consequence of fundamental properties of quantum field theory. But when trying to construct it this way one faces quite a few subtleties, and to the best of my knowledge a complete derivation from first principles has never been given. Instead, one usually postulates the existence of a closed associative operator product algebra as a *separate input*, called the *bootstrap* hypothesis or requirement. Note, however, that operator product expansions do *not* converge in the operator norm, but only weakly, i.e. when applied to a vector in $\mathcal{H}$. Thus while one may regard the collection of all fields as a vector space (over some function field), the operator product does not really make this space into an associative algebra in the mathematical sense. [12]

---

[12] The theory of vertex operator algebras ([Frenkel et al. 1988]) provides a means for formulating operator products in a purely algebraic setting, without having to resort to arguments based on complex analysis, e.g. contour deformation or analytic continuation. The basic structures of a vertex operator algebra are an infinite set of products and the action of the derivative ∂; the commutator with respect to the first of these products yields the Lie algebra structure of the space $\mathcal{W}_o$ of zero modes. The variables z etc. are treated as formal variables, which implies e.g. that $(z-w)^{-1}$ is interpreted as $z^{-1}(1-w/z)^{-1} = z^{-1}\sum_{n\geq 0}(w/z)^n$. Thus $1/(z-w) + 1/(w-z) = z^{-1}\sum_{n\in\mathbb{Z}}(w/z)^n$, which in terms of complex analysis corresponds to the delta function on the unit circle. The relation with the radial ordering prescription is that upon re-interpreting the variables as complex numbers, the formal power series occurring in the vertex operator formulation become convergent series precisely if the relevant numbers are radially ordered in the appropriate manner.

Operator products of primary fields ϕ can be written in the form

$$\phi_A(z,\bar{z})\,\phi_B(w,\bar{w}) = \sum_C C_{AB}{}^C \,(z-w)^{\Delta_C-\Delta_A-\Delta_B}(\bar{z}-\bar{w})^{\bar{\Delta}_C-\bar{\Delta}_A-\bar{\Delta}_B}\phi_C(w,\bar{w}) + \cdots,$$
$$(37)$$

where the ellipsis stands for contributions of descendants $\varphi_C \in [\phi_C]$. The numbers $C_{AB}{}^C$ introduced in (37) are called the *operator product coefficients*. The corresponding coefficients which involve descendants are completely fixed by the $\mathcal{W}$-symmetry in terms of those of the primaries (and of their conformal weights and the values of central charges). For instance, the coefficient with which the descendant $\partial^n \varphi$ of a quasi-primary field φ appears equals the coefficient for φ multiplied by $(n!)^{-1}\prod_{l=0}^{n-1}(l+\Delta_A-\Delta_B+\Delta_\varphi)/(l+2\Delta_\varphi)$. On the other hand, the relation between the operator product coefficients for quasi-primaries and those for primaries has been worked out in some detail only for $\mathcal{W} = \mathcal{V}ir$. Nevertheless, at least in principle it is possible to solve the theory completely, i.e. to compute all correlation functions, by expressing them in terms of the coefficients $C_{AB}{}^C$ involving only primaries and of the structure constants of $\mathcal{W}$. The determination of the numbers $C_{AB}{}^C$ is therefore one of the main goals in conformal field theory.

1.7 Correlation Functions and Chiral Blocks

Operator product expansions can only be valid when applied to vectors in $\mathcal{H}$. In fact, because the presence of additional fields typically affects manipulations with integration contours, strictly speaking they even only hold when matrix elements with respect to vectors in $\mathcal{H}$ are taken, e.g. for vacuum matrix elements

$$\mathcal{G}(\{z_j\},\{\bar{z}_j\}) \equiv \langle \varphi_1(z_1,\bar{z}_1)\,\varphi_2(z_2,\bar{z}_2)\cdots\varphi_p(z_p,\bar{z}_p)\rangle$$
$$:= (v_0\,|\,\varphi_1(z_1,\bar{z}_1)\,\varphi_2(z_2,\bar{z}_2)\cdots\varphi_p(z_p,\bar{z}_p)\,v_0)\,. \qquad (38)$$

Such a vacuum expectation value is called the (p-point) *correlation function* of the fields $\varphi_1, \varphi_2, \ldots, \varphi_p$. Given the operator product algebra, one can in principle evaluate any correlation function by expressing successively all products as linear combinations of fields until one ends up with a linear combination of fields acting on v_0, and then using the fact (see below) that the only field with non-vanishing one-point function is the identity field $\mathbf{1}$. In practice, this procedure only works in simple cases such as free field theories. However, some general properties of correlators can be derived without a detailed knowledge of the operator products. They are implied by the *Ward identities* for $\mathcal{W}$, which are obtained as follows. Consider a correlation function of the form $\langle W^i(z)\,\varphi_1(z_1,\bar{z}_1)\,\varphi_2(z_2,\bar{z}_2)\cdots\rangle$, multiplied with some power z^n, and perform a z-integration over a contour $\mathcal{C}$ that encircles all of the 'insertion points' z_i. By deforming $\mathcal{C}$ to infinity one learns that for $n > -\Delta_i$ this integrated correlator vanishes; on the other hand, after deforming the contour into a union of contours each of which encircles precisely one of the insertion points z_j, one can insert the operator product $W^i(z)\varphi_1(z_j,\bar{z}_j)$ and perform the integration for each of these contours separately. These manipulations amount to a linear differential equation for the correlation function

$\langle \varphi_1(z_1, \bar{z}_1) \, \varphi_2(z_1, \bar{z}_1) \cdots \rangle$. In particular, in the case of $W^i(z) = T(z)$, the operator product (34) leads to the *projective* Ward identities

$$\sum_{i=1}^{p} z_i^n \left(z_i \frac{\partial}{\partial z_i} + (n+1)\Delta_i \right) \mathcal{G}(\{z_j\}, \{\bar{z}_j\}) = 0 \quad \text{for} \quad n \in \{0, \pm 1\} \qquad (39)$$

for correlation functions of primary fields. When $W = Vir$ and the number p of insertions is small, the projective Ward identities are particularly effective:

- The relation (39) with $n = -1$ tells us that one-point functions are constant. Thus $\langle \phi_A \rangle = \lim_{z \to 0} \langle \phi_A(z) \rangle = (v_0 \, | \, v_A)$, which is non-zero only if $\phi_A = \phi_0 = 1$.
- In the case of two-point functions, the general solution of the $n = -1$ identity is $\mathcal{G} = \mathcal{G}(z_1 - z_2)$; $n = 0$ then gives $\mathcal{G} \propto (z_1 - z_2)^{-\Delta_1 - \Delta_2}$, and finally $n = 1$ shows that $\Delta_1 = \Delta_2$. When the chiral symmetry algebra W is maximally extended, the constant of proportionality (which is not fixed by the Ward identities, since they are linear), is zero unless ϕ_2 is the *conjugate field* of ϕ_1, which carries the complex conjugate representation of W. Thus $\langle \phi_A \phi_B \rangle \propto \delta_{A,B+}$, where $\phi_{A+} = (\phi_A)^+$.
- Similarly, three-point functions are fixed up to a multiplicative constant:

$$\mathcal{G} \propto (z_1 - z_2)^{\Delta_3 - \Delta_1 - \Delta_2}(z_2 - z_3)^{\Delta_1 - \Delta_2 - \Delta_3}(z_3 - z_1)^{\Delta_2 - \Delta_3 - \Delta_1} . \qquad (40)$$

- For four-point functions the situation is already more involved (they will be studied in more detail in Sect. 2.7 and Sect. 4.4). The projective Ward identities only imply that every four-point function can be written as an arbitrary function of a certain variable w, multiplied with definite powers (which are linear combinations of the conformal weights) of the coordinate differences $z_i - z_j$. This variable w is a cross-ratio; e.g. one may choose $w = (z_2 - z_1)(z_3 - z_4)/(z_3 - z_1)(z_2 - z_4)$, which is obtained by applying to all insertion points the projective transformation $z \mapsto (z_2 - z_1)(z - z_4)/(z - z_1)(z_2 - z_4)$, upon which z_1, z_2, z_3, z_4 are mapped to ∞, 1, w, 0. In other words, without loss of generality we can set three of the insertion points to ∞, 1, 0, i.e. consider correlators of the form

$$\mathcal{G}(z, \bar{z}) \equiv \mathcal{G}_{ABCD}(z, \bar{z}) = \langle \phi_A(\infty, \infty) \, \phi_B(1, 1) \, \phi_C(z, \bar{z}) \, \phi_D(0, 0) \rangle . \qquad (41)$$

(In the case of three-point functions one can analogously map the insertion points to the fixed values ∞, 1, 0, which explains why in that case the dependence on the z_i is completely fixed. More generally, any p-point function with $p \geq 3$ can be written as a function of $p - 3$ cross-ratios, multiplied by powers of the $z_i - z_j$.)

The Ward identities involve only one chiral half of the theory, so that when studying them one can ignore the presence of the antichiral half. Of course, there are analogous identities involving the antichiral algebra, too. As a consequence, the correlation function of a *two*-dimensional theory that is obtained by combining its two chiral halves can be expressed as a linear combination of the products of independent solutions of the chiral and antichiral Ward identities,

$$\mathcal{G}(z, \bar{z}) = \sum_{I=1}^{M} \sum_{\bar{I}=1}^{\bar{M}} a_{I\bar{I}} \, \mathcal{F}_I(z) \, \overline{\mathcal{F}}_{\bar{I}}(\bar{z}) . \qquad (42)$$

The individual solutions $\mathcal{F}_I$ and $\overline{\mathcal{F}}_I$ are called the *chiral blocks* (or the conformal blocks) of $\mathcal{G}$. For $p > 3$ the sum (4.9) contains, in general, more then one term (see also Sect. 2.7), and hence does *not* factorize into a chiral and an antichiral part. This means in particular that (except for the observables $W^i(z)$) such a factorization is not possible for the fields $\varphi = \varphi(z, \bar{z})$. But still, as also indicated by the form (4.9) of correlators, one can regard primary fields ϕ as combinations

$$\phi_A(z, \bar{z}) = \sum_q c_q \, \varpi_q(z) \, \bar{\varpi}_{\bar{q}}(\bar{z}) \tag{43}$$

of suitable chiral objects ϖ_q and $\bar{\varpi}_{\bar{q}}$, which are called *chiral vertex operators*. In order for this decomposition to make sense, for each primary ϕ_A the label q must be understood as standing not only for A, but also for two additional primary field labels B, C [13] such that $\varpi_q \equiv \varpi_{A;B,C}$ constitutes a map $\mathcal{H}_B \to \mathcal{H}_C$ which intertwines the action of the chiral algebra, i.e. $\varpi_{A;B,C} \circ R_B(W_n^i) = R_C(W_n^i) \circ \varpi_{A;B,C}$. In terms of operator products, the restriction to a specific *source* $\mathcal{H}_B$ and *range* $\mathcal{H}_C$ corresponds to considering only the terms involving the family $[\phi_C]$ in the operator product $\phi_A \phi_B$; correspondingly, the coefficients $c_q \equiv c_{A;B,C}$ in (43) are in fact nothing but the operator product coefficients $C_{AB}{}^C$.

By construction, the chiral blocks can be interpreted as the vacuum expectation values of suitable chiral vertex operators, e.g. $\mathcal{F}_I = \langle \varpi_{0;A,A} \, \varpi_{A;B,I} \, \varpi_{I;C,D+} \, \varpi_{D+;D,0} \rangle$ for the four-point function $\mathcal{G}_{ABCD}$. Chiral blocks are generically not functions, but are multi-valued.[14] They are single-valued on a suitable multiple covering of the complex plane (and hence can be regarded as sections of some bundle). Analytic continuation connects the different sheets of that covering. Correspondingly, the exchange of two chiral vertex operators is governed by a representation of the braid group rather than the permutation group.

The requirement that the chiral and antichiral blocks combine to single-valued correlation functions of the two-dimensional theory yields algebraic equations for the linear coefficients $a_{I\bar{I}}$ in (4.9). In principle, these equations can be solved to obtain these coefficients up to over-all normalization; the latter is left undetermined because the Ward identities are *linear* differential equations. For the three-point functions there is only a single chiral block (when $\mathcal{W}$ is maximal) and hence only a single coefficient $a \equiv a_{ABC}$. Comparison with the operator product algebra (37) shows that $a_{ABC} = \mathcal{C}_{ABC} := \sum_D \mathcal{C}_{AB}{}^D d_{DC}$, where $d_{AB} := (z_1 - z_2)^{2\Delta_A} (\bar{z}_1 - \bar{z}_2)^{2\bar{\Delta}_A} \langle \phi_A(z_1) \phi_B(z_2) \rangle \propto \delta_{A,B+}$ plays the rôle of a metric in the space of fields. In short, the three-point functions of primary fields are essentially the operator product coefficients.

---

[13] as well as possibly some multiplicity label, compare Sect. 2.2.

[14] Sometimes the blocks are also called *holomorphic* respectively anti-holomorphic blocks. Thus in this context the term 'holomorphic' is used in a rather loose sense; even the full correlation functions are in general only *mero*morphic.

2   Fusion Rules, Duality and Modular Transformations

2.1   Fusion Rules

In the first lecture we learned that upon forming radially ordered products and when considered inside correlation functions, the fields φ of a conformal field theory realize a closed associative operator product algebra. Unfortunately, in practice this structure looks extremely complicated. However, a large amount of information about the operator products is already contained in a much more transparent structure, the so-called *fusion rules*. Roughly speaking, the fusion rules constitute the basis-independent contents of the operator product algebra, i.e. count the number $N_{AB}{}^C$ of times that the family $[\phi_C]$ of the primary field ϕ_C appears in the operator product of primaries ϕ_A and ϕ_B. In other words, they tell how many distinct couplings among primary fields, respectively $\mathcal{W}$-families, are possible. To encode this information, one associates to each primary field ϕ_A an abstract object Φ_A, and introduces an abstract multiplication '$\star$' by writing

$$\Phi_A \star \Phi_B = \sum_C N_{AB}{}^C \Phi_C \,. \tag{44}$$

The integers $N_{AB}{}^C$ are known as the *fusion rule coefficients*. Note that the product (44) is *not* isomorphic to the ordinary tensor product of modules of $\mathcal{W}$. [15]

The $\mathcal{W}$-family $[\phi_C]$ occurs in the product of ϕ_A with ϕ_B iff $N_{AB}{}^C$ is non-zero. This might suggest that $N_{AB}{}^C$ can take only the values 0 or 1, corresponding to the alternative whether $[\phi_C]$ appears in $\phi_A(z)\phi_B(w)$ or not. But one and the same family $[\phi_C]$ may couple in several distinct ways to ϕ_A and ϕ_B so that values $N_{AB}{}^C > 1$ are possible as well. Accordingly, the fields φ_C appearing in the expansion (37) are not necessarily all distinct, i.e. it may happen that $\varphi_{C_1} = \varphi_{C_2}$ for $C_1 \neq C_2$; nevertheless it is not allowed just to add up the corresponding operator product coefficients, because the relative values of coefficients involving different members of a family $[\phi_k]$ are fixed by the Ward identities of $\mathcal{W}$.

It turns out to be convenient to formalize the general properties of fusion rules which are implied by the principles of conformal field theory that I outlined in lecture 1. To this end one regards the objects Φ_A and numbers $N_{AB}{}^C$ as the basis elements and structure constants of a ring over the integers $\mathbb{Z}$ or of an algebra over the complex numbers $\mathbb{C}$. These structures are called the *fusion ring*, respectively the *fusion algebra*, of the theory; their defining properties are:

(F1) they are commutative and associative, and they have a unit element (namely, Φ_0, the abstract object that is associated to the identity primary field);

(F2) there is a distinguished basis (namely the one consisting of the objects Φ_A) in which the structure constants are non-negative and which contains the unit element;

(F3) the evaluation at the unit element provides an involutive automorphism, called the *conjugation* and denoted by $\Phi_A \mapsto (\Phi_A)^+$.

---

[15] This follows directly from the observation that upon forming tensor products central charges add up, whereas the fusion product yields an object which appears in the same theory and hence has the same values of central charges as the original fields.

In terms of the structure constants $N_{AB}{}^C$, these properties read as follows.
(N1) $N_{BA}{}^C = N_{AB}{}^C$, $\sum_D N_{AB}{}^D N_{DC}{}^B = \sum_D N_{BC}{}^D N_{DA}{}^B$ and $N_{oB}{}^C = \delta_B{}^C$; (N2) $N_{BA}{}^C \geq 0$;
(N3) $C_{AB} \equiv N_{AB}{}^o = \delta_{A,B^+}$ for some order-two permutation $A \mapsto A^+$ of the index set (such that $(\Phi_A)^+ = \Phi_{A^+}$), and $(A^+)^+ = A$ as well as $N_{A^+B^+}{}^{C^+} = N_{AB}{}^C$.
It then also follows that $N_{ABC} := N_{AB}{}^{C^+}$ is totally symmetric.

Note that in the identification of the Φ_A as a basis it is implicit that for all A, B the number $\sum_C N_{AB}{}^C$ is finite; conformal field theories which satisfy this requirement are called *quasi-rational*. A *rational* conformal field theory, i.e. one with only a finite number of sectors, is also quasi-rational; while this is not manifest in the definition, it follows easily from the very existence of a fusion ring. It is worth stressing that (quasi-) rationality is *not* a fundamental property of conformal field theories. However, in practice it is often indispensable, since it allows one to perform many calculations explicitly. This manifests itself for the first time when one studies the representation theory of the fusion ring. (Whether the concept of a fusion ring is still applicable in non-quasi-rational theories is not known.) Accordingly, I will from now on restrict my attention to rational conformal field theories only, unless stated otherwise.

The fusion ring of a rational conformal field theory is a finite-dimensional commutative associative ring. As a consequence, each of its irreducible representations is one-dimensional, and every finite-dimensional representation is isomorphic to the direct sum of such irreducible representations. In particular, the adjoint representation π_{ad}, defined by $\pi_{\mathrm{ad}}(\Phi_A) := N_A$, where N_A denotes the matrix with entries $(N_A)_B{}^C = N_{AB}{}^C$, must be isomorphic to the direct sum of one-dimensional irreducible representations (in fact, each inequivalent one-dimensional representation appears in this sum precisely once). In other words, there exists a unitary matrix S which 'diagonalizes the fusion rules' in the sense that – simultaneously for all values of the label A – the matrix

$$D_A := S^{-1} N_A S \tag{45}$$

is diagonal, and also the relations $CS = S^* = SC$ are valid.

It should be noted that the row and column labels of the matrix S that is introduced this way are a priori on a rather distinct footing: the row index labels the elements Φ_A of the distinguished basis, while the column index counts the inequivalent one-dimensional irreducible representations π_B. While the two sets of labels have the same order (so that, as I already did above, one can use the same symbols for either type of labels), in general there does not exist a canonical bijection between them. As it turns out, however, for those fusion rings which occur in conformal field theory such a canonical bijection does exist, and moreover, when implementing this bijection the diagonalizing matrix S possesses the highly non-trivial property of being symmetric, and also satisfies $S_{Ao} > 0$ (in particular $S_{Ao} \in \mathbb{R}$) for all values of the label A. It then follows that the one-dimensional irreducible representations π_A obey

$$\pi_A(\Phi_B) \equiv (D_B)_A{}^A = S_{BA}/S_{oA} . \tag{46}$$

As a consequence, the relation (45) amounts to an expression [16]

$$N_{AB}^{\ \ C} = \sum_D \frac{S_{AD} S_{BD} (S^{-1})_{CD}}{S_{0D}} \tag{47}$$

of the fusion rule coefficients in terms of the matrix S, and also that $S^2 = C$, so that (47) can be rewritten more symmetrically as $N_{ABC} = \sum_D S_{AD} S_{BD} S_{CD}/S_{0D}$.

2.2 Duality

The associativity of the operator product algebra implies that the fusion rules are associative as well. I will now use the same information to deduce properties of the correlation functions. Let us study the four-point functions $\mathcal{G} = \langle \phi_A \phi_B \phi_C \phi_D \rangle$ of primary fields. Applying the operator product algebra on the first two and on the last two fields, one has $\langle \phi_A \phi_B \phi_C \phi_D \rangle = \langle (\phi_A \phi_B)(\phi_C \phi_D) \rangle \equiv \langle \Re(\Re(\phi_A \phi_B) \Re(\phi_C \phi_D)) \rangle$, which pictorially amounts to

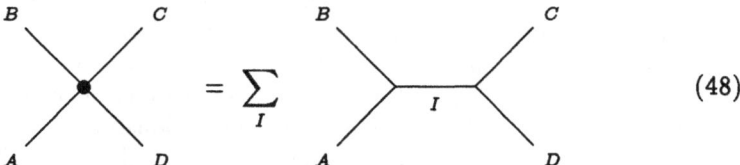

$$\tag{48}$$

But using associativity of the operator product algebra, one could also form products in a different order, resulting in pictures like

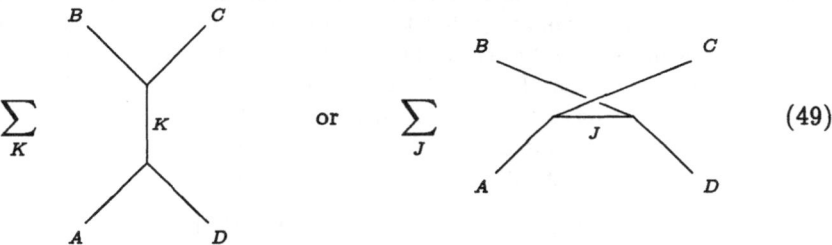

$$\tag{49}$$

Of course, because of the radial ordering prescription each of these choices is only valid for a definite order of the absolute values of the insertion points z_i of the fields $\phi_A, \dots, \phi_D$, and hence the different pictures describe functions which coincide only upon an appropriate analytic continuation. Now when the two chiral halves of a two-dimensional theory are combined, so that $\bar{z}_i$ is identified as the complex conjugate of z_i, one must require that every correlation function is single-valued on the Riemann sphere when the dependence on both z_j and $\bar{z}_j$ is accounted for; thus the three descriptions above for the four-point correlators $\mathcal{G}$ must yield one and the same function. Together with (4.9) this yields the

---

[16] This is not yet the Verlinde ([Verlinde 1988]) formula, because at this point there is no connection to modular transformations yet – this interpretation must be postponed until Sect. 2.4.

crossing symmetry relations for the four-point functions, which when choosing the insertion points as in formula (41) read

$$\mathcal{G}_{ABCD}(z,\bar{z}) = \mathcal{G}_{BCDA}(1-z,1-\bar{z}) = z^{-2\Delta_C}\bar{z}^{-2\Delta_C}\mathcal{G}_{ACBD}(z^{-1},\bar{z}^{-1}). \quad (50)$$

On the other hand, each of the pictures describes the chiral blocks in a different 'channel', and hence combining the two chiral halves amounts to three different decompositions of the four-point function as a sum over products of chiral blocks (for more details, see Sect. 2.7). In a quasi-rational theory the sums are finite, and as a consequence the different systems of chiral blocks are linearly related:

$$\hat{\mathcal{F}}_{ABCD,K}(z_1,z_2,z_3,z_4) = \sum_I \mathsf{F}_{KI}{\scriptstyle\begin{bmatrix}BC\\AD\end{bmatrix}} \mathcal{F}_{ABCD,I}(z_2,z_3,z_4,z_1),$$

$$\check{\mathcal{F}}_{ABCD,J}(z_1,z_2,z_3,z_4) = \sum_I \mathsf{B}_{JI}{\scriptstyle\begin{bmatrix}BC\\AD\end{bmatrix}} \mathcal{F}_{ABCD,I}(z_1,z_3,z_2,z_4). \qquad (51)$$

The coefficients $\mathsf{F}{\scriptstyle\begin{bmatrix}BC\\AD\end{bmatrix}}$ and $\mathsf{B}{\scriptstyle\begin{bmatrix}BC\\AD\end{bmatrix}}$ that appear in these linear relations are called the *fusing* and *braiding matrices*, and the corresponding manipulations with chiral blocks as visualized in the pictures above are referred to as fusing and braiding transformations, or generically as *duality* transformations. Strictly speaking, the indices I and J that I used to label the various blocks in a given channel generically do not just count the intermediate primary fields, but for the case of fusion rule coefficients larger than 1 also the various possible couplings among their families, e.g. $I \equiv (\alpha, I, \beta)$ with $\alpha \in \{1, 2, , ..., \mathsf{N}_{AB}{}^I\}$, $\beta \in \{1, 2, , ..., \mathsf{N}_{IC}{}^D\}$. For notational simplicity, in (51) and also below I suppress all multiplicity indices. [17]

The fusing and braiding matrices obey a number of compatibility relations, which can be derived by considering suitable duality transformations of higher p-point functions and demanding that the result does not depend on the individual set of transformations, but only on the initial and final configurations of (external and intermediate) fields. These relations can all be reduced to those obtained for five-point functions and which involve five and six different configurations, respectively ([Moore and Seiberg 1990]); the latter are called the *polynomial equations*, or more specifically, the *pentagon* and *hexagon* equations and read

$$\mathsf{F}_{HI}{\scriptstyle\begin{bmatrix}AB\\EF\end{bmatrix}} \mathsf{F}_{JF}{\scriptstyle\begin{bmatrix}CD\\EG\end{bmatrix}} = \sum_K \mathsf{F}_{KF}{\scriptstyle\begin{bmatrix}BD\\IG\end{bmatrix}} \mathsf{F}_{JI}{\scriptstyle\begin{bmatrix}AK\\EG\end{bmatrix}} \mathsf{F}_{HK}{\scriptstyle\begin{bmatrix}AB\\JD\end{bmatrix}},$$

$$\sum_K \mathsf{B}_{KI}{\scriptstyle\begin{bmatrix}AB\\DE\end{bmatrix}} \mathsf{B}_{GB}{\scriptstyle\begin{bmatrix}AC\\KF\end{bmatrix}} \mathsf{B}_{HK}{\scriptstyle\begin{bmatrix}BC\\DG\end{bmatrix}} = \sum_L \mathsf{B}_{LB}{\scriptstyle\begin{bmatrix}BC\\IF\end{bmatrix}} \mathsf{B}_{HI}{\scriptstyle\begin{bmatrix}AC\\DL\end{bmatrix}} \mathsf{B}_{GL}{\scriptstyle\begin{bmatrix}AB\\HF\end{bmatrix}} \qquad (52)$$

(note that the required summations do *not* amount to matrix multiplication of the F and B matrices). When also correlation functions on higher genus Riemann surfaces are considered, then there arises one additional independent relation, obtained from manipulating one-point blocks on the torus. In principle one may start from the fusion rules to set up the system (52) of polynomial equations and then solve this system for the F and B matrices, but already for a small number of sectors this becomes very hard (see ([Fuchs et al. 1995a]) for examples).

---

[17] Those readers who want to see them are referred to e.g. ([Fuchs et al. 1995a]).
  The relation with the notation used there is $\mathsf{F}_{\alpha I\beta,\gamma J\delta}{\scriptstyle\begin{bmatrix}AB\\CD\end{bmatrix}} = F^{(ABC)D}_{\alpha I\beta,\gamma J\delta}$ and
$\mathsf{B}_{\alpha I\beta,\gamma J\delta}{\scriptstyle\begin{bmatrix}AB\\CD\end{bmatrix}} = \sum_K \sum_{\lambda,\mu,\nu} ((F^{(BAC)D})^{-1})_{\alpha I\beta,\lambda K\nu} R^{(AB)K}_{\lambda,\mu} F^{(ABC)D}_{\mu K\nu,\gamma J\delta}.$

2.3   Counting States: Characters

The structure of an irreducible highest weight module $\mathcal{H}_A$ of $\mathcal{W}$ is rather complicated. But a lot of information is already contained in its *weight system*, i.e. in the collection of weights λ of the vectors in a basis of $\mathcal{H}_A$. To keep track of the weight system of a module, one sums over formal exponentials e^λ of its weights, counting their multiplicities, or what is equivalent (writing $e^\lambda = (v_\lambda \,|\, e^{W_\circ} v_\lambda)$ with v_λ the orthonormal basis vectors of $\mathcal{H}_A$), takes the trace over the module of formal exponentials of elements $W_\circ$ of $\mathcal{W}_\circ$. The so obtained quantity [18]

$$\chi_A(W_\circ) := \mathrm{tr}_{\mathcal{H}_A} e^{W_\circ} \tag{53}$$

is called the *character* of the irreducible $\mathcal{W}$-module $\mathcal{H}_A$. (Compare the characters of irreducible modules of Lie algebras, see Sect. 3.6).

It is often convenient to regard χ_A as a function of complex variables, which are the coefficients in an expansion of a generic element $W_\circ$ of $\mathcal{W}_\circ$ in some suitable basis, and for some purposes it is sufficient to specialize to characters at vanishing values of some of these variables. A very specific, but nevertheless most important, case is provided by the *Virasoro-specialized* character, for which $W_\circ$ is taken to have only a component in the direction of $L_0 - \frac{1}{24} C$:

$$\chi_A(\tau) := \mathrm{tr}_{\mathcal{H}_A} e^{2\pi i \tau (L_0 - C/24)} . \tag{54}$$

It is a function of one complex variable τ and converges in the upper half-plane. By construction, $\chi_A(\tau) \cdot \exp(-2\pi i \tau (\Delta_A - c/24))$ is a power series in $q = \exp(2\pi i \tau)$, with the coefficient of q^n the dimension of the subspace of grade n.

The $\mathcal{V}ir$-specialized character $\chi_A(\tau)$ can be interpreted as a chiral block for the '0-point correlation function' (or vacuum-to-vacuum amplitude) on a torus with modular parameter τ, which in the diagrammatic description used in (48) corresponds to a one-loop graph without any external lines. By identifying $L_0 - C/24$ as a Hamiltonian, $\chi_A(\tau)$ can also be regarded as a partition function,

$$\chi_A(\tau) = \mathrm{tr}_{\mathcal{H}_A} e^{-\beta H} , \tag{55}$$

in a thermal state of complex inverse temperature $\beta = -2\pi i \tau$.

Once the characters are known, the computation of the 0-point function on the torus for the full two-dimensional theory with symmetry algebra $\mathcal{W} \oplus \overline{\mathcal{W}}$

---

[18] That this construction makes sense has its origin in the fact that for any set of numbers a_n indexed by $\mathbb{Z}$, the formal Laurent series $f(q) := \sum_n a_n q^n$ in some indeterminate q contains the same information as the a_n. Here this recipe is generalized to the case where multiplicities are labelled by weights rather than by integers; accordingly one associates to each weight λ a formal variable e^λ on which one imposes the usual properties of exponentials, i.e. $e^\lambda e^\mu = e^{\lambda + \mu}$ and $e^0 = 1$. In more mathematical terms, the basic observation is that the weights form an abelian group L under vector addition. The formal exponentials constitute a basis for the group algebra $\mathbb{C}L$ of L. When considered as elements of $\mathbb{C}L$, one can add up these exponentials, and hence can consider χ_A as an element of this group algebra, respectively, in the case of infinite-dimensional modules, of some completion of the group algebra.

amounts to specifying non-negative integers Z_{AB} which tell how often the chiral block corresponding to the irreducible $\mathcal{W}$-representation labelled by A gets combined with the antichiral block for the irreducible $\overline{\mathcal{W}}$-representation labelled by B. The total *partition function* of the conformal field theory is then given by

$$Z \equiv \mathcal{Z}(\tau, \bar{\tau})|_{\tau^* = \tau} := \sum_{A,B} \chi_A(\tau) Z_{AB} \chi_B(\bar{\tau})|_{\tau^* = \tau}, \tag{56}$$

where we have to identify $\bar{\tau}$ with the complex conjugate of τ. To qualify as a partition function of a physical theory, the matrix Z clearly must fulfill a number of consistency requirements. By construction, its entries Z_{AB} are non-negative integers; moreover, the vacuum sector $\mathcal{H}_0$ must be unique, and hence one needs $Z_{00} = 1$. Further properties arise from modular invariance, to be discussed next.

2.4 Modularity

The space of characters of the unitary irreducible highest weight modules of the chiral algebra $\mathcal{W}$ of a rational conformal field theory carries a unitary representation of the group $\mathrm{SL}_2(\mathbb{Z})$. (The elements of $\mathrm{SL}_2(\mathbb{Z})$ are 2×2-matrices with integral entries and determinant one.) To show this crucial fact of life would go much beyond the scope of this review, and in fact I am not aware of any rigorous proof. A matrix $M \in \mathrm{SL}_2(\mathbb{Z})$ acts on the characters by a change of parameters; in the particular case of Virasoro-specialized characters, the action is given by

$$\tau \mapsto M\tau := \frac{a\tau + b}{c\tau + d} \quad \text{for } M = \begin{pmatrix} a & b \\ c & d \end{pmatrix} \tag{57}$$

$(a, b, c, d \in \mathbb{Z},\ ad - bc = 1)$, i.e. $M \cdot \chi(\tau) := \chi(M\tau)$. Note that here the elements M and $-M$ of $\mathrm{SL}_2(\mathbb{Z})$ act in the same way, so that the Virasoro-specialized characters actually carry a representation of the quotient group $\mathrm{PSL}_2(\mathbb{Z}) = \mathrm{SL}_2(\mathbb{Z})/\{\mathbb{1}, -\mathbb{1}\}$. This group is known as the *modular group* of the torus, and correspondingly the mapping (57) is called a *modular transformation*.

The parameter τ which (for convergence of $\chi(\tau)$) takes values in the upper complex half-plane can indeed be interpreted geometrically as parametrizing a torus, namely the one obtained by identifying the opposite edges of the quadrangle with corners $0, 1, \tau$ and $1 + \tau$ in the complex plane. The complex structure on this torus depends on τ only up to modular transformations.

The group $\mathrm{SL}_2(\mathbb{Z})$ is freely generated by two elements S and T modulo the relations $S^2 = (ST)^3$ and $S^4 = \mathbb{1}$. For $\mathrm{PSL}_2(\mathbb{Z})$ these get supplemented by the relation $S^2 = \mathbb{1}$; in the general case, $S^2 = C$ is the matrix for the conjugation $A \mapsto A^+$, i.e. $C_{AB} = \delta_{A,B^+}$. S and T are represented on the parameter τ as

$$T: \ \tau \mapsto \tau + 1 \quad \text{and} \quad S: \ \tau \mapsto -1/\tau. \tag{58}$$

(In terms of the temperature $\beta^{-1} = (-2\pi i \tau)^{-1}$, S exchanges the low and high temperature regimes – or, thinking of β as a coupling constant, the strong and weak coupling regions.) The corresponding matrices that act on the space of characters as $\chi_A(\tau + 1) = \sum_B T_{AB} \chi_B(\tau)$ and $\chi_A(-1/\tau) = \sum_B S_{AB} \chi_B(\tau)$

are, correspondingly, referred to as the (modular) S-matrix and T-matrix of the conformal field theory. From the definition (54) of Virasoro-specialized characters it follows immediately that the T-matrix is diagonal and unitary, with entries

$$T_{AA} = \exp(2\pi i \Delta_A - c/24) \,. \tag{59}$$

In contrast, the explicit form of the S-matrix, which turns out to be a symmetric and unitary matrix, is much more difficult to obtain; in fact it is by no means obvious that the modular inversion S closes on the space of characters. In particular, knowing only the Virasoro-specialized characters is typically not sufficient to determine S (e.g., conjugate fields ϕ_A and ϕ_{A^+} have identical Virasoro-specialized characters); rather one must use the full characters.

It is by no means accidental that in the definition (45) I used the same letter S for the matrix that diagonalizes the fusion rules as for the matrix that implements $\tau \mapsto -1/\tau$. Namely, for a modular invariant rational conformal field theory this diagonalizing matrix (when normalizations are chosen in such a way that it is symmetric) coincides with the modular inversion S as introduced here. With this identification, the identity (47) expressing the fusion rule coefficients through S is known as the *Verlinde formula* ([Verlinde 1988]). [19]

In order to construct a conformal field theory consistently on a Riemann surface of genus larger than zero, the partition function (56) must be invariant under the modular transformations of that surface, in particular under (57) in the case of the torus which has genus 1. I should point out that, as a consequence, modular invariance is a basic property of any conformal field theory model that is relevant to string theory, but that it is *not* a fundamental property of conformal field theory as such, and there are situations where it is absent. (For instance, in the case of the critical Ising model modular invariance restricts the number of sectors to three, which have the same highest weight with respect to the chiral and the antichiral Virasoro algebra, as listed in table (127) below. But for many purposes one must also consider other sectors, which correspond to the disorder parameter and to a chiral or antichiral free fermion.) Also, in general more requirements than just modular invariance are necessary for the existence and consistency of a conformal field theory. It is therefore not surprising that there exist modular invariant partition functions which cannot correspond to any conformal field theory (for some examples see ([Fuchs et al. 1995b])).

The requirement that (56) is invariant under the modular transformations (57) is equivalent to demanding that the matrix $Z = (Z_{AB})$ satisfies

$$[Z, S] = [Z, T] = 0 \,, \tag{60}$$

---

[19] The existing derivations of the Verlinde formula range from formal manipulations with chiral blocks (see for instance exercise 3.6 in ([Moore and Seiberg 1990])), localization of path integrals and surgery manipulations on three-manifolds to highly non-trivial arguments in algebraic geometry (for reviews and further references, I refer to ([Beauville 1994]) and to ([Schottenloher 1995])), as well as to the web page http://www.desy.de/~jfuchs/Vfcb.html). Typically the more rigorous the arguments are, the more restricted is the range of theories to which they apply; e.g. the algebraic geometry derivation applies only to the case of (most) WZW theories.

i.e. commutes with both the S- and the T-matrix. An immediate consequence of $[Z, T] = 0$ is that the 'conformal spin' $\Delta - \bar{\Delta}$ must be an integer, so that each field is 'local' with respect to all other fields. As a consequence, at any genus all correlation functions of the two-dimensional theory are single-valued functions.

There always exists a straightforward solution to the constraints (60), namely $Z = \mathbb{1}$. This is called the *diagonal* or *A-type* modular invariant. When $\mathcal{W}$ is maximal, then this is in fact ([Moore and Seiberg 1989]) the only solution, up to possibly a permutation which is an automorphism of the fusion rules, i.e. $Z_{AB} = \delta_{A,\pi B}$ with $N_{\pi A, \pi B}{}^{\pi C} = N_{AB}{}^{C}$. However, often one does not know the maximal chiral algebra, so that it makes sense to analyze the constraint (60) also in the case of non-maximal $\mathcal{W}$ (but with the theory still being rational). As it turns out, (60) constitutes a powerful restriction, and its implementation is a highly non-trivial task. So far the solutions have been classified only for very specific types of theories, e.g. for the WZW theories (see Sect. 4.1) based on $A_1^{(1)}$, the Virasoro minimal models ([Cappelli et al. 1987]), for WZW theories based on $A_2^{(1)}$ ([Gannon 1994]), and for $N = 2$ superconformal minimal models ([Gannon 1996]).

2.5   Free Bosons

The simplest examples of conformal field theories are those which describe massless free bosons or fermions. Here I will sketch the bosonic case. The classical action for a massless free boson X living on a D-dimensional space-time manifold of metric $g_{\mu\nu}$ reads $S_X \propto \int d^D x \sqrt{\det g}\, g^{\mu\nu}\, \partial_\mu X\, \partial_\nu X$. Variation of S_X with respect to $g_{\mu\nu}$ yields the energy-momentum tensor

$$T_{\mu\nu} = -\partial_\mu X \partial_\nu X + \tfrac{1}{2} g_{\mu\nu} \sum_{\sigma,\tau} g^{\sigma\tau}\, \partial_\sigma X\, \partial_\tau X\,, \tag{61}$$

which is conserved ($\sum_\mu \partial^\mu T_{\mu\nu} = 0$) and has trace $\sum_\mu T_\mu^\mu \propto 1 - D/2$. Thus in $D = 2$ dimensions $T_{\mu\nu}$ is traceless; in complex coordinates, this means $T_{z\bar{z}} = 0$, so that conservation reduces to $\partial_{\bar{z}} T_{zz} = 0 = \partial_z T_{\bar{z}\bar{z}}$, or in other words, $T \equiv T_{zz} = T(z)$ and $\bar{T} \equiv T_{\bar{z}\bar{z}} = \bar{T}(\bar{z})$, as is needed for a conformal theory. As T has conformal weight $\Delta = 2$, at the classical level the boson X has scaling dimension zero. Correspondingly, X is *not* a proper conformal field in the sense of Sect. 1.5; it can be written as $X(z,\bar{z}) = X(z) + \bar{X}(\bar{z})$ where $X(z)$ has an expansion

$$X(z) = X_0 - \mathrm{i}\, P \ln(z) + \mathrm{i} \sum_{n \in \mathbb{Z} \setminus \{0\}} \tfrac{1}{n} J_n\, z^{-n}\,. \tag{62}$$

Thus X is not single-valued on the complex plane. But the derivatives ∂X and $\bar{\partial} \overline{X}$ (which appear in the action S_X) are, and so are (suitably normal ordered) exponentials $:\!e^{\mathrm{i}qX}\!:$. The canonical commutation relations for the free boson X yield Heisenberg relations for its modes: $[X_0, J_n] = 0$ for $n \neq 0$ and

$$[X_0, P] = \mathrm{i}\,, \qquad [J_m, J_n] = m\, \delta_{m+n,0}\,. \tag{63}$$

The chiral algebra $\mathcal{W}$ of this theory is the semi-direct sum of the Virasoro algebra generated by $T(z)$ and a *Heisenberg algebra* generated by the Virasoro-primary field $J = \sum_{n \in \mathbb{Z}} J_n z^{-n-1}$ ($J_0 \equiv P$). In terms of X these fields read

$$J(z) = i\,\partial X(z), \qquad T(z) = -\tfrac{1}{2}\,{:}\partial X(z)\,\partial X(z){:}\; = \tfrac{1}{2}\,{:}J^2(z){:}\;. \qquad (64)$$

Using the two-point function $\langle X(z)\,X(w)\rangle \simeq -\ln(z-w)$ and Wick's rule for calculating correlators of normal ordered products of free fields, one checks that the conformal central charge has the value $c = 1$ and that J has conformal weight $\Delta_J = 1$. The subalgebra $\mathcal{W}_0$ is spanned by L_0 and the zero mode $J_0 \equiv P$ of the field J. The primary fields of this theory are $\phi_q = {:}e^{iqX}{:}$; they are labelled by charges, i.e. J_0-eigenvalues, $q \in \mathbb{R}$, and their fusion rules read $\Phi_p \star \Phi_q = \Phi_{p+q}$.

To be precise, these statements refer to the situation that the modes of the classical boson X, in particular the zero mode X_0, are allowed to take values on the whole real line $\mathbb{R}$. In contrast, when the boson is *compactified* on a circle of radius R, in the sense that one identifies the values X_0 and $X_0 + 2\pi R$ of the zero mode, and when in addition $R^2 = 2\mathcal{N}$ is an even integer, then there are two additional (Virasoro- and J-primary) fields in $\mathcal{W}$ which have conformal dimension $\Delta = \mathcal{N}$. Moreover, in this case the theory is rational, and with a suitable normalization of the current J the possible charge values are $q \in \{0, 1, \ldots, 2\mathcal{N}-1\}$. The conformal dimension of a primary field of charge q is then $\Delta_q = q^2/4\mathcal{N}$. The fusion rules read $\Phi_p \star \Phi_q = \Phi_{p+q \bmod 2\mathcal{N}}$, and the modular S-matrix has entries

$$S_{pq} = \exp(-\pi i\,pq/\mathcal{N})/\sqrt{2\mathcal{N}}\,. \qquad (65)$$

Similarly, when $R^2 = 2r/s$ (r, s coprime) is any other rational number, the theory is still rational and $\mathcal{W}$ contains additional fields of conformal weight $\Delta = \mathcal{N} := rs$ in $\mathcal{W}$. But in the general case $\mathcal{W}$ contains fields of still higher weight, and one deals with a non-diagonal extension of the theory that has squared radius $2\mathcal{N}$. However, the theories with radius R and radius $2/R$, which look most different when formulated in terms of a classical action, are actually one and the same conformal field theory; in particular for $R^2 = 2/\mathcal{N}$ one recovers a diagonal theory.

All the theories just described have $c = 1$. In fact, any other unitary rational $c = 1$ conformal field theory can be formulated in terms of a free boson as well. Namely, each of those theories can be obtained as a so-called *orbifold* of the theory of a free boson on a circle. In general, forming an orbifold of a conformal field theory means that one restricts the observables to those invariant under some (discrete) group Z of automorphisms. In the case at hand, Z must be a discrete subgroup of O(3), which leads to three isolated rational theories for which Z is non-abelian, and to a continuous family of theories for which $Z = \mathbb{Z}_2$, with the non-trivial element corresponding to the identification of X with $-X$. For rational square radius, the latter $\mathbb{Z}_2$-orbifolds are again rational theories, but for non-rational values of R^2 they are not even quasi-rational.

2.6 Simple Currents

Each fusion ring contains a distinguished basis element, the unit element Φ_0. This element clearly has the property to be invertible within the ring, i.e. in mathematical terms, it is a *unit* of the ring. But a fusion ring may contain also further basis elements which are units in this sense. These are called *simple currents* of the ring, or of the conformal field theory, and they turn out to be of considerable importance. In the theory of a free boson at rational square radius that was just described, in fact each basis element Φ_p is a simple current, with inverse $(\Phi_p)^{-1} = \Phi_{2N-p}$. Another important example for a simple current is the field which implements the GSO projection in superstring theory. In this subsection I sketch the most intriguing properties of simple currents; for more details see e.g. ([Schellekens and Yankielowicz 1990]) and ([Fuchs 1994]).

A simple current J can be equivalently characterized as a primary field for which the fusion product with the conjugate field just yields the unit element,

$$ J \star J^+ = \Phi_0 , \qquad (66) $$

or for which the fusion rules are *simple* in the sense that for each A the fusion product $J \star \Phi_A$ belongs again to the distinguished basis (it is then simply written as $\Phi_{JA} \equiv \Phi_{J\star A}$). In terms of the fusion rule coefficients this means that $\sum_C N_{JB}{}^C = 1$ for all B. [20] Due to the associativity of the fusion product, the product of two simple currents is again a simple current. Simple currents thus form an abelian group under the fusion product, i.e. a sub*group* of the fusion ring (this group has been termed the *center* of the theory). A rational theory can of course accommodate only finitely many simple currents. This implies that simple currents are unipotent; the smallest positive integer $N \equiv N_J$ such that $J^N = \Phi_0$ is called the *order* of J (here $J^2 \equiv J \star J$, etc.). Furthermore, any simple current organizes the primary fields into orbits $[\Phi_A] := \{\Phi_A, \Phi_{JA}, \Phi_{J^2A}, \dots, \Phi_{J^{N-1}A}\}$; the size $N_A := \|[\Phi_A]\|$ of any orbit $[\Phi_A]$ is a divisor of the order $N = \|[\Phi_0]\|$ of J.

A crucial result about simple currents is that in a unitary theory, S-matrix elements involving fields on the same simple current orbit differ only by a phase:

$$ S_{J^pA, J^qB} = e^{2\pi i p Q_J(B)} e^{2\pi i q Q_J(A)} e^{2\pi i p q Q_J(J)} \cdot S_{AB} . \qquad (67) $$

Here $Q_J(A)$ is the *monodromy charge* of ϕ_A with respect to J, defined as the combination

$$ Q_J(A) := \Delta_J + \Delta_A - \Delta_{JA} \mod \mathbb{Z} \qquad (68) $$

---

[20] The term 'current' is chosen ([Schellekens and Yankielowicz 1989]) in anticipation of the fact that these fields can be used for an extension of the chiral algebra, and (also when the *conformal* dimension is not equal to 1) such a field is often called a current. The qualification 'simple' refers to the behavior with respect to fusion rules (and to correlation functions as well, see ([Fuchs and Gepner 1987])), but also fits with the fact that in algebraic quantum field theory a superselection sector with *statistical* dimension equal to 1 is called a simple sector.

of conformal weights (in a rational theory, $Q_J(A)$ is a rational number). The monodromy charge is additive under the operator product, i.e. for any C with $N_{AB}^C \neq 0$ one has $Q_J(A) = Q_J(B) + Q_J(C)$ mod $\mathbb{Z}$. Based on (67), one can show that for any simple current which satisfies $N_J \Delta_J \in \mathbb{Z}$, the matrix Z with entries

$$Z_{AB} = \frac{N}{N_A} \sum_{n=0}^{N_A-1} \delta_{B,J^nA}\, \delta^{(\mathbb{Z})}(Q_J(A) + \tfrac{n}{2} Q_J(J)) \tag{69}$$

provides a modular invariant. Here $\delta^{(\mathbb{Z})}$ denotes the function with $\delta^{(\mathbb{Z})}(x) = 1$ for $x \in \mathbb{Z}$ and $\delta^{(\mathbb{Z})}(x) = 0$ otherwise. Note that all non-zero entries of Z are between fields on the same simple current orbit. If the center is the cyclic group $\{\Phi_0, J, J^2, \ldots, J^{N-1}\} \cong \mathbb{Z}_N$ generated by a simple current J, then (69) is the only modular invariant of the theory with this property. When the center is not cyclic, then there are further possibilities for simple current extensions which are parametrized by the so-called discrete torsion ([Kreuzer and Schellekens 1994]).

When Δ_J is integral, the presence of $\delta^{(\mathbb{Z})}$ in (69) means that only fields with vanishing monodromy charge occur, and the modular invariant reduces to

$$\mathcal{Z} = \sum_{A:\,Q_J(A)=0} \frac{N}{N_A} \left| \sum_{n=0}^{N_A-1} \chi_{J^nA} \right|^2 . \tag{70}$$

Invariants of this type are called *integer spin simple current extensions*. They are interpreted as diagonal invariants of a theory with an extended chiral algebra $\mathcal{W}$, with the additional fields in $\mathcal{W}$ being the simple currents J^n. Note that typically several irreducible modules of the original chiral algebra $\mathcal{W}_{\text{unext.}}$ combine to a single irreducible module of $\mathcal{W}$; according to (70), the characters of the irreducible $\mathcal{W}$-modules are proportional to the orbit sum $\sum_n \chi_{J^nA}$. Moreover, not every irreducible $\mathcal{W}_{\text{unext.}}$-module will be contained in a module of $\mathcal{W}$, and this is precisely encoded in the requirement of vanishing monodromy charge.

The orbit sums of characters of the original 'unextended' theory appear in (70) with multiplicities N/N_A. Orbits A with $N/N_A > 1$ are called *fixed points* of the simple current J. To interpret these multiplicities, one must recall (see Sect. 2.4) that for maximally extended $\mathcal{W}$ each (equivalence class of) irreducible $\mathcal{W}$-module appears precisely once. Thus a multiplicity in front of the complete square indicates that there exist several non-isomorphic $\mathcal{W}$-modules which reduce to one and the same module of $\mathcal{W}_{\text{unext.}}$, or in other words, that such a term corresponds to several distinct primary fields in the conformal field theory with the enlarged chiral algebra. However, this prescription is unambiguous only for $N/N_A = 2$ or 3. For larger values it can also happen that one must include (part of) the prefactor N/N_A into the extended character; as this can only be done for complete squares, the number of distinct possible interpretations is then equal to the number of ways that N/N_A can be written as a sum of squares. For example, when $N/N_A = 5$ one either deals with five inequivalent $\mathcal{W}$-modules which have ($\mathcal{W}_{\text{unext.}}$-specialized) character $\sum_{n=0}^{N_A-1} \chi_{J^nA}$, or else with one such module plus another $\mathcal{W}$-module whose character equals twice that sum.

Given an invariant of the type (70), it is in general a highly non-trivial task to investigate whether a fully consistent conformal field theory can be associated to that modular invariant. In particular one would like to compute the relevant $SL_2(\mathbb{Z})$-representation. While the T-matrix for the extended theory is just the restriction of the original T-matrix to allowed orbits, for the S-matrix one only gets some constraints involving the S-matrix of the original theory, and determining those elements which involve two fixed points proves to be an intricate problem. However, recently a general class of solutions to this *fixed point resolution* problem was constructed ([Fuchs et al. 1996c, Bantay 1996]); this construction yields in particular a general closed expression for the S-matrix.

2.7   Operator Product Algebra from Fusion Rules

The results of Sect. 1.7 show that the operator product coefficients of primary fields are the normalizations of the corresponding three-point functions, but these normalizations cannot be determined from the Ward identities. With the help of four-point functions, on the other hand, one *can* compute the operator product coefficients. Indeed, for maximal $\mathcal{W}$ the coefficients a_{II} that appear in the decomposition (4.9) of four-point correlators into three-point functions are of the form $a_{II} = a_I \delta_{I,\pi I}$ for some permutation π. Moreover, provided that the chiral blocks are correctly normalized, the a_I are just the products $C_{AB}{}^I C_{ICD}$ of the relevant operator product coefficients. As a consequence, one can regard a conformal field theory as completely solved once all four-point functions of its primaries are known. These correlators, in turn, can be deduced from the fusion rules of the theory by using general analytic properties of the correlators to construct linear differential equations which they must satisfy.[21] The independent solutions of these differential equations are the chiral blocks, which transform into each other under duality transformations (see (51)). Having obtained explicit formulæ for the blocks, one can therefore determine the duality matrices F and B and express the coefficients a_I, and hence the operator product coefficients, through the entries of these matrices ([Blok and Yankielowicz 1989, Fuchs 1989]).

To achieve this result in practice, one just exploits the basic properties of (rational or quasi-rational, genus zero) conformal field theories. First, one can invoke projective invariance to work with four-point correlators $\mathcal{G}(z,\bar{z})$ of the special form (41). Next, the closure of the operator product algebra implies that $\mathcal{G}$ can be written as a sum of products of three-point functions (compare the picture (48)), and the contributions from each 'intermediate' family $[\phi_I]$ combine to a block of definite analytic behavior. Furthermore, by the fact that the symmetry algebra is the direct sum $\mathcal{W} \oplus \overline{\mathcal{W}}$ of two chiral halves, the z- and $\bar{z}$-dependence factorize, which amounts to the decomposition (4.9). The number

---

[21] Often these differential equations also follow from the presence of null vectors in the Verma modules of $\mathcal{W}$. An example is provided by the Knizhnik–Zamolodchikov equations which will be discussed in Sect. 4.4.

M ($= \bar{M}$ for maximally extended $\mathcal{W}$) of blocks in (4.9) is determined as

$$M = \sum_I \mathrm{N}_{AB}{}^I \mathrm{N}_{ICD}, \qquad (71)$$

with I labelling the intermediate families $[\phi_I]$, through the coefficients $\mathrm{N}_{AB}{}^C$.

To determine the chiral blocks, one observes that according to the operator product expansion (37), $\phi_i(z,\bar{z})\phi_j(w,\bar{w})$ depends on z and w as $(z-w)^\alpha$, with α determined by the conformal dimensions of the fields involved. More precisely,

$$\mathcal{F}_I(z) \propto z^{-\Delta_C - \Delta_D + \tilde{\Delta}_I^{(CD)}} \quad \text{for } z \to 0. \qquad (72)$$

Here Δ_C is the conformal dimension of the primary field ϕ_C, while $\tilde{\Delta}_I^{(CD)} = \Delta_I^{(CD)} + \mu_I^{(CD)}$ is the conformal dimension of that field $\varphi_I \in [\phi_I]$, of grade $\mu_I^{(CD)}$, in the family of the primary $\phi_I \equiv \phi_I^{(CD)}$ which gives the leading contribution to the coupling between ϕ_C, ϕ_D and $[\phi_I]$. The result (4.10) corresponds to the decomposition of $\mathcal{G}$ into three-point blocks as indicated in figure (48). Analogously, decomposing as in picture (49) yields chiral blocks $\hat{\mathcal{F}}_K$ and $\check{\mathcal{F}}_J$ which behave as

$$\hat{\mathcal{F}}_K \propto (1-z)^{-\Delta_C - \Delta_B + \tilde{\Delta}_K^{(CB)}} \ \text{for } z \to 1, \quad \check{\mathcal{F}}_J \propto (z^{-1})^{\Delta_C - \Delta_A + \tilde{\Delta}_J^{(CA)}} \ \text{for } z \to \infty \quad (73)$$

with $\tilde{\Delta}_J^{(CB)}$ and $\tilde{\Delta}_K^{(CA)}$ defined analogously as $\tilde{\Delta}_I^{(CD)}$. As already described in Sect. 2.2, the systems $\{\mathcal{F}_I\}$, $\{\hat{\mathcal{F}}_K\}$ and $\{\check{\mathcal{F}}_J\}$ of chiral blocks transform into each other upon analytic continuation. On the other hand, since they correspond to distinct couplings they must be algebraically independent. According to elementary results from the theory of ordinary linear differential equations (see e.g. ([Yoshida 1987])) it then follows that the chiral blocks $\mathcal{F}_I$ constitute the M independent solutions of an Mth order differential equation.

The construction of the differential equation is essentially equivalent to solving a *Riemann monodromy problem*, i.e. determine a collection of functions with prescribed monodromy. The Riemann monodromy problem always possesses a solution, provided that one allows for the possible presence of apparent singularities, around which the monodromy is trivial, in the differential equation.

3 Kac–Moody Algebras

3.1 Cartan Matrices

Kac–Moody algebras constitute a certain class of (generically infinite-dimensional) Lie algebras. Roughly, they are those Lie algebras which possess both a Cartan matrix (implying the existence of a triangular decomposition) and a Killing form. They include all finite-dimensional simple Lie algebras, and I will start by summarizing some pertinent features of those well-known algebras.

A finite-dimensional simple Lie algebra $\mathfrak{g} = \mathfrak{g}(A)$ is characterized by its *Cartan matrix* A; the basic Lie bracket relations of $\mathfrak{g}$ can be expressed through A as

$$
\boxed{A} \quad
\begin{aligned}
&[H^\alpha, H^\beta] = 0 \\
&[H^\alpha, E^\beta] = (\alpha^\vee, \beta)\, E^\beta \\
&[E^\alpha, E^{-\alpha}] = H^\alpha \\
&[E^\alpha, E^\beta] = e_{\alpha,\beta}\, E^{\alpha+\beta} \ \text{ for } \alpha \neq -\beta
\end{aligned}
\qquad
\boxed{B} \quad
\begin{aligned}
&[H^i, H^j] = 0 \\
&[H^i, E^j_\pm] = \pm A^{ji} E^j_\pm \\
&[E^i_+, E^j_-] = \delta_{ij}\, H^j \\
&(\mathrm{ad}_{E^i_\pm})^{1-A^{ji}} E^j_\pm = 0
\end{aligned}
\qquad (74)
$$

in a so-called Cartan–Weyl basis and Chevalley–Serre basis, respectively. Let me explain some of the notation that is used here. When $\mathfrak{g}$ has rank r, A is an $r \times r$-matrix, and $\mathfrak{g}$ has a basis $\{E^\alpha\} \cup \{H^i \,|\, i = 1, 2, ..., r\}$ obeying the relations (74 A) (in addition to antisymmetry and the Jacobi identity, the two defining properties of a Lie bracket). The r generators H^i span an abelian Lie algebra, called the *Cartan subalgebra* $\mathfrak{g}_0$ of $\mathfrak{g}$; the maps $\mathrm{ad}_{H^i} \colon x \mapsto [H^i, x]$ are diagonalizable, and to each *root*, i.e. vector $\alpha = (\alpha^i)$ of eigenvalues of these maps, there corresponds a generator E^α of $\mathfrak{g}$, and one writes $H^\alpha := (\alpha^\vee, H)$ with $\alpha^\vee := 2\alpha/(\alpha, \alpha)$. Among the roots there is a special subset $\{\alpha^{(i)} \,|\, i = 1, 2, ..., r\}$, called the *simple roots*, through which all roots can be expressed as integral linear combinations with either only positive or only negative coefficients; the respective roots are called positive and negative roots. Writing $E^{\pm\alpha^{(i)}} =: E^i_\pm$, $\mathfrak{g} = \mathfrak{g}(A)$ is generated algebraically [22] by $3r$ generators $\{E^i_\pm, H^i \,|\, i = 1, 2, ..., r\}$ (called the Chevalley generators of $\mathfrak{g}$) modulo the relations (74 B). The relations in the last line of (74 B) are called the Serre relations (the symbol $(\mathrm{ad}_x)^n$ is a shorthand for $\mathrm{ad}_x \circ \mathrm{ad}_x \circ \cdots \circ \mathrm{ad}_x$ (n factors), e.g. $(\mathrm{ad}_x)^3(y) \equiv [x, [x, [x, y]]]$), and the full set (74 B) is known as the Chevalley–Serre relations. Cartan matrices are conveniently encoded into *Dynkin diagrams*: to each possible value of the label i one associates a node or vertex, and for $i \neq j$ the nodes i and j are connected by $A^{ij} A^{ji}$ lines.

Of course, not any arbitrary square matrix qualifies as a Cartan matrix. In order that $\mathfrak{g}(A)$ is finite-dimensional and simple, the matrix A must satisfy

$$
A^{ij} \in \mathbb{Z}, \qquad A^{ii} = 2, \qquad A^{ij} \leq 0 \text{ for } i \neq j, \qquad A^{ij} = 0 \Leftrightarrow A^{ji} = 0, \qquad (75)
$$

as well as

$$
\det A > 0, \qquad (76)
$$

and must be indecomposable (i.e. must not be rearrangeable to block-matrix form by any simultaneous permutation of its rows and columns; decomposable matrices obeying (75) and (76) describe direct sums of simple Lie algebras).

---

[22] By '(algebraically) generated' – as opposed to '(linearly) spanned' – one means that the elements of the Lie algebra $\mathfrak{g}$ are obtained as arbitrary linear combinations of arbitrary (multiple) Lie brackets of the basic generators. Such a characterization of the Lie algebra $\mathfrak{g}$, not to be confused with the description of $\mathfrak{g}$ as the linear span of a basis, is called a *presentation* of $\mathfrak{g}$ by generators modulo relations.

The analysis of these conditions leads to the following classification of all finite-dimensional simple Lie algebras:

$$A_r \, (r \geq 1), \quad B_r \, (r \geq 2), \quad C_r \, (r \geq 3), \quad D_r \, (r \geq 4), \quad E_6, \quad E_7, \quad E_8, \quad F_4, \quad G_2. \qquad (77)$$

Thus there are four infinite series, the *classical* Lie algebras $A_r, \ldots, D_r$, as well as the five *exceptional* simple Lie algebras $E_6, \ldots, G_2$. In particular there is a unique simple Lie algebra of rank one, namely $A_1 \cong \mathfrak{sl}(2)$, with Cartan matrix $A(A_1) = 2$, and there are three simple Lie algebras of rank two, namely $A_2 \cong \mathfrak{sl}(3)$, $B_2 \equiv C_2 \cong \mathfrak{so}(5)$ and G_2, with Cartan matrices

$$A(A_2) = \begin{pmatrix} 2 & -1 \\ -1 & 2 \end{pmatrix}, \quad A(B_2) = \begin{pmatrix} 2 & -2 \\ -1 & 2 \end{pmatrix}, \quad A(G_2) = \begin{pmatrix} 2 & -3 \\ -1 & 2 \end{pmatrix}. \qquad (78)$$

By construction, the space spanned by the g-roots α is the space $\mathfrak{g}_\circ^\star$ dual to the Cartan subalgebra $\mathfrak{g}_\circ$. $\mathfrak{g}_\circ^\star$, called the *weight space* of $\mathfrak{g}$, is an r-dimensional vector space with euclidean scalar product $(\cdot, \cdot)$; the simple roots, and likewise the simple *coroots* $\alpha^{(i)\vee} := 2\alpha^{(i)}/(\alpha^{(i)}, \alpha^{(i)})$, form a basis of $\mathfrak{g}_\circ^\star$. The entries of the Cartan matrix encode the non-orthonormality of these bases; more precisely,

$$A^{ij} = (\alpha^{(i)\vee}, \alpha^{(j)}) \equiv 2 \, (\alpha^{(i)}, \alpha^{(j)})/(\alpha^{(j)}, \alpha^{(j)}). \qquad (79)$$

Further properties of the root system, i.e. the collection of all roots of $\mathfrak{g}$, are the following. The only multiple of a root α that is again a root is $-\alpha$. Also, for any pair α, β of roots, the linear combination $\alpha + m\beta$ is a root for a set of consecutive integers m, but not for any other value of m; for pairs of simple roots, this *root string* can be directly read off the Serre relations. For the lengths of roots at most two different values are possible; the relative length squared of a longer and a shorter root is given by $\max\{|A^{ij}| \mid i \neq j\}$. In particular, when $A^{ij} \in \{0, -1\}$ for $i \neq j$ (in which case $\mathfrak{g}$ is said to be *simply laced*), then all roots have the same length. Also, when different root lengths do occur, then this already happens among the simple roots, and each of the two sets of simple roots of equal length corresponds to a connected sub-diagram of the Dynkin diagram. (Also, in order that the Dynkin diagram specifies the Lie algebra $\mathfrak{g}$ uniquely, one supplements the multiple link which connects these two sub-diagrams by an arrow pointing from the longer to the shorter simple roots.)

The *Killing form* of $\mathfrak{g}$ associates to any two elements x and y of $\mathfrak{g}$ the number

$$\kappa(x, y) \equiv (x|y) := \mathrm{tr}(\mathrm{ad}_x \circ \mathrm{ad}_y). \qquad (80)$$

It is bilinear, symmetric, and satisfies $\kappa(x, [y, z]) = \kappa([x, y], z)$. Conversely, any map with these properties equals the Killing form up to a multiplicative constant. When restricted to the Cartan subalgebra $\mathfrak{g}_\circ$, the Killing form induces by duality an inner product on the weight space $\mathfrak{g}_\circ^\star$, which, when suitably normalized, coincides with the euclidean product on $\mathfrak{g}_\circ^\star$ that was already used above.

Another important basis of the weight space $\mathfrak{g}_\circ^\star$ consists of the *fundamental weights* $\Lambda_{(i)}$, which are defined by the requirement that

$$(\Lambda_{(i)}, \alpha^{(j)\vee}) = \delta_i^{\ j} \qquad (81)$$

for $i, j = 1, 2, \ldots, r$. This basis plays a special rôle in the representation theory of $\mathfrak{g}$, or more precisely, for its highest weight modules (the concept of highest weight modules was already discussed in the context of chiral algebras in Sect. 1.3, and will be dealt with in more detail in Sect. 3.5 below). Namely, among the irreducible highest weight modules $\mathcal{H}_\Lambda$ of $\mathfrak{g}$, precisely those are finite-dimensional whose highest weight Λ is *dominant integral*, i.e. is a non-negative integral linear combination of the fundamental weights $\Lambda_{(i)}$. Moreover, each such module can be obtained by forming tensor products of just a few special highest weight modules, for all of which the highest weight is a fundamental weight.

The *Weyl group* W of the Lie algebra $\mathfrak{g}$ is a finite group which is generated by the reflections r_i, $i \in \{1, 2, \ldots, r\}$, in the weight space $\mathfrak{g}_0^\star$ of $\mathfrak{g}$ about the hyperplanes through the origin which are perpendicular to the simple roots $\alpha^{(i)}$. On the fundamental weights $\Lambda_{(j)}$ these reflections act as

$$r_i: \quad \Lambda_{(j)} \mapsto \Lambda_{(j)} - \delta_{i,j} \sum_k A^{ik} \Lambda_{(k)}. \tag{82}$$

Each element $w \in W$ can be written (non-uniquely) as a product of the reflections r_i. The minimal number of r_i that is needed to form such a *Weyl word* for w is called the *length* $\ell(w)$ of w, and $(-1)^{\ell(w)} =: \mathrm{sign}(w)$ is called the *sign* of w.

3.2 Symmetrizable Kac–Moody Algebras

A considerable portion of the description above directly generalizes to a much larger class of Lie algebras. Recall that the finite-dimensional simple Lie algebras are obtained when imposing indecomposability as well as the properties (75) and (76) of A; in particular the rank of A is equal to r. One arrives at more general Lie algebras $\mathfrak{g}$ when one stipulates again that $\mathfrak{g} = \mathfrak{g}(A)$ is algebraically generated by Chevalley generators $E_\pm^i$, H^i modulo the relations (74 B), with the matrix A still satisfying (75), but relaxes the condition (76) on the determinant of A. The algebras obtained this way are known as *Kac–Moody* Lie algebras, and A as a *generalized* Cartan matrix. All Kac–Moody algebras, except for the simple ones described above, are infinite-dimensional (for affine Kac–Moody algebras, this follows from their realization that I will present in the next subsection).

Remarkably, even removing the condition (76) altogether does not destroy much of the power and beauty of these Lie algebras and their representation theory. But to keep some specific nice properties, it is necessary to restrict oneself to a (still comprehensive) subclass, the so-called *symmetrizable* Kac–Moody algebras. These are characterized by the existence of a non-degenerate diagonal matrix D such that the matrix DA is symmetric. A symmetrizable Kac–Moody algebra is simple (that is, it does not possess any non-trivial ideal, i.e. subalgebra $\mathfrak{h}$ such that $[\mathfrak{h}, \mathfrak{g}] \subseteq \mathfrak{h}$, and has dimension larger than 1), if an only if $\det A \neq 0$.

Every symmetrizable Kac–Moody algebra possesses a bilinear symmetric invariant form $(\cdot | \cdot)$ – in the finite-dimensional case this is the Killing form (80). However, when $\mathfrak{g}$ is strictly defined as above, this form is degenerate as soon as the symmetrized Cartan matrix has a zero eigenvalue. To make the form non-degenerate, one needs to enlarge the Cartan subalgebra $\mathfrak{g}_0$ of $\mathfrak{g}$ by so-called outer *derivations* D. More precisely, to every zero eigenvalue of A there is associated an independent central element $K^a \in \mathfrak{g}$ which is a suitable linear combination of the generators H^i. For each K^a one must introduce an independent derivation D^a; this can be done in such a way that $(K^a|D^b) = \delta^{ab}$ and $(K^a|K^b) = 0 = (D^a|D^b)$, and then $(\cdot|\cdot)$ is non-degenerate. It is usually this enlarged Lie algebra that is referred to as the Kac–Moody algebra $\mathfrak{g} = \mathfrak{g}(A)$ associated to A.

When (76) is not removed completely, but only relaxed to the requirement

$$\det A_{\{i\}} > 0 \quad \text{for all } i = 0, 1, ..., r \tag{83}$$

for the matrices $A_{\{i\}}$ that are obtained from A by deleting the ith row and ith column (and when A is indecomposable), then A is called an *affine* Cartan matrix, and $\mathfrak{g} = \mathfrak{g}(A)$ an *affine Lie algebra*. In (83) I changed the labelling convention for the entries of A: they now take values in $\{0, 1, ..., r\}$, so that A is an $(r+1) \times (r+1)$-matrix; note that the validity of (83) implies that the rank of A is at least r. One can also characterize an affine Cartan matrix by the requirement that the symmetrized Cartan matrix DA is positive semidefinite, but not positive definite. (If DA is positive definite, then $\mathfrak{g}(A)$ is a finite-dimensional simple Lie algebra.) The center of an affine Lie algebra is one-dimensional, and correspondingly one includes into the definition of $\mathfrak{g}$ an outer derivation D.

Once the classification of simple Lie algebras up to some rank is known, classifying the affine Lie algebras with that rank is straightforward. For instance, just like for $r = 1$ requiring $\det A > 0$ immediately leads to the three simple rank-two Cartan matrices (78), the requirement that A has one positive and one zero eigenvalue yields two rank-one affine algebras, denoted by $A_1^{(1)}$ and $A_1^{(2)}$:

$$A(A_1^{(1)}) = \begin{pmatrix} 2 & -2 \\ -2 & 2 \end{pmatrix}, \qquad A(A_1^{(2)}) = \begin{pmatrix} 2 & -4 \\ -1 & 2 \end{pmatrix}. \tag{84}$$

For $r > 1$, the crucial observation is that by deleting the ith row and ith column ($i \in \{0, 1, ..., r\}$ arbitrary) from an affine Cartan matrix, one must produce the Cartan matrix of a direct sum of finite-dimensional simple Lie algebras. This leads to the following classification. There are seven infinite series of affine Lie algebras, and in addition nine isolated cases; they are denoted by

$$A_r^{(1)} \, (r \geq 2), \; B_r^{(1)} \, (r \geq 3), \; C_r^{(1)} \, (r \geq 2), \; D_r^{(1)} \, (r \geq 4), \; B_r^{(2)} \, (r \geq 3), \; C_r^{(2)} \, (r \geq 2), \; \tilde{B}_r^{(2)} (r \geq 2),$$
$$A_1^{(1)}, \; E_6^{(1)}, \; E_7^{(1)}, \; E_8^{(1)}, \; F_4^{(1)}, \; G_2^{(1)}, \; A_1^{(2)}, \; F_4^{(2)}, \; G_2^{(3)}. \tag{85}$$

The affine algebras are thus all denoted by symbols $X_r^{(\ell)}$, with X_r the symbol for one of the simple algebras (77) and $\ell \in \{1, 2, 3\}$.[23] The algebras with $\ell = 1$ are called *untwisted* affine algebras, while those with $\ell = 2, 3$ are the *twisted* ones.

---

[23] For $\ell = 2, 3$ several different conventions are in use. The notation adopted here is taken from ([Fuchs 1992b]); a rather different one is used in ([Kac 1990]).

3.3   Affine Lie Algebras as Centrally Extended Loop Algebras

The description of a Kac–Moody algebra $\mathfrak{g}$ in terms of generators $\{E_\pm^i,\, H^i\}$ and relations (74 B) encodes the structure of $\mathfrak{g}$ in a very compact form, but it is in fact not too transparent. (For instance it is difficult to deduce the root system, and it even requires some effort to decide whether $\mathfrak{g}$ is finite-dimensional.) But *affine* Kac–Moody algebras possess a specific realization which for many purposes is much more convenient and which arises naturally in conformal field theory.

Consider the space $\bar{\mathfrak{g}}_{\mathrm{loop}}$ of analytic mappings from the circle S^1 to some Lie algebra $\bar{\mathfrak{g}}$. (For future convenience I denote the algebra to start with by $\bar{\mathfrak{g}}$ rather than $\mathfrak{g}$.) Fourier analysis shows that when S^1 is regarded as the unit circle in the complex plane with coordinate $z = e^{2\pi i t}$, then a (topological[24]) basis of the function space $\bar{\mathfrak{g}}_{\mathrm{loop}}$ is given by $\{T_n^a \mid a = 1, 2, \ldots, \dim\bar{\mathfrak{g}};\ n \in \mathbb{Z}\}$, where

$$T_n^a := \bar{T}^a \otimes z^n \equiv \bar{T}^a \otimes e^{2\pi i n t} \tag{86}$$

with $z \in S^1$ and $\{\bar{T}^a \mid a = 1, 2, \ldots, \dim\bar{\mathfrak{g}}\}$ a basis of $\bar{\mathfrak{g}}$. Moreover, $\bar{\mathfrak{g}}_{\mathrm{loop}}$ inherits a natural bracket with respect to which it becomes a Lie algebra itself, namely

$$[T_m^a, T_n^b] := [\bar{T}^a, \bar{T}^b] \otimes (z^m z^n) = \sum_{c=1}^{\dim\bar{\mathfrak{g}}} f^{ab}{}_c\, \bar{T}^c \otimes z^{m+n} = \sum_{c=1}^{\dim\bar{\mathfrak{g}}} f^{ab}{}_c\, T_{m+n}^c\,. \tag{87}$$

Accordingly, $\bar{\mathfrak{g}}_{\mathrm{loop}}$ is called the *loop algebra* over $\bar{\mathfrak{g}}$. The subset of the loop algebra that is spanned by the generators T_0^a is a Lie subalgebra, called the zero mode subalgebra of $\bar{\mathfrak{g}}_{\mathrm{loop}}$; it is isomorphic to $\bar{\mathfrak{g}}$.

Now let $\bar{\mathfrak{g}} = X_r$ be a finite-dimensional simple Lie algebra. Its loop algebra $\bar{\mathfrak{g}}_{\mathrm{loop}}$, which is infinite-dimensional, turns out to be closely related to the affine algebra $\mathfrak{g} = X_r^{(1)}$.[25] However, unlike affine Lie algebras, $\bar{\mathfrak{g}}_{\mathrm{loop}}$ has a trivial center (and also does not possess any interesting representations). But starting from $\bar{\mathfrak{g}}_{\mathrm{loop}}$ one can obtain a Lie algebra which does possess a center, by a canonical construction known as *central extension*. Namely, by simply adjoining ℓ additional generators K^j, $j = 1, 2, \ldots, \ell$, to a basis $\{T^a\}$ of an arbitrary Lie algebra $\mathfrak{g}$, one constructs an algebra $\hat{\mathfrak{g}}$ whose dimension exceeds the dimension of $\mathfrak{g}$ by ℓ and for which the K^j are central, as follows. One keeps the original values $f^{ab}{}_c$ of those structure constants which involve only the generators $\hat{T}^a$ and imposes the relations $[K^i, K^j] = 0$ and $[\hat{T}^a, K^j] = 0$ for all i, j and all a. Here by $\hat{T}^a$ I denote the image of the generator T^a of $\mathfrak{g}$ in the extended algebra $\hat{\mathfrak{g}}$. The general form of the brackets among these generators $\hat{T}^a$ reads $[\hat{T}^a, \hat{T}^b] = \sum_c f^{ab}{}_c\, \hat{T}^c + \sum_{i=1}^{\ell} f^{ab}{}_i\, K^i$. The new additional structure constants $f^{ab}{}_i$ are not arbitrary, but are restricted

---

[24] A topological or analytic basis, sometimes also called Hilbert space basis, of a vector space V is characterized by the property that the closure in the topology of V of the set of all finite linear combinations of basis vectors is the whole space V. In contrast, an ordinary vector space basis of V has the property that every vector in V is a finite linear combination of basis vectors.

[25] The notation in the lists (77) and (85) is chosen in such a way that $\mathfrak{g} = X_r^{(1)}$ with $X \in \{A, B, C, D, E, F, G\}$ precisely corresponds to the simple Lie algebra $\bar{\mathfrak{g}} = X_r$.

by the Jacobi identity. Clearly, one should count only those extensions as genuine central extensions for which $\hat{\mathfrak{g}}$ is not just the direct sum of $\mathfrak{g}$ and an abelian Lie algebra; such a direct sum is certainly obtained when $f^{ab}{}_i \equiv 0$, but this is typically not the only choice for which this happens.

It is in general a difficult cohomological question whether a given Lie algebra allows for non-trivial central extensions or not. But it is quite straightforward to see that finite-dimensional simple Lie algebras $\bar{\mathfrak{g}} = X_r$ do not possess non-trivial central extensions, whereas their loop algebras $\bar{\mathfrak{g}}_{\text{loop}}$ possess indeed a unique non-trivial extension by a single central generator K, with brackets

$$[K, \hat{T}_n^a] = 0 \,, \qquad [\hat{T}_m^a, \hat{T}_n^b] = \sum_{c=1}^{\dim \bar{\mathfrak{g}}} f^{ab}{}_c \, \hat{T}_{m+n}^c + m\,\delta_{m+n,0}\,\bar{\kappa}^{ab}\,K \,; \qquad (88)$$

here $\bar{\kappa}^{ab} = \bar{\kappa}(\bar{T}^a, \bar{T}^b)$, with $\bar{\kappa}$ the Killing form (80) of $\bar{\mathfrak{g}}$. Moreover, the untwisted affine Lie algebra $\mathfrak{g} = X_r^{(1)}$ is obtained from this centrally extended loop algebra $\widehat{\bar{\mathfrak{g}}_{\text{loop}}}$ by including a derivation D, i.e. a generator with Lie brackets

$$[D, \hat{T}_m^a] = m\,\hat{T}_m^a \,, \qquad [D, K] = 0 \,. \qquad (89)$$

(Thus D measures the mode number m and hence provides a $\mathbb{Z}$-grading, which is of precisely the same form as that of the Virasoro operator $-L_0$ in the case of chiral algebras $\mathcal{W}$, cf. equation (7).) In short, the vector space structure of $\mathfrak{g}$ is

$$\mathfrak{g} = \hat{\mathfrak{g}} \oplus \mathbb{C}D = \mathbb{C}\left[z, z^{-1}\right] \otimes_{\mathbb{C}} \bar{\mathfrak{g}} \oplus \mathbb{C}K \oplus \mathbb{C}D \,. \qquad (90)$$

Just like for loop algebras, the zero modes $\hat{T}_0^a$ of $\mathfrak{g}$ span a subalgebra which is isomorphic to the simple algebra $\bar{\mathfrak{g}}$; this is called the *horizontal* subalgebra of $\mathfrak{g}$.

Upon choosing a Cartan–Weyl basis $\{\bar{T}^a\} = \{H^i \,|\, i = 1, 2, \ldots, r\} \cup \{E^{\bar{\alpha}}\}$ of $\bar{\mathfrak{g}} \equiv \bar{\mathfrak{g}}(\bar{A})$, the defining Lie bracket relations (74 A) of $\mathfrak{g}$ take the form

$$[H_m^i, H_n^j] = m\,(\bar{D}\bar{A})^{ij}\,\delta_{m+n,0}\,K \,, \qquad [H_m^i, E_n^{\bar{\alpha}}] = \bar{\alpha}^i\,E_{m+n}^{\bar{\alpha}} \,,$$

$$[E_m^{\bar{\alpha}}, E_n^{\bar{\beta}}] = e_{\bar{\alpha},\bar{\beta}}\,E_{m+n}^{\bar{\alpha}+\bar{\beta}} \qquad \text{for } \bar{\alpha} + \bar{\beta} \text{ a } \bar{\mathfrak{g}}\text{-root}, \qquad (91)$$

$$[E_n^{\bar{\alpha}}, E_{-n}^{-\bar{\alpha}}] = \sum_{i=1}^{r} \bar{\alpha}_i H_0^i + nK \,.$$

Here $m, n \in \mathbb{Z}$, $i, j \in \{1, 2, \ldots, r\}$, and $\bar{\alpha}, \bar{\beta}$ are arbitrary $\bar{\mathfrak{g}}$-roots.

So far only the untwisted affine algebras have been obtained. But along similar lines one can realize the twisted ones as well. The only modification is to give up the single-valuedness of the maps f from S^1 to $\bar{\mathfrak{g}}$ that were used in the loop construction above. More precisely, the *twisted* algebras are obtained when instead of $f(e^{2\pi i}z) = f(z)$ one imposes the *twisted* boundary conditions

$$x \otimes f(e^{2\pi i}z) = \omega(x) \otimes f(z) \qquad (92)$$

for all $x \in \bar{\mathfrak{g}}$, where ω is an automorphism of the horizontal subalgebra $\bar{\mathfrak{g}}$ of finite order N. In other words, f is now no longer a function on the circle S^1, but rather on an N-fold covering of S^1. However, when performing this construction for two automorphisms ω and ω' for which $\omega^{-1}\omega'$ is an inner automorphism of $\bar{\mathfrak{g}}$, one obtains two realizations of one and the same abstract Lie algebra; in particular, when ω is inner, then one gets a realization of an untwisted algebra which differs from the realization (91). As a consequence, the twisted affine Lie algebras correspond in fact to equivalence classes of outer modulo inner automorphisms of $\bar{\mathfrak{g}}$; in each such class there is a distinguished representative which is induced by a symmetry of the Dynkin diagram of $\bar{\mathfrak{g}}$.

3.4   The Triangular Decomposition of Affine Lie Algebras

For any affine Lie algebra $\mathfrak{g}$ one can determine a triangular decomposition of $\mathfrak{g}$ that is analogous to (12) for chiral algebras. One first observes that a Cartan subalgebra $\mathfrak{g}_0$ of $\mathfrak{g}$ is spanned by $\{K, D, H_0^i \,|\, i = 1, 2, \dots, r\}$. The roots (i.e., vectors of eigenvalues) with respect to (H_0, K, D) are given by

$$\alpha = (\bar{\alpha}, 0, n) \quad \text{for } \bar{\alpha} \in \bar{\Phi},\ n \in \mathbb{Z} \quad \text{and} \quad \alpha = (0, 0, n) \quad \text{for } n \in \mathbb{Z} \backslash \{0\}, \qquad (93)$$

where $\bar{\Phi} = \{\bar{\alpha}\}$ denotes the root system of $\bar{\mathfrak{g}}$. These roots correspond to the generators $E_n^{\bar{\alpha}}$ with $n \in \mathbb{Z}$ and to H_n^j with $n \neq 0$, respectively. While the roots $(\bar{\alpha}, 0, n)$ are non-degenerate, i.e. appear with multiplicity one, the roots $(0, 0, n)$ do not depend on the label j of H_n^j and hence have multiplicity r. The degenerate roots are all integral multiples of $\delta := (0, 0, 1)$. Also note that the roots of the derived algebra $[\mathfrak{g}, \mathfrak{g}] =: \hat{\mathfrak{g}}$ are all infinitely degenerate, because in $\hat{\mathfrak{g}}$ there is no Cartan subalgebra generator that is able to distinguish between different labels n. Thus another rôle of the derivation D is to avoid such infinite multiplicities.

Having made a choice for distinguishing between positive and negative roots of $\bar{\mathfrak{g}}$, a system of positive roots ($\alpha > 0$) of $\mathfrak{g}$ can be chosen to consist of those roots $(\bar{\alpha}, 0, n)$ for which either $n > 0$ and $\bar{\alpha} \in \bar{\Phi}$ is an arbitrary $\bar{\mathfrak{g}}$-root or zero, or else $n = 0$ and $\bar{\alpha}$ is a positive $\bar{\mathfrak{g}}$-root. The remaining roots are then the negative ones ($\alpha < 0$). Together with $\mathfrak{g}_0$, the subalgebras $\mathfrak{g}_+ := \mathrm{span}\{E^\alpha \,|\, \alpha > 0\}$ and $\mathfrak{g}_- := \mathrm{span}\{E^{-\alpha} \,|\, \alpha > 0\}$ then provide a triangular decomposition

$$\mathfrak{g} = \mathfrak{g}_+ \oplus \mathfrak{g}_0 \oplus \mathfrak{g}_- \qquad (94)$$

(direct sum of vector spaces) of $\mathfrak{g}$, i.e. satisfy $[\mathfrak{g}_\pm, \mathfrak{g}_0 \oplus \mathfrak{g}_\pm] \subseteq \mathfrak{g}_\pm$ analogous to the relation (13) for chiral symmetry algebras.

The simple roots $\alpha^{(i)}$ of $\mathfrak{g}$ provide a basis of the root space such that in the expansion $\alpha = \sum_{i=0}^r b_i\, \alpha^{(i)}$ one has, for all $i = 0, 1, \dots, r$, $b_i \in \mathbb{Z}_{\geq 0}$ if $\alpha > 0$ and $b_i \in \mathbb{Z}_{\leq 0}$ if $\alpha < 0$. With the above choice of positive roots, the simple roots are

$$\alpha^{(i)} = (\bar{\alpha}^{(i)}, 0, 0) \equiv \bar{\alpha}^{(i)} \quad \text{for } i = 1, 2, \dots, r \quad \text{and} \quad \alpha^{(0)} = (-\bar{\theta}, 0, 1) = \delta - \bar{\theta}, \qquad (95)$$

where $\bar{\alpha}^{(i)}$ are the simple roots of the horizontal subalgebra $\bar{\mathfrak{g}}$ and $\bar{\theta}$ is the highest root of $\bar{\mathfrak{g}}$. For each $i \in \{0, 1, \dots, r\}$ the generator $E^{\alpha^{(i)}}$ corresponding to a simple root is precisely the generator that in the Chevalley–Serre formulation is denoted by E_+^i. The relation to the $E_n^{\bar{\alpha}}$-notation is thus

$$E_+^i = E_0^{\bar{\alpha}^{(i)}} \quad \text{for } i = 1, 2, \dots, r, \qquad E_+^0 = E_1^{-\bar{\theta}}. \qquad (96)$$

3.5   Representation Theory

Every Kac–Moody algebra $\mathfrak{g}$ has a triangular decomposition $\mathfrak{g} = \mathfrak{g}_+ \oplus \mathfrak{g}_0 \oplus \mathfrak{g}_-$ similar to the one described above. Correspondingly a particularly interesting class of modules (representation spaces) V of $\mathfrak{g}$ are those which have a basis on which the whole Cartan subalgebra $\mathfrak{g}_0$ acts diagonally, and hence possess a decomposition $V = \bigoplus_\lambda V_{(\lambda)}$ into subspaces such that

$$R(H^i)\, v_\lambda = \lambda^i \cdot v_\lambda \qquad (97)$$

for all $v_\lambda \in V_{(\lambda)}$. The vectors $\lambda \equiv (\lambda^i)_i$ are called the *weights* of the module V; they are elements of the weight space $\mathfrak{g}_0^*$, and the numbers λ^i are their components in the basis furnished by the fundamental weights $\Lambda_{(i)}$.

   A *maximal* weight Λ satisfies by definition $R(E^\alpha)\, v_\Lambda = 0$ for all positive roots α and all $v_\Lambda \in V_{(\Lambda)}$. A module which has exactly one weight Λ with this property is called a *highest weight module* and is denoted by V_Λ. All other elements of V_Λ can be obtained by applying step operators for negative roots to v_Λ, i.e. every $v \in V_\Lambda$ is contained in $\mathfrak{U}(\mathfrak{g}_-)\, v_\Lambda$, where $\mathfrak{U}$ denotes the enveloping algebra.

   A Kac–Moody algebra $\mathfrak{g}$ is algebraically generated by its subalgebras $\mathfrak{g}_{(i)}$ that are associated to the simple $\mathfrak{g}$-roots, i.e. $\mathfrak{g}_{(i)} = \mathrm{span}\{E_\pm^i, H^i\}$ (together with the derivations, but these do not concern us here). From (74 B) one reads off that $\mathfrak{g}_{(i)} \cong \mathfrak{sl}(2)$ for each i; one can therefore reduce many issues in the representation theory of $\mathfrak{g}$ to that of the simple Lie algebra $\mathfrak{sl}(2)$. A particularly important application is the following. All non-trivial $\mathfrak{g}$-modules are infinite-dimensional, but in favorable circumstances the subspaces into which an *irreducible* highest weight module $\mathcal{H}_\Lambda$ decomposes with respect to the subalgebras $\mathfrak{g}_{(i)}$ are all finite-dimensional. (Loosely speaking, such $\mathfrak{g}$-modules are 'less infinite-dimensional' than others.) This happens precisely when the highest weight $\Lambda = \sum_i \Lambda^i \Lambda_{(i)}$ of $\mathcal{H}_\Lambda$ is *dominant integral*, that is, a positive integral linear combination, i.e.

$$\Lambda^i \equiv (\Lambda, \alpha^{(i)\vee}) \in \mathbb{Z}_{\geq 0} \quad \text{for all } i \qquad (98)$$

(i.e. for $i = 0, 1, \ldots, r$ in the case of affine Lie algebras), of fundamental weights.

   For affine $\mathfrak{g}$ it is convenient to write $\mathfrak{g}$-weights (cf. (93)) as triples $\lambda = (\bar\lambda, k, m)$ with k the eigenvalue of the central element K and m the *grade*, i.e. the eigenvalue of the derivation D. Then for untwisted $\mathfrak{g}$, the fundamental weights are

$$\Lambda_{(i)} = (\bar\Lambda_{(i)}, \tfrac{1}{2}(\bar\theta, \bar\theta) a_i^\vee, 0) \text{ for } i = 1, 2, \ldots, r \quad \text{and} \quad \Lambda_{(0)} = (0, \tfrac{1}{2}(\bar\theta, \bar\theta), 0). \quad (99)$$

Here $a_i^\vee$ are the dual Coxeter labels, defined as the coefficients of $\bar\theta^\vee \equiv 2\bar\theta/(\bar\theta, \bar\theta)$ in the basis of simple coroots of $\bar{\mathfrak{g}}$. Also, the property (98) means, first, that the horizontal part $\bar\Lambda$ of Λ is a dominant integral $\bar{\mathfrak{g}}$-weight, and second, that

$$k^\vee := 2k/(\bar\theta, \bar\theta), \qquad (100)$$

called the *level* of $\mathcal{H}_\Lambda$, is a non-negative integer; and third, there is the bound

$$0 \leq \sum_{i=1}^r a_i^\vee \bar\Lambda^i \equiv (\bar\Lambda, \bar\theta^\vee) \leq k^\vee. \qquad (101)$$

The precise structure of an irreducible highest weight module $\mathcal{H}_\Lambda$ strongly depends on whether the highest weight Λ is dominant integral or not. For generic Λ already the Verma module $\mathcal{V}_\Lambda$, which is isomorphic to $\mathfrak{U}(\mathfrak{g}_-)\,v_\Lambda$, without any additional relations (compare equation (18)), is irreducible. In contrast, for dominant integral Λ the Verma module contains null vectors. The entire information on the null vector structure of $\mathcal{V}_\Lambda$ is encoded in the weight Λ: all null vectors can be obtained by acting with $\mathfrak{U}(\mathfrak{g}_-)$ on the so-called *primitive* null vectors

$$v^{(i)} := (E^{-\alpha^{(i)}})^{\Lambda^i+1} v_\Lambda , \tag{102}$$

where i can take all possible values, i.e. $i \in \{0, 1, \ldots, r\}$ for affine algebras. The irreducible module $\mathcal{H}_\Lambda$ is obtained from the Verma module $\mathcal{V}_\Lambda$ by taking the quotient with respect to the submodule spanned by all null vectors. For affine $\mathfrak{g}$, the quotienting by the null vectors $v^{(i)}$ with $i = 1, 2, \ldots, r$ can be implemented by just restricting to irreducible rather than Verma modules with respect to the horizontal subalgebra $\bar{\mathfrak{g}}$ of $\mathfrak{g}$; also, one has $v^{(0)} = (E^{-\alpha^{(0)}})^{\Lambda^0+1} v_\Lambda = (E^{\bar\theta}_{-1})^{k^\vee-(\bar\theta^\vee,\bar\Lambda)+1} v_\Lambda$.

3.6   Characters

Precisely as in the case of the chiral algebras $\mathcal{W}$ that arise in conformal field theory (see Sect. 2.3), one can encode the information contained in the weight system of a module V of a Kac–Moody algebra $\mathfrak{g}$ in its *character*

$$\chi_V := \textstyle\sum_\lambda \mathrm{mult}_V(\lambda)\, \mathrm{e}^\lambda . \tag{103}$$

Since the weights λ are linear functions on $\mathfrak{g}_0$, the formal exponential e^λ – and hence also the character χ_V – can be regarded as a function on $\mathfrak{g}_0$, i.e. $\mathrm{e}^\lambda \colon \mathfrak{g}_0 \to \mathbb{C}$, $h \mapsto \mathrm{e}^\lambda(h) := \exp(\lambda(h))$, where 'exp' is the ordinary exponential function. Moreover, via the invariant bilinear form the Cartan subalgebra $\mathfrak{g}_0$ can be identified with its dual space, i.e. with the weight space $\mathfrak{g}_0^\star$. Correspondingly, by setting

$$\mathrm{e}^\lambda : \quad \mu \mapsto \mathrm{e}^\lambda(\mu) := \exp[(\lambda,\mu)] \quad \text{for all } \mu \in \mathfrak{g}_0^\star , \tag{104}$$

where $(\cdot\,,\cdot)$ is the inner product in weight space, one can interpret the e^λ and the character χ_V also as functions from the weight space to the complex numbers.

One of the most intriguing results in the representation theory of symmetrizable Kac–Moody algebras is the *Weyl–Kac character formula* which states that the character χ_Λ of an irreducible $\mathfrak{g}$-module $\mathcal{H}_\Lambda$ with highest weight Λ satisfies

$$\chi_\Lambda(\mu) = \frac{\sum_{w\in W} \mathrm{sign}(w)\, \exp[(w(\Lambda+\rho),\mu)]}{\sum_{w\in W} \mathrm{sign}(w)\, \exp[(w(\rho),\mu)]} . \tag{105}$$

Here W is the Weyl group of $\mathfrak{g}$ which, just like in the finite-dimensional case, is the group generated by reflections about the hyperplanes perpendicular to the simple roots, $\mathrm{sign}(w) \in \{\pm 1\}$ is the sign of a Weyl group element w, and $\rho = \sum_i \Lambda_{(i)}$ is the *Weyl vector* of $\mathfrak{g}$. The formula (105) is a consequence of a deep interplay between the structure of the Weyl group and the relation between irreducible and Verma modules. The character $\mathcal{X}_\Lambda$ of a Verma module $\mathcal{V}_\Lambda$ is easily obtained with the help of the Poincaré–Birkhoff–Witt theorem, which tells how to construct bases of universal enveloping algebras; from the form of such a basis it follows that each generator E^α associated to a positive $\mathfrak{g}$-root α contributes a geometric series in $e^{-\alpha}$ to $\mathcal{X}_\Lambda$, and by summing these series one finds that

$$\mathcal{X}_\Lambda = e^\Lambda \cdot \prod_{\alpha>0}(1 - e^{-\alpha})^{-\mathrm{mult}(\alpha)}. \tag{106}$$

Moreover, irreducible characters are invariant under the Weyl group W, $\chi_\Lambda(w(\mu))$ $= \chi_\Lambda(\mu)$, while Verma characters acquire a factor of $\mathrm{sign}(w)$. When combining these results with a linear relation that relates irreducible and Verma characters and with basic properties of W, one obtains the character formula in the form

$$\chi_\Lambda = \frac{\sum_{w \in W} \mathrm{sign}(w)\, e^{w(\Lambda+\rho)-\rho}}{\prod_{\alpha>0}(1 - e^{-\alpha})^{\mathrm{mult}(\alpha)}}. \tag{107}$$

Furthermore, for $\Lambda = 0$, $\mathcal{H}_\Lambda$ is just the trivial one-dimensional module so that it has character $\chi_0 \equiv 1$, and hence (107) implies the *denominator identity*

$$\prod_{\alpha>0}(1 - e^{-\alpha})^{\mathrm{mult}(\alpha)} = \sum_{w \in W} \mathrm{sign}(w)\, e^{w(\rho)-\rho} \tag{108}$$

(which is the source of many combinatorial identities). Upon re-inserting this result into (107) one arrives at the character formula in the form (105).

With the help of Poisson resummation, the character formula allows one to analyze the behavior of characters under modular transformations. One finds that the set of characters of irreducible highest weight modules at fixed level of every untwisted (and of some twisted) affine Lie algebra spans a module for a unitary representation of $\mathrm{SL}_2(\mathbb{Z})$ – compare the Kac–Peterson formula (119) below. As observed after equation (89), the derivation $D \in \mathfrak{g}_0$ of an affine Kac–Moody algebras $\mathfrak{g}$ behaves just like $-L_0$. Thus the restriction of the character to the span $\mathbb{C}D$ of D yields precisely the Virasoro-specialized character that I already introduced in (54); however, the Virasoro-specialized characters are usually not sufficient to determine the representation matrices of $\mathrm{SL}_2(\mathbb{Z})$.

The character formula is valid for arbitrary symmetrizable Kac–Moody algebras. Untwisted affine Lie algebras are nevertheless distinguished, as the formula is then often easy to evaluate. It becomes particularly simple when $\mathfrak{g}$ is simply laced and at level 1. In this case the Virasoro-specialized character $\mathcal{X}_\Lambda(\tau)$ is, up to an over-all power of $q = \exp(2\pi i\tau)$, just the $-r$ th power of Euler's function $\varphi(q) = \prod_{n=1}^{\infty}(1 - q^n)$ multiplied with a theta function for the $\bar{\mathfrak{g}}$-root lattice shifted by the highest weight. E.g. for $A_{N-1}^{(1)}$ at level one the formula reads

$$\mathcal{X}_{\Lambda_{(j)}}(\tau) = (q^{1/24}\varphi(q))^{-(N-1)}\, q^{-j^2/2N} \sum_{\substack{j_1,j_2,\dots,j_N \in \mathbb{Z} \\ j_1+j_2+\dots+j_N=j}} q^{(j_1^2+j_2^2+\dots+j_N^2)/2}. \tag{109}$$

for $j = 1, 2, \dots, N-1$.

Let me return to the relation between Verma and irreducible modules. The maximal submodule V_{max} that must be divided out from $\mathcal{V}_\Lambda$ in order to obtain the irreducible module $\mathcal{H}_\Lambda$ is *not* the direct sum of the Verma modules that are 'headed' by the primitive null vectors. Rather, in $V_{\max}$ the latter can overlap (moreover, in general there also exist proper submodules which are *not* Verma modules). However, when Λ is an integrable weight, the overlap between two Verma submodules $\mathcal{V}_{\mu_1}$ and $\mathcal{V}_{\mu_2}$ of $\mathcal{V}_\Lambda$ is a submodule of both $\mathcal{V}_{\mu_1}$ and $\mathcal{V}_{\mu_2}$ and is again obtained from primitive null vectors (now in $\mathcal{V}_{\mu_1}$ respectively $\mathcal{V}_{\mu_2}$). Moreover, the precise structure is governed by the Weyl group of $\mathfrak{g}$. Based on this connection one can construct a semi-infinite complex of $\mathfrak{g}$-modules and $\mathfrak{g}$-module homomorphisms, the so-called *BGG resolution* of $\mathcal{H}_\Lambda$, in which each module except for the one at the right-most place, which is $\mathcal{H}_\Lambda$, is a direct sum of Verma modules. (The structure looks rather simple, but the proof, which works even for non-symmetrizable Kac–Moody algebras, is most complicated.)

4   WZW Theories and Coset Theories

4.1   WZW Theories

The results of Sect. 3 show in particular that an untwisted affine Lie algebra $\mathfrak{g}$ enjoys all properties that are needed for a chiral algebra of conformal field theory, except that it does not contain the Virasoro algebra. The latter shortcoming is remedied by just prescribing the bracket relations $[L_m, K] = 0 = [C, \hat{T}_n^a]$ and

$$[L_m, \hat{T}_n^a] = -n\,\hat{T}_{m+n}^a \tag{110}$$

among the generators of $\mathfrak{g}$ and $\mathcal{V}$ir. This way one arrives at the semi-direct sum of $\mathcal{V}$ir and the centrally extended loop algebra $\hat{\mathfrak{g}} = [\mathfrak{g}, \mathfrak{g}]$ and identifies the Virasoro generator L_0 with minus the derivation D of the affine algebra $\mathfrak{g}$.

While this observation does not automatically imply that one can associate a conformal field theory to any untwisted affine Lie algebra $\mathfrak{g}$, indeed there exists a class of conformal field theories with such a chiral algebra $\mathfrak{g} \oplus \mathcal{V}$ir, namely the so-called *WZW* (Wess–Zumino–Witten) *theories* ([Knizhnik and Zamolodchikov 1984]). These theories are characterized by the fact that their chiral symmetry algebra consists of the energy-momentum tensor $T(z)$ and the *current*

$$J(z) = \sum_{m \in \mathbb{Z}} z^{-m-1}\,\hat{T}_m^a \tag{111}$$

whose modes L_m and $\hat{T}_n^a$ satisfy the relations (3), (88) and (110), together with the requirement that the Virasoro generators L_m are of the *Sugawara* form, i.e.

$$L_m = \frac{1}{2(k^\vee + g^\vee)} \sum_{n \in \mathbb{Z}} \bar{\kappa}_{ab} :\hat{T}_{m+n}^a \hat{T}_{-n}^b: \,. \tag{112}$$

Here the colons $:\ :$ denote a normal ordering prescription (compare equation (30)), $k^\vee$ is the level of the relevant $\mathfrak{g}$-modules, and $g^\vee$ is the dual Coxeter number of the horizontal subalgebra $\bar{\mathfrak{g}}$ of $\mathfrak{g}$, i.e., the (properly normalized) eigenvalue of the quadratic Casimir operator in the adjoint representation of $\bar{\mathfrak{g}}$. The Sugawara formula (112) implies that WZW theories are governed by the representation theory of $\mathfrak{g}$, and there is no need to study the Virasoro algebra independently of $\mathfrak{g}$. Moreover, it follows from the representation theory of $\mathfrak{g}$ summarized in Sect. 3.5 that many quantities of interest of a WZW theory can be studied entirely in terms of the finite-dimensional algebra $\bar{\mathfrak{g}}$ and of the level $k^\vee$. E.g., because of the relation (112) the Virasoro central charge c can be expressed as

$$c(\mathfrak{g}, k^\vee) = \frac{k^\vee \dim \bar{\mathfrak{g}}}{k^\vee + g^\vee}. \tag{113}$$

Note that in a quantum field theory the factor $(k^\vee + g^\vee)^{-1}$ makes sense only if $k^\vee$ can be regarded as a $\mathbb{C}$-number; in other words, it is necessary that all $\mathfrak{g}$-representations appearing in a WZW theory have one and the same value of the level. For unitary theories, $k^\vee$ is a non-negative integer.

Sometimes one uses the term WZW theory also to refer to a somewhat more general class of models, namely those where the affine algebra $\mathfrak{g}$ is replaced by a *current algebra*, i.e. by a direct sum of untwisted affine algebras and of Heisenberg algebras. (The Lie bracket relations of the Heisenberg algebra, which generates the observables $\mathcal{W}$ for the theory of a free boson, and which is often also called the $\hat{\mathfrak{u}}(1)$ current algebra, were displayed in (63). Note that this Lie algebra is *not* a Kac–Moody algebra.) Those models for which $\mathfrak{g}$ is just an untwisted affine algebra, and hence $\bar{\mathfrak{g}}$ is simple, are then called 'simple' WZW theories.

4.2 WZW Primaries

The primary fields of a unitary WZW theory with diagonal modular invariant are in one-to-one correspondence with the integrable highest weights Λ of an untwisted affine Lie algebra $\mathfrak{g}$ at level $k^\vee$, i.e. with the dominant integral $\bar{\mathfrak{g}}$-weights $\bar{\Lambda}$ that satisfy (101). Only finitely many weights of $\bar{\mathfrak{g}}$ fulfill the condition (101); WZW theories are therefore rational conformal field theories. WZW primary fields $\phi = \phi_\Lambda(z)$ can be characterized by the properties that (compare also (27))

$$[\hat{T}_n^a, \phi_\Lambda(0)] = 0 \quad \text{for } n > 0 \qquad \text{and} \qquad [\hat{T}_0^a, \phi_\Lambda(0)] = \bar{T}_{(\Lambda)}^a \phi_\Lambda(0), \tag{114}$$

where $\bar{T}_{(\Lambda)}^a$ is a shorthand for the representation matrix $\bar{R}_\Lambda(\bar{T}^a)$. In terms of operator products, these relations correspond to

$$\underline{J^a(z)\,\phi_\Lambda}(w,\bar{w}) = (z-w)^{-1}\,\bar{T}_{(\Lambda)}^a \phi_\Lambda(w,\bar{w}). \tag{115}$$

The operator product of two current fields reads

$$\underline{J^a(z)\,J^b}(w) = (z-w)^{-2}\bar{\kappa}^{ab}K - (z-w)^{-1}\sum_c f^{ab}{}_c\,J^c(w); \tag{116}$$

thus while J^a is a primary field of the Virasoro algebra, it is not primary with respect to $\mathfrak{g}$, but rather a descendant of the identity field $\mathbf{1}$ at grade one.

The conformal dimension of a ($\mathfrak{g}$-) primary field with highest weight Λ is

$$\Delta_\Lambda = \frac{(\bar{\Lambda}, \bar{\Lambda} + 2\bar{\rho})}{2\,(k^\vee + g^\vee)}, \tag{117}$$

with $\bar{\rho}$ the Weyl vector of $\bar{g}$. Note that $(\bar{\Lambda}, \bar{\Lambda} + 2\bar{\rho})$ is the quadratic Casimir eigenvalue of the irreducible $\bar{g}$-module that has highest weight $\bar{\Lambda}$. Conjugate primary fields have highest weights which are 'charge-conjugate' with respect to $\bar{g}$, and hence identical conformal dimensions, as they should.

The Ward identity corresponding to the zero modes of the current reads

$$\sum_{i=1}^{p} R_i(\bar{x}) \langle \varphi_{\lambda_1} \varphi_{\lambda_2} \cdots \varphi_{\lambda_p} \rangle = 0, \tag{118}$$

where φ_{λ_j} are arbitrary conformal fields of $\bar{g}$-weight λ_j and $\bar{x} = \sum_a \xi_a \hat{T}_0^a$ is any element of $\bar{g}$. By taking $\bar{x}$ in the Cartan subalgebra of $\bar{g}$, the formula (118) tells us that $\left(\sum_{j=1}^{p} \lambda_j \right) \langle \varphi_{\lambda_1} \cdots \varphi_{\lambda_p} \rangle = 0$. Thus a correlation function can be non-vanishing only if it is '$\bar{g}$-neutral', i.e. if $\sum \lambda_j = 0$. In particular, $\langle \phi_{\Lambda_1} \cdots \phi_{\Lambda_p} \rangle \equiv 0$ for each correlation function which only involves (non-trivial) fields ϕ_Λ which are primary not only with respect to $\mathcal{V}ir$, but also with respect to g. In order to study non-vanishing correlation functions it is therefore necessary to deal also with correlators which involve secondary fields. It is, however, sufficient to allow just for 'horizontal' g-descendants, which are obtained from the primaries ϕ_Λ by acting only with zero mode currents $\hat{T}_0^a$ and which are still $\mathcal{V}ir$-primary. (An analogous complication arises whenever the zero mode subalgebra $\mathcal{W}^0$ of the chiral symmetry algebra is non-abelian.)

4.3   Modularity, Fusion Rules and WZW Simple Currents

As already mentioned in Sect. 3.6, the irreducible characters of a WZW theory span a module for a unitary representation of $\mathrm{SL}_2(\mathbb{Z})$. The explicit form of the modular T-matrix is obtained by inserting (117) into the general formula (59). The modular S-matrix is given by the *Kac–Peterson formula*

$$S_{\Lambda,\Lambda'} = \mathcal{N} \sum_{w \in \overline{W}} \mathrm{sign}(w) \, \exp[-\tfrac{2\pi i}{k^\vee + g^\vee} (w(\bar{\Lambda} + \bar{\rho}), \bar{\Lambda}' + \bar{\rho})]. \tag{119}$$

Here the summation is over the Weyl group $\overline{W}$ of the horizontal subalgebra $\bar{g}$ of g; the normalization $\mathcal{N}$ follows from the requirement that S must be unitary.

Using these explicit formulæ, it has in particular been possible to construct a large number of modular invariant combinations of characters. However, so far the classification of physical modular invariants has been completed only for $\bar{g} = \mathfrak{sl}(2)$, where an *A-D-E* pattern emerges ([Cappelli et al. 1987]), and ([Gannon 1994]) for $\bar{g} = \mathfrak{sl}(3)$. In particular, despite much progress (see e.g. ([Gannon 1995])) it is not yet known whether already all 'exceptional' modular invariants, which are similar to the E_7-type invariant for $\mathfrak{sl}(2)$, have been found.

Via the Verlinde formula (47), the Kac–Peterson formula can be employed to compute the fusion rules of WZW theories. Moreover, comparison with the character formula (105) shows that $S_{\Lambda,\Lambda'}$ is essentially the character of an irreducible module of the horizontal subalgebra $\bar{g}$ evaluated at a specific rational element of the weight space $\bar{g}_0^*$. By combining these observations with the unitarity of the modular S-matrix and with knowledge about the tensor products of $\bar{\mathfrak{g}}$-modules in a clever manner, one finds the following *Kac–Walton formula*

$$\mathrm{N}_{\Lambda\Lambda'}{}^{\Lambda''} = \sum_{w \in W} \mathrm{sign}(w)\, \mathrm{mult}_{\bar{\Lambda}'}(\hat{w}(\bar{\Lambda}'' + \bar{\rho}) - \bar{\rho} - \bar{\Lambda}) \tag{120}$$

for the fusion rule coefficients $\mathrm{N}_{\Lambda\Lambda'}{}^{\Lambda''}$ of WZW theories. Here $\bar{\rho}$ is the Weyl vector of the horizontal subalgebra $\bar{\mathfrak{g}}$, W is the Weyl group of $\mathfrak{g}$, $\hat{w}$ denotes the (affine) action of $w \in W$ that is induced on the weight space of $\bar{\mathfrak{g}}$, and $\mathrm{mult}_{\bar{\Lambda}}(\bar{\mu})$ is the multiplicity with which the weight $\bar{\mu}$ arises for the irreducible highest weight module $\mathcal{H}_{\bar{\Lambda}}$ of $\bar{\mathfrak{g}}$. Since W is an infinite group, (120) would not be of practical use if one really had to perform the sum over all of W, even though only a finite number of Weyl group elements gives a non-zero contribution. Fortunately, this is not necessary, because one can express the relevant elements of W through a finite algorithm as products of the (finitely many) reflections which correspond to simple $\mathfrak{g}$-roots (see e.g. Appendix A of ([Fuchs et al. 1995b])).

All simple currents of WZW theories are known ([Fuchs 1991]). Except for an isolated case appearing for $E_8^{(1)}$ at level two, they are, besides the identity field, the primary fields with highest weights $k^\vee \Lambda_{(i)}$, where $k^\vee$ is the level and $\Lambda_{(i)}$ is a fundamental weight for which the Coxeter label a_i – the coefficient of $\bar{\alpha}^{(i)}$ in the expansion of $\bar{\theta}$ with respect to the basis of simple $\bar{\mathfrak{g}}$-roots – has the value $a_i = 1$. The action of such a simple current J on the highest weights of the primary fields corresponds to a symmetry of the Dynkin diagram of $\mathfrak{g}$, and the monodromy charge $\mathrm{Q}_\mathrm{J}(\Lambda)$ is proportional to the conjugacy class of $\bar{\Lambda}$.

4.4   The Knizhnik–Zamolodchikov Equation

Recall that one obtains irreducible highest weight modules of $\mathcal{W}$ by 'setting to zero' the null vectors in the corresponding Verma module. In more field theoretical terms, this means that application of certain elements of the universal enveloping algebra $\mathfrak{U}(\mathcal{W}_-)$ to the highest weight vector yields a vector which is not contained in the space of physical states. Inserting such an operator into a correlation function must therefore give zero. Identities for correlators that are obtained this way are called *null vector equations*. In WZW theories, where $\mathcal{W} \supset \mathfrak{g}$, one clearly gets null vector equations by acting with the enveloping algebra of $\mathfrak{g}_-$, corresponding to the null vectors (102). These *Gepner–Witten* equations are purely algebraic relations (quite different from the null vector differential equations that arise for $\mathcal{W} = \mathcal{V}$ir when $c < 1$) and will not be treated here.

But in WZW theories the Sugawara relation (112) provides an additional source for null vectors. The associated null vector equations are known as *Knizhnik–Zamolodchikov equations*. I will formulate them for the four-point functions

$$\mathcal{G}(z,\bar{z}) = \langle \phi_{\Lambda_1}(\infty,\infty)\, \phi_{\Lambda_2}(1,1)\, \phi_{\Lambda_3}(z,\bar{z})\, \phi_{\Lambda_4}(0,0) \rangle. \tag{121}$$

As already observed, due to the Ward identity (118) such correlators vanish when the fields ϕ_Λ are primary with respect to $\mathfrak{g}$, and to get non-trivial correlators one must also allow for ($\mathcal{V}ir$-primary) horizontal descendants which arise from the ϕ_Λ by application of the zero modes $\hat{T}_0^a$ of the current; I will denote such descendants of ϕ_Λ by ϕ_Λ^λ. By $\bar{\mathfrak{g}}$-symmetry, one can then expand (121) as

$$\mathcal{G}_{\Lambda_1\Lambda_2\Lambda_3\Lambda_4}(z,\bar{z}) \equiv \mathcal{G}_{\Lambda_1\Lambda_2\Lambda_3\Lambda_4}^{\lambda_1\lambda_2\lambda_3\lambda_4}(z,\bar{z}) = \sum_{i=1}^{\tilde{M}} \mathcal{G}_{\Lambda_1\Lambda_2\Lambda_3\Lambda_4;i}(z,\bar{z}) \cdot (\mathcal{T}_i)^{\bar{\lambda}_1\bar{\lambda}_2\bar{\lambda}_3\bar{\lambda}_4}, \quad (122)$$

where $\{\mathcal{T}_i \,|\, i = 1, 2, ..., \tilde{M}\}$ is a basis of invariant $\bar{\mathfrak{g}}$-tensors $\mathcal{T}_i$ for the tensor product representation $\bar{R}_{\bar{\Lambda}_1} \otimes \bar{R}_{\bar{\Lambda}_2} \otimes \bar{R}_{\bar{\Lambda}_3} \otimes \bar{R}_{\bar{\Lambda}_4}$ of $\bar{\mathfrak{g}}$. In this description, the specific choice of horizontal descendants is encoded in the corresponding components $(\mathcal{T}_i)^{\bar{\lambda}_1\bar{\lambda}_2\bar{\lambda}_3\bar{\lambda}_4}$ of the invariant tensors, while the 'reduced' functions $\mathcal{G}_{\Lambda_1\Lambda_2\Lambda_3\Lambda_4;i}$ only depend on the primary fields. Also, the number $\tilde{M}$ of independent invariant tensors is the $\bar{\mathfrak{g}}$-analogue of the number $M = \sum \mathrm{N}_{AB}{}^{I} \mathrm{N}_{ICD}$ (cf. (71)) of chiral blocks appearing in (4.9), i.e. one has $\tilde{M} = \sum_{\bar{\mu}} \bar{\mathrm{N}}_{\bar{\Lambda}_1\bar{\Lambda}_2}{}^{\bar{\mu}} \bar{\mathrm{N}}_{\bar{\mu}\bar{\Lambda}_3}{}^{\bar{\Lambda}_4}$, where $\bar{\mathrm{N}}_{\bar{\Lambda}\bar{\Lambda}'}{}^{\bar{\Lambda}''}$ are the tensor product coefficients which appear in the decomposition $\bar{R}_{\bar{\Lambda}} \otimes \bar{R}_{\bar{\Lambda}'} = \bigoplus_{\bar{\Lambda}''} \bar{\mathrm{N}}_{\bar{\Lambda}\bar{\Lambda}'}{}^{\bar{\Lambda}''} \bar{R}_{\bar{\Lambda}''}$ of tensor products of $\bar{\mathfrak{g}}$-representations into irreducible subrepresentations. (For any value of the level, the fusion rule coefficients $\mathrm{N}_{\Lambda\Lambda'}{}^{\Lambda''} = \mathrm{N}_{\Lambda\Lambda'}{}^{\Lambda''}(k^\vee)$ are majorized by the $\bar{\mathrm{N}}_{\Lambda\Lambda'}{}^{\Lambda''}$, and $\mathrm{N}_{\Lambda\Lambda'}{}^{\Lambda''}(k^\vee) = \bar{\mathrm{N}}_{\Lambda\Lambda'}{}^{\Lambda''}$ for $k^\vee$ large enough. Hence one has $\tilde{M} \geq M(k^\vee)$ for all $k^\vee$, and $\lim_{k^\vee \to \infty} M(k^\vee) = \tilde{M}$.)

With the expansion (122), the Knizhnik–Zamolodchikov equation constitutes an ordinary first order matrix differential equation, reading

$$-\tfrac{1}{2}(k^\vee + g^\vee)\,\partial\mathcal{G}_{\Lambda_1\Lambda_2\Lambda_3\Lambda_4;i}(z,\bar{z}) = \sum_{j=1}^{\tilde{M}} \left(\frac{P_{ij}}{z} + \frac{Q_{ij}}{z-1} \right) \mathcal{G}_{\Lambda_1\Lambda_2\Lambda_3\Lambda_4;j}(z,\bar{z}), \quad (123)$$

where the matrices P and Q are given by $P = \sum_{a,b=1}^{\dim \bar{\mathfrak{g}}} \bar{\kappa}_{ab} \bar{R}_{\bar{\Lambda}_3}(\bar{T}^a) \otimes \bar{R}_{\bar{\Lambda}_4}(\bar{T}^b)$ and $Q = \sum_{a,b=1}^{\dim \bar{\mathfrak{g}}} \bar{\kappa}_{ab} \bar{R}_{\bar{\Lambda}_3}(\bar{T}^a) \otimes \bar{R}_{\bar{\Lambda}_2}(\bar{T}^b)$. When the basis of invariant tensors is chosen in such a way that they correspond precisely to the various irreducible subrepresentations in the tensor product $\bar{R}_{\bar{\Lambda}_3} \otimes \bar{R}_{\bar{\Lambda}_4}$, then the matrix P is diagonal.

Decoupling of the system (123) leads to linear differential equations of order $\tilde{M}$ for the components $\mathcal{G}_{\Lambda_1\Lambda_2\Lambda_3\Lambda_4;i}$. The solutions to these equations can be given in terms of contour integrals which generalize the contour integral formula for hypergeometric functions. This is in principle straightforward, but in practice it is rather tedious, and accordingly it has been fully worked out only for the case of $\bar{\mathfrak{g}} = \mathfrak{sl}(2)$ and for some specific isolated four-point functions for which the order $\tilde{M}$ of the differential equation is small.

4.5   Coset Conformal Field Theories

The well-established representation theory of affine Lie algebras makes WZW theories amenable to detailed study. The *coset construction* exploits the representation theory of affine algebras to investigate also more complicated models. The basic idea is to consider embeddings $\mathfrak{h} \hookrightarrow \mathfrak{g}$, where both Lie algebras are current algebras at some level(s) $k^\vee$, and to analyze the difference of the two Virasoro algebras $\mathcal{V}ir_\mathfrak{g}$ and $\mathcal{V}ir_\mathfrak{h}$ (which are obtained via the Sugawara formula (112) for $\mathfrak{g}$ and $\mathfrak{h}$, respectively), i.e. the Lie algebra with generators

$$L_m^{\mathfrak{g}/\mathfrak{h}} := L_m^{\mathfrak{g}} - L_m^{\mathfrak{h}}. \quad (124)$$

When the embedding of $\mathfrak{h}$ into $\mathfrak{g}$ is induced by an embedding $\bar{\mathfrak{h}} \hookrightarrow \bar{\mathfrak{g}}$ of the respective horizontal subalgebras, then these combinations commute with every generator J_n^a of $\mathfrak{h}$, $[L_m^{\mathfrak{g}/\mathfrak{h}}, \hat{T}_n^a] = 0$, implying that $[L_m^{\mathfrak{g}/\mathfrak{h}}, L_n^{\mathfrak{g}/\mathfrak{h}}] = [L_m^{\mathfrak{g}}, L_n^{\mathfrak{g}}] - [L_m^{\mathfrak{h}}, L_n^{\mathfrak{h}}]$. As a consequence the generators $L_n^{\mathfrak{g}/\mathfrak{h}}$ span a Virasoro algebra with central element $C_{\mathfrak{g}/\mathfrak{h}} = C_{\mathfrak{g}} - C_{\mathfrak{h}}$, called the *coset Virasoro algebra*. It is, however, still an open problem whether for *any* embedding $\bar{\mathfrak{h}} \hookrightarrow \bar{\mathfrak{g}}$ this prescription of a Virasoro algebra can be complemented in such a way that one arrives at a consistent conformal field theory – called the *coset theory* and briefly denoted by '$(\mathfrak{g}/\mathfrak{h})_{k^{\vee}}$' – and if so, whether that theory is unique. To decide these questions, one must in particular construct the (maximal) chiral algebra $\mathcal{W}_{\mathfrak{g}/\mathfrak{h}}$ of $(\mathfrak{g}/\mathfrak{h})_{k^{\vee}}$, as well as the spectrum of the theory, i.e. tell which modules – of the coset chiral algebra $\mathcal{W}_{\mathfrak{g}/\mathfrak{h}}$, or at least of the coset Virasoro algebra $\mathcal{V}ir_{\mathfrak{g}/\mathfrak{h}}$ – appear.

As it turns out, to obtain the spectrum of the coset theory is a somewhat delicate issue. While by construction the generators of the coset Virasoro algebra act on the chiral state space of the WZW theory based on $\mathfrak{g}$, i.e. on the direct sum $\mathcal{H}_{\mathfrak{g}} = \bigoplus_\Lambda \mathcal{H}_\Lambda$ of all inequivalent irreducible highest weight modules $\mathcal{H}_\Lambda$ of $\mathfrak{g}$ at level $k^{\vee}$, $\mathcal{H}_{\mathfrak{g}}$ is *not* the state space $\mathcal{H}_{\mathfrak{g}/\mathfrak{h}}$ of the coset theory. The crucial observation is that due to $[\mathfrak{h}, \mathcal{V}ir_{\mathfrak{g}/\mathfrak{h}}] = 0$, retaining the full state space $\mathcal{H}_{\mathfrak{g}}$ would imply that the coset theory would possess (infinitely many) primary fields of zero conformal dimension other than the identity field, namely all the currents $J^a(z)$ of the subalgebra $\mathfrak{h}$ and all their $\mathfrak{h}$-descendants. In a field theoretic language this means that – in the full theory, not just accidentally in some approximation to it – the vacuum is not unique. This disaster is avoided by requiring that these fields act trivially on $\mathcal{H}_{\mathfrak{g}/\mathfrak{h}}$, i.e. by imposing the gauge principle that $\mathcal{H}_{\mathfrak{g}}$-vectors which differ only by the action of $\mathfrak{U}(\mathfrak{h})$ merely provide different descriptions of the same physical situation and hence represent one and the same vector in $\mathcal{H}_{\mathfrak{g}/\mathfrak{h}}$. In short, the elements of $\mathcal{H}_{\mathfrak{g}/\mathfrak{h}}$ are $\mathfrak{U}(\mathfrak{h})$-*orbits* of vectors in $\mathcal{H}_{\mathfrak{g}}$ rather than individual vectors of $\mathcal{H}_{\mathfrak{g}}$. Concretely, the restriction to $\mathfrak{U}(\mathfrak{h})$-orbits is implemented by decomposing the $\mathfrak{g}$-modules $\mathcal{H}_\Lambda$ into $\mathfrak{h}$-modules $\mathcal{H}_{\Lambda'}$ as [26]

$$\mathcal{H}_\Lambda = \bigoplus_{\Lambda'} \mathcal{H}_{(\Lambda;\Lambda')} \otimes \mathcal{H}_{\Lambda'}, \qquad (125)$$

and the *branching spaces* $\mathcal{H}_{(\Lambda;\Lambda')}$ appearing here are natural candidates for the irreducible modules of the coset theory. In terms of characters, this corresponds to regarding the *branching functions* $b_{(\Lambda;\Lambda')}$ which appear in the decomposition

$$\chi_\Lambda^{\mathfrak{g}}(\tau) = \sum_{\Lambda'} b_{(\Lambda;\Lambda')}(\tau) \chi_{\Lambda'}^{\mathfrak{h}}(\tau) \qquad (126)$$

of the characters $\chi_\Lambda^{\mathfrak{g}}$ of $\mathfrak{g}$ with respect to the characters $\chi_{\Lambda'}^{\mathfrak{h}}$ of $\mathfrak{h}$ as the characters of the coset theory. It follows immediately from (126) that branching functions have a definite behavior under modular transformations, namely the same as the irreducible characters of the tensor product $\mathfrak{g} \oplus \mathfrak{h}^*$ of the WZW theory based on

---

[26] Here and below I denote quantities referring to the subalgebra $\mathfrak{h}$ by the same symbol as the corresponding quantities for $\mathfrak{g}$, but with a prime added. In particular Λ and Λ' stand for integrable highest weights of $\mathfrak{g}$ and $\mathfrak{h}$, respectively.

g and a putative conformal field theory which behaves just like the WZW theory based on the affine Lie algebra $\mathfrak{h}$ except that, as indicated by the notation '*', it carries the complex conjugate representation of $SL_2(\mathbb{Z})$.

4.6   Field Identification

As it turns out, even after fixing the gauge symmetry $\mathfrak{h}$ there still remain severe problems; they can be attributed to the fact that in general the redundancy symmetry of the coset theory is larger than the obvious gauge symmetry $\mathfrak{h}$. It is instructive to observe this in the example of the critical Ising model, which can be realized with $\mathfrak{h} = A_1^{(1)}$ at level 2 embedded into $\mathfrak{g} = A_1^{(1)} \oplus A_1^{(1)}$ at levels 1. This model is a rational theory with $c = \frac{1}{2}$ (more generally, by taking levels $m+2$, 1 and $m+1$, respectively, one obtains the whole minimal series (21) of $\mathcal{V}$ir); hence its spectrum follows already from the representation theory of the Virasoro algebra. Namely, there are three primary fields, and comparing their conformal dimensions $\Delta = \bar{\Delta}$ with those of the WZW primaries one finds that each of them can be realized by two distinct pairs $(\Lambda; \Lambda')$ of weights:

Field		Δ	$(\Lambda; \Lambda')$
identity field	1	0	$(0, 0; 0)$, $(1, 1; 2)$
order parameter σ		1/16	$(0, 1; 1)$, $(1, 0; 1)$
energy operator ψ		1/2	$(1, 1; 0)$, $(0, 0; 2)$

(127)

Thus each field expected from the representation theory of $\mathcal{V}$ir seems to appear twice. On the other hand, half of the possible pairs $(\Lambda; \Lambda')$, namely those with $\Lambda_1 + \Lambda_2 + \Lambda' \in 2\mathbb{Z}+1$, do not correspond to any field of the theory at all; inspection shows that these pairs possess vanishing branching functions.

This situation generalizes for arbitrary coset theories. First, typically there are selection rules, i.e. the branching functions for certain pairs $(\Lambda; \Lambda')$ vanish. This observation is in itself not too disturbing, as one just has to make sure to find all selection rules. [27] However, along with each selection rule there also comes a redundancy, i.e. non-vanishing branching functions for distinct pairs $(\Lambda; \Lambda')$ turn out to be identical. In particular the putative vacuum sector seems to occur several times. By the same argument which forced us to divide out the action of the subalgebra $\mathfrak{h}$, one learns that this degeneracy cannot be interpreted as the multiple appearance of a corresponding primary field in the spectrum

---

[27] But still, while empirically for most coset theories all selection rules come from conjugacy class selection rules for the embedding $\bar{\mathfrak{h}} \hookrightarrow \bar{\mathfrak{g}}$, so far no prescription for enumerating all selection rules for every coset theory is known. Another possible reason for the vanishing of branching functions is the occurrence of additional null states in the Verma modules. This indeed happens for conformal embeddings, for which by definition the coset central charge vanishes, and presumably also for the closely related maverick ([Dunbar and Joshi 1993]) cosets. For all conformal embeddings, the level of $\mathfrak{g}$ is $k^\vee = 1$, while $k^\vee = 2$ for all known maverick cosets.

(in particular, it would then not be possible to obtain the required modular transformation properties). Rather, the correct interpretation is that a primary field of the coset theory is not associated to an individual pair $(\Lambda; \Lambda')$, but rather to an appropriate equivalence class $[(\Lambda; \Lambda')]$ of such pairs. This prescription has been termed *field identification* (which is a bit of a misnomer, since it is pairs of labels rather than conformal fields that get identified).

Both selection rules and field identification can be described conveniently via the concept of the *identification group* G_{id} ([Schellekens and Yankielowicz 1990]). The elements $(J; J')$ of G_{id} are simple currents of the tensor product theory $\mathfrak{g} \oplus \mathfrak{h}^*$ of zero conformal dimension $\Delta_J - \Delta_{J'}$; they act on pairs $(\Lambda; \Lambda')$ via the fusion product. The selection rules are equivalent to the vanishing of the *monodromy charges* $Q_{(J;J')}((\Lambda; \Lambda')) := Q_J(\Lambda) - Q_{J'}(\Lambda')$ of any allowed branching function with respect to all $(J; J') \in G_{id}$; here $Q_J(\Lambda)$ is the combination $Q_J(\Lambda) = \Delta_\Lambda + \Delta_J - \Delta_{J \star \Lambda}$ (see equation (68)) of conformal weights. Moreover, the equivalence classes in the field identification are precisely the orbits

$$[(\Lambda; \Lambda')] := \{ (\Upsilon; \Upsilon') \mid \Upsilon = J\Lambda, \ \Upsilon' = J'\Lambda' \text{ for some } (J; J') \in G_{id} \} \qquad (128)$$

of G_{id}. Provided that all G_{id}-orbits have a common length, taking one branching function out of each G_{id}-orbit and combining them diagonally yields a modular invariant spectrum. In this case the modular S-matrix is given by

$$S^{\mathfrak{g}/\mathfrak{h}}_{[(\Lambda; \Lambda')], [(\Upsilon; \Upsilon')]} = S_{\Lambda, \Upsilon} (S'_{\Lambda', \Upsilon'})^* , \qquad (129)$$

where $(\Lambda; \Lambda')$ and $(\Upsilon; \Upsilon')$ are arbitrary representatives of the orbits $[(\Lambda; \Lambda')]$ and $[(\Upsilon; \Upsilon')]$, respectively, i.e. by restricting the S-matrix $S \otimes (S')^*$ of $\mathfrak{g} \oplus \mathfrak{h}^*$ to orbits.

4.7   Fixed Points

The next degree of generality, and difficulty, is reached when G_{id}-orbits with different sizes appear. (Of course, all sizes are divisors of the size of the orbit through $(0; 0)$, on which G_{id} acts freely, i.e. of the order $N = |G_{id}|$ of G_{id}.) Orbits of non-maximal size are referred to as *fixed points*. As soon as fixed points are present, taking precisely one representative out of each orbit $[(\Lambda; \Lambda')]$ does not give a modular invariant spectrum. This is a severe problem, as it seems to prevent one from gaining control over the coset theory $(\mathfrak{g}/\mathfrak{h})_{k^\vee}$ by completely understanding it in terms of the underlying WZW theories based on $\mathfrak{g}$ and $\mathfrak{h}$. The solution to this problem [28] shows, however, that $(\mathfrak{g}/\mathfrak{h})_{k^\vee}$ is in fact still fully controlled by the affine algebras $\mathfrak{g}$ and $\mathfrak{h}$, provided that one implements additional novel structures associated to Kac–Moody algebras, the so-called *twining characters* and *orbit Lie algebras*. The basic observation is that the coset modules $\mathcal{H}_{(\Lambda; \Lambda')}$ for fixed points are *not* irreducible and hence a *fixed point resolution* into irreducible subspaces must be performed. This results in particular in a formula which expresses the S-matrix of $(\mathfrak{g}/\mathfrak{h})_{k^\vee}$ in terms of the group characters

---

[28] See ([Fuchs et al. 1996b]) or, for a condensed exposition, ([Fuchs et al. 1996d]).

of stabilizer subgroups of G_{id} and of certain modular transformation matrices $S^{[J;J']}$ which are products of S-matrices of the relevant orbit Lie algebras.

In closing this last regular subsection, I would like to reiterate the remark made earlier that while in principle it is desirable to work with observable quantities only, the use of non-observable objects often proves to be convenient. The situation at hand provides an outstanding example for such an approach, in that extremely non-trivial effects can be detected without even knowing precisely what the observables $W_{\mathfrak{g}/\mathfrak{h}}$ (except for the subalgebra $\mathcal{V}\mathrm{ir}_{\mathfrak{g}/\mathfrak{h}}$) look like.

5 Addenda

5.1 Omissions

Lack of time forced me to omit a number of topics which should definitely occur in a decent introduction to conformal field theory and to Kac–Moody algebras. To partially compensate this shortcoming, I give a brief list of some of these issues and point out a few relevant references. Concerning citations, I decided to sacrifice historical accuracy for up-to-dateness, so I usually refer either to reviews or to the most recent publications I am aware of, which can serve as a guide to further literature. (This remark applies likewise to most of the references that I already mentioned in the main text.)

Sect. 1: more on the representation theory of the Virasoro algebra ([Ganchev and Petkova 1993, Schottenloher 1995, Schellekens 1996]); classification and properties of chiral algebras ([Bouwknegt and Schoutens 1993, Watts 1997]); vertex operator algebras ([Gebert and Nicolai 1995, Kac 1996]); more on chiral blocks and chiral vertex operators ([Alvarez-Gaumé et al. 1990, Moore and Seiberg 1990, Cremmer et al. 1994]); conformal field theory and higher genus Riemann surfaces ([Moore and Seiberg 1990, Bantay and Vecsernyés 1995]); the C^*-algebraic approach to conformal field theory ([Fredenhagen et al. 1992, Wassermann 1996, Gabbiani and Fröhlich 1993, Böckenhauer 1996]); 'logarithmic' conformal theories ([Gaberdiel and Kausch 1996, Flohr 1996]).

Sect. 2: more on duality, crossing symmetry, and polynomial equations [Alvarez-Gaumé et al. 1990, Moore and Seiberg 1990]; the representation theory and the classification of fusion rings [Fröhlich and Kerler 1993, Fuchs 1994, Eholzer 1995]; the rôle of quantum groups, or more general, quantum symmetries ([Alvarez-Gaumé et al. 1990, Schomerus 1995, Fuchs et al. 1995a, Recknagel 1996]); classification of modular invariants ([Kreuzer and Schellekens 1994, Gannon 1996, Fuchs et al. 1996a, Gannon et al. 1996]); free fermion [Goddard and Olive 1986, Fuchs 1992b] and orbifold ([Bantay 1994, Kac and Todorov 1996, Borisov et al. 1997]) conformal field theories; Galois symmetry ([Buffenoir et al. 1995, Fuchs et al. 1996a, Fuchs and Schweigert 1996]); more on the classification of $c = 1$ and related conformal field theories ([Furlan et al. 1992, Flohr 1993]) and on the Coulomb gas description of $c < 1$ theories ([Furlan et al. 1989, Schellekens 1996]).

Sect. 3: more on Lie algebras with triangular decomposition [Kac 1990, Fulton and Harris 1991, Moody and Pianzola 1995, Fuchs and Schweigert 1997]; realizations of affine Lie algebra representations via free fermions or via vertex operators ([Goddard and Olive 1986, Kac 1990, Dolan et al. 1996, Gebert and Nicolai 1996]), and through more complicated free field constructions ([de Boer and Feher 1996]); character formulæ and combinatorial identities ([Kac 1990, Borcherds 1995]); generalized Kac–Moody algebras and their rôle in string theory ([Gebert and Nicolai 1995, Harvey and Moore 1996]).

Sect. 4: the affine-Virasoro construction which generalizes the Sugawara formula ([Halpern et al. 1996, Halpern and Obers 1996]); Lagrangian formulations of WZW and coset theories ([Hori 1996, Rhedin 1996, Schweigert 1996]); Knizhnik–Zamolodchikov–Bernard equations ([Felder and Wieczerkowski 1996, Ivanov 1996]); WZW theories which have fractional level ([Mathieu et al. 1992, Furlan et al. 1993, Petersen et al. 1996]) or are based on non-reductive Lie algebras ([Figueroa-O'Farrill and Stanciu 1996, Arfaei and Parvizi 1996]); Hamiltonian reduction of WZW theories ([Fehér et al. 1992, Ganchev and Petkova 1993, Boresch et al. 1995]); the relevance of Knizhnik–Zamolodchikov equations to moduli spaces of flat connections, to quasi-Hopf algebras and to integrable systems ([Ivanov 1996, Alekseev et al. 1996]).

5.2 Outlook

It is often claimed (implicitly or explicitly) that in conformal field theory all interesting problems are already solved. This statement is not entirely correct (as is e.g. indicated by the large number of recent publications quoted above); some examples of open questions are the following.

- A complete understanding of the rôle of modular invariance, in particular in view of the presence of similar structures in arbitrary rational two-dimensional theories (cf. e.g. ([Rehren 1990, Bantay and Vecsernyés 1995, Müger 1996]));
- a mathematically rigorous proof of the Verlinde formula for general conformal field theories, not just for (diagonal) WZW theories;
- structural information on fusion rules of non-quasirational theories (e.g. existence of vector space bases rather than only topological bases), and the possible description of such theories as deformations or limits of quasirational theories;
- the quantum field-theoretic interpretation of non-unitary, and in particular of logarithmic conformal field theories;
- the classification of modular invariants for all WZW theories – as well as deeper insight into its guiding principles (note that even the A-D-E pattern that arises for $\bar{\mathfrak{g}} = \mathfrak{sl}(2)$ essentially emerges in a technical rather than a conceptual manner);
- the rôle of 'apparent singularities' that appear in the differential equations for chiral blocks ([Fuchs 1992a]) as possible parameters in the classification of rational conformal field theories;
- simple current and Galois symmetries of braiding and fusing matrices, and a complete analysis of the behavior of fusing matrices under 'tetrahedral' transformations in the presence of fusion rule coefficients larger than one;
- a more substantial classification of modular fusion rings;
- an efficient algorithm for solving the polynomial equations for prescribed fu-

sion rules;

- a description of conformal field theory on surfaces with boundaries ([Cardy 1984, Pradisi et al. 1996]) at the same level of sophistication as for closed surfaces;
- a complete description of the space of physical states for arbitrary coset theories, in particular a characterization and enumeration of all maverick cosets.

Acknowledgements. I am grateful to K. Blaubär, B. Haßler, C. Schweigert and P. Vecsernyés for suggesting various improvements to the manuscript.

References

Alekseev, A.Yu., Recknagel, A., Schomerus, V. (1996): preprint hep-th/9610066
Alvarez-Gaumé, L., Gómez, C., Sierra, G. (1990): in: *The Physics and Mathematics of Strings*, L. Brink et al., eds. (World Scientific, Singapore), p. 16
Arfaei, H., Parvizi, S. (1996): Mod. Phys. Lett. A **11**, 1289
Bantay, P. (1994): Int. J. Mod. Phys. A **9**, 1443
Bantay, P. (1996): preprint hep-th/9611124
Bantay, P., Vecsernyés, P. (1995): preprint hep-th/9506186
Beauville, A. (1994): preprint alg-geom/9404001
Belavin, A.A., Polyakov, A.M., Zamolodchikov, A.B. (1984): Nucl. Phys. B **241**, 333
Blok, B., Yankielowicz, S. (1989): Nucl. Phys. B **321**, 717
Blumenhagen, R., Flohr, M., Kliem, A., Nahm, W., Recknagel, A., Varnhagen, R. (1991): Nucl. Phys. B **361**, 255
Böckenhauer, J. (1996): Rev. Math. Phys. **8**, 925
Borcherds, R.E. (1995): Invent. math. **120**, 161
Boresch, A., Landsteiner, K., Lerche, W., Sevrin, A. (1995): Nucl. Phys. B **436**, 609
Borisov, L., Halpern, M.B., Schweigert, C. (1997): preprint hep-th/9701061, to appear in Int. J. Mod. Phys. A
Bouwknegt, P., Schoutens, K. (1993): Phys. Rep. **223**, 183
Buffenoir, E., Coste, A., Lascoux, J., Degiovanni, P., Buhot, A. (1995): Ann. Inst. Poincaré **63**, 41
Cappelli, A., Itzykson, C., Zuber, J.-B. (1987): Commun. Math. Phys. **113**, 1
Cardy, J.L. (1984): Nucl. Phys. B **240** [FS12] (1984) 514
Cremmer, E., Gervais, J.-L., Roussel, J.-F. (1994): Nucl. Phys. B **413**, 244
de Boer, J., Feher, L., (1996): Mod. Phys. Lett. A **11**, 1999
Dolan, L., Goddard, P., Montague, P. (1996): Commun. Math. Phys. **179**, 61
Dunbar, D.C., Joshi, K.G. (1993): Mod. Phys. Lett. A **8**, 2803
Eholzer, W. (1995): J. Math. Phys. **172**, 623
Fehér, L., O'Raifeartaigh, L., Ruelle, P., Tsutsui, I., Wipf, A. (1992): Phys.Rep. **222**, 1
Felder, G., Wieczerkowski, C. (1996): Commun. Math. Phys. **176**, 163
Figueroa-O'Farrill, J., Stanciu, S. (1996): Nucl. Phys. B **458**, 137
Flohr, M. (1993): Commun. Math. Phys. **157**, 179
Flohr, M. (1996): preprint hep-th/9605151
Fredenhagen, K., Rehren, K.-H., Schroer, B. (1992): Rev. Math. Phys. [spec.iss.], 111
Frenkel, I.B., Lepowsky, J., Meurman, A. (1988): *Vertex Operator Algebras and the Monster* (Academic Press, New York)
Fröhlich, J., Kerler, T. (1993): *Quantum Groups, Quantum Categories and Quantum Field Theory* [Lecture Notes in Mathematics 1542] (Springer Verlag, Berlin)

Fuchs, J. (1989): Nucl. Phys. B **328**, 585
Fuchs, J. (1991): Commun. Math. Phys. **136**, 345
Fuchs, J. (1992a): Nucl. Phys. B **386**, 343
Fuchs, J. (1992b): *Affine Lie Algebras and Quantum Groups* (Cambridge University Press, Cambridge)
Fuchs, J. (1994): Fortschr. Phys. **42**, 1
Fuchs, J., Ganchev, A.Ch., Vecsernyés, P. (1995a): Int. J. Mod. Phys. A **10**, 3431
Fuchs, J., Gepner, D. (1987): Nucl. Phys. B **294**, 30
Fuchs, J., Schellekens, A.N., Schweigert, C. (1995b): Nucl. Phys. B **437**, 667
Fuchs, J., Schellekens, A.N., Schweigert, C. (1996a): Commun. Math. Phys. **176**, 447
Fuchs, J., Schellekens, A.N., Schweigert, C. (1996b): Nucl. Phys. B **461**, 371
Fuchs, J., Schellekens, A.N., Schweigert, C. (1996c): Nucl. Phys. B **473**, 323
Fuchs, J., Schellekens, A.N., Schweigert, C. (1996d): preprint hep-th/9612093, to appear in the Proceedings of the XXIth International Colloquium on *Group Theoretical Methods in Physics*, H.-D. Doebner and W. Scherer, eds.
Fuchs, J., Schweigert, C. (1996): preprint hep-th/9609124, to appear in Commun. Math. Phys.
Fuchs, J., Schweigert, C. (1997): *Symmetries, Lie Algebras and Representations* (Cambridge University Press, Cambridge)
Fuks, D.B. (1986): *Cohomology of Infinite-dimensional Lie Algebras* (Consultants Bureau, New York)
Fulton, W., Harris, J. (1991): *Representation Theory, a First Course* (Springer Verlag, New York)
Furlan, P., Ganchev, A., Paunov, R., Petkova, V.B. (1993): Nucl. Phys. B **394**, 665
Furlan, P., Paunov, R., Todorov, I.T. (1992): Fortschr. Phys. **40**, 211
Furlan, P., Sotkov, G., Todorov, I.T. (1989): Riv. Nuovo Cim. **12**, 1
Gabbiani, F., Fröhlich, J. (1993): Commun. Math. Phys. **155**, 569
Gaberdiel, M.R., Kausch, H. (1996): Nucl. Phys. B **477**, 293
Ganchev, A.Ch., Petkova, V.B. (1993): Phys. Lett. B **318**, 77
Gannon, T. (1994): Commun. Math. Phys. **161**, 233
Gannon, T. (1995): preprint q-alg/9510026
Gannon, T. (1996): preprint hep-th/9608063
Gannon, T., Ruelle, P., Walton, M.A. (1996): Commun. Math. Phys. **179**, 121
Gebert, R.W., Nicolai, H. (1995): Commun. Math. Phys. **172**, 571
Gebert, R.W., Nicolai, H. (1996): preprint hep-th/9608014
Goddard, P., Olive, D.I. (1986): Int. J. Mod. Phys. A **1**, 303
Haag, R. (1992): *Local Quantum Physics* (Springer Verlag, Berlin)
Halpern, M.B., Kiritsis, E.B., Obers, N.A., Clubok, K. (1996): Phys. Rep. **265**, 1
Halpern, M.B., Obers, N.A. (1996): Int. J. Mod. Phys. A **11**, 4837
Harvey, J.A., Moore, G. (1996): Nucl. Phys. B **463**, 315
Hori, K. (1996): Commun. Math. Phys. **182**, 1
Ivanov, D. (1996): preprint hep-th/9610207
Kac, V.G. (1990): *Infinite-dimensional Lie Algebras*, third edition (Cambridge University Press, Cambridge)
Kac, V.G. (1996): in: *New Trends in Quantum Field Theory*, A.Ch. Ganchev, R. Kerner and I. Todorov, eds. (Heron Press, Sofia), p. 261
Kac, V.G., Todorov, I.T. (1996): preprint hep-th/9612078
Knizhnik, V.G., Zamolodchikov, A.B. (1984): Nucl. Phys. B **247**, 83
Kreuzer, M., Schellekens, A.N. (1994): Nucl. Phys. B **411**, 97

Mathieu, P., Senechal, P., Walton, M.A. (1992): Int. J. Mod. Phys. A **7** Suppl. 1B, 731

Moody, R.V., Pianzola, A. (1995): *Lie Algebras With Triangular Decomposition* (John Wiley, New York)

Moore, G., Seiberg, N. (1989): Nucl. Phys. B **313**, 16

Moore, G., Seiberg, N. (1990): in: *Physics, Geometry, and Topology*, H.C. Lee, ed. (Plenum, New York), p. 263

Müger, M. (1996): preprint hep-th/9606175

Petersen, J.L., Rasmussen, J., Yu, M. (1996): Nucl. Phys. B **481**, 577

Pradisi, G., Sagnotti, A., Stanev, Ya.S. (1996): Phys. Lett. B **381**, 97

Recknagel, A. (1996): preprint hep-th/9612178

Rehren, K.-H. (1990): in: *The Algebraic Theory of Superselection Sectors. Introduction and Recent Results*, D. Kastler, ed. (World Scientific, Singapore), p. 333

Rhedin, H. (1996): Phys. Lett. B **373**, 76

Schellekens, A.N. (1996): Fortschr. Phys. **44**, 605

Schellekens, A.N., Yankielowicz, S. (1989): Nucl. Phys. B **327**, 673

Schellekens, A.N., Yankielowicz, S. (1990): Int. J. Mod. Phys. A **5**, 2903

Schomerus, V. (1995): Commun. Math. Phys. **169**, 193

Schottenloher, M. (1995): preprint gk-mp-9510/21, available as http://www.mathematik.uni-muenchen.de/~gkadmin/pr/gk_mp_9510_21.ps.gz

Schweigert, C. (1996): preprint hep-th/9611092, to appear in Nucl. Phys. B

Streater, R.F., Wightman, A.S. (1989): *PCT, Spin and Statistics, and All That* (Addison-Wesley, Reading)

Verlinde, E. (1988): Nucl. Phys. B **300** [FS22] (1988) 360

Wassermann, A. (1996): preprint DPMMS Cambridge, available as http://www.dpmms.cam.ac.uk/home/emu/ajw/bourbaki9.ps

Watts, G.M.T. (1997): contribution to these Proceedings

Yoshida, M. (1987): *Fuchsian Differential Equations* (Vieweg Verlag, Braunschweig)

W-Algebras and Their Representations

Gerard M.T. Watts

Department of Mathematics, King's College London,
Strand, London, WC2R 2LS, U.K.

Abstract. An introduction is given to the basic concepts, construction and representation theory of W-algebras.

1 Introduction to W-Algebras

A conformal field theory describes a set of fields which depend on the coordinates $(z, \bar{z})$ on the complex plane, and the space on which these fields act. Every conformal field theory has a subset of fields which are independent of $\bar{z}$, and this is called its chiral algebra.

It would be nice to classify all possible chiral algebras; known examples are affine Lie algebras, the Virasoro algebra and the super-Virasoro algebra. What can be more general than these?

In order to answer this question, we have to have a framework in which to discuss this problem, and some associated notation. One can take different approaches, treating conformal field theory as e.g. a D–module [42], a Vertex operator algebra [42] or the approach we shall follow, a Meromorphic cft [38]. Of these three, the last is the most similar to the physicists' methods for performing calculations in conformal field theory, and so we shall use this, if only because it allows us to prove some results which otherwise are a bit mysterious.

2 Meromorphic Conformal Field Theory

Meromorphic conformal field theory is an abstraction and formalisation of the properties of conformal field theories where the only singularities which occur in the operator product expansions are poles. It is not the only form of conformal field theory which may be formulated rigorously, but it has been done so.

The structure of meromorphic conformal field theory (mcft) is developed in [38], and we shall only present a brief summary here. A mcft consists of a Hilbert space $\mathcal{H}$ and a vertex operator map from a dense subspace $\mathcal{F}$ of $\mathcal{H}$ into the space of fields. There are two distinguished states: the vacuum $|0\rangle$ and the 'conformal state' $|L\rangle$, whose vertex operator is the stress-energy tensor of the theory, and whose modes form a copy of the Virasoro algebra.

The vertex operator V is a map $V\colon \mathcal{H} \times \mathbb{C} \to \mathrm{End}(\mathcal{H})$ which has to satisfy the following conditions:

[1] $V(|\phi\rangle, z)|0\rangle = e^{zL_{-1}}|\phi\rangle$

[2] $\langle\phi_1|V(|\psi\rangle, z)|\phi_2\rangle$ is a meromorphic function

[3] $\langle\phi_1|V(|\psi\rangle, z)V(|\chi\rangle, z')|\phi_2\rangle$ is a holomorphic function for $|z| > |z'|$.

[4] $\langle\phi_1|V(|\psi\rangle, z)V(|\chi\rangle, z')|\phi_2\rangle = \epsilon\langle\phi_1|V(|\chi\rangle, z')V(|\psi\rangle, z)|\phi_2\rangle$ by analytic continuation, where $\epsilon = 1$ unless both ψ and χ are fermionic, in which case $\epsilon = -1$

The distinguished state $|L\rangle$ corresponds to the Stress-Energy tensor,

$$V(|L\rangle, z) \equiv L(z) \equiv \sum_m L_m z^{-m-2}, \qquad (1)$$

where the modes L_m satisfy the Virasoro algebra,

$$[L_m, L_n] = \frac{c}{12}m(m^2 - 1)\delta_{m,-n} + (m - n)L_{m+n}, \qquad (2)$$

where c is a central element called the central charge. The subset L_{-1}, L_0, L_1 generate an $su(1, 1)$ subalgebra which is the Lie algebra of the global Möbius transformations and which leave $|0\rangle$ invariant.

These axioms allow us to prove the operator product expansion, that is

$$V(|\psi\rangle, z)V(|\chi\rangle, z') = V(V(|\psi\rangle, z - z')|\chi\rangle, z'), \qquad (3)$$

and to show that $L_{-1} \equiv \partial/\partial z$, that is

$$V(L_{-1}|\psi\rangle, z) = \frac{\partial}{\partial z}V(|\psi\rangle, z). \qquad (4)$$

For a state $|\psi\rangle$ of definite L_0 eigenvalue h, we introduce the mode expansion

$$V(|\psi\rangle, z) = \psi(z) = \sum_m \psi_m z^{-m-h}, \qquad (5)$$

so that the operator product expansion (3) of two fields $\psi(z)$ and $\phi(z')$ of weights h and h' becomes

$$\psi(z)\phi(z') = \sum_m (z - z')^{-m}V(\psi_{-h-m}|\phi\rangle, z'), \qquad (6)$$

This now allows us to give a definition of the normal ordered product of $\psi(z)$ and $\phi(z')$ as

$$(\psi\phi)(z) = V(\psi_{-h}\phi_{-h'}|0\rangle, z), \qquad (7)$$

with modes

$$(\psi\phi)_m = \sum_{n \leq -h} \psi_n\phi_{m-n} + \sum_{n > -h} \phi_{m-n}\psi_n. \qquad (8)$$

This is a non-commutative, non-associative operation. There are many other normal ordering prescriptions, which all differ by finite local fields, for example the prescription of Nahm [45] which includes a projection onto the subspace annihilated by L_1.

How does this all work in practice? Let's consider the simplest non-trivial example, being the operator product of $L(z)$ with $L(z')$. Since these fields are both of weight 2, we can immediately use (6) to give

$$L(z)L(z') = \sum_m (z - z')^{-m} V(L_{m-2} L_{-2} |0\rangle, z') . \qquad (9)$$

As we shall only be interested in the singular part, we only need to consider the states

$$L_{m-2} L_{-2} |0\rangle , \qquad (10)$$

for $m > 0$. We use the Virasoro commutation relations to find

$$L_{m-2} L_{-2} |0\rangle = \left(L_{-2} L_{m-2} + \frac{c}{2} \delta_{m,4} + m L_{m-4} \right) |0\rangle , \qquad (11)$$

m	1	2	3	4	≥ 5
$L_{m-2} L_{-2} \|0\rangle$	$L_{-1} L_{-2} \|0\rangle$	$2 L_{-2} \|0\rangle$	0	$(c/2) \|0\rangle$	0

$$(12)$$

Combining these results, and remembering that $L_{-1} \equiv \partial/\partial z$, we get

$$L(z) L(z') = \frac{c/2}{(z - z')^4} + \frac{2 L(z')}{(z - z')^2} + \frac{L'(z')}{z - z'} + \text{regular terms} . \qquad (13)$$

Using the standard method for obtaining the commutator of two modes from the double contour integral of the singular part of the operator product expansion, we recover the Virasoro algebra (2).

Taking Wick contractions may be much faster for working out operator product expansions of expressions involving only free fields, but for the more complicated algebras we will have, the method outlined above is the only practical method, and is the method by which we shall work out all operator product expansions.

3   What is a W Algebra?

We are now in a position to define a W-algebra.

[1] *A W-algebra is a meromorphic conformal field theory.*
[2] *There is a distinguished set of fields $W^a(z)$ which are primary with respect to the Virasoro algebra and have conformal weight h^a*

This means that

$$L(z) W^a(z') = \frac{h^a W^a(z')}{(z - z')^2} + \frac{W^{a\prime}(z')}{z - z'} + \text{regular terms} , \qquad (14)$$

or equivalently

$$[L_m , W_n^a] = (m(h^a - 1) - n) W_{m+n}^a . \qquad (15)$$

[3] *The Hilbert space $\mathcal{H}$ is spanned by states of the form*

$$W_{-m_1}^{a_1} W_{-m_2}^{a_2} \cdots \cdots W_{-m_s}^{a_s} \, L_{-n_1} \cdots \cdots L_{-n_y} |0\rangle \,, \qquad (16)$$

which are ordered, in the sense that we consider all fields of the same type first, $a_i \geq a_{i+1}$, and that the modes are always increasing for fields of the same type, i.e. $n_j \geq n_{j+1} \geq 2$ and $m_j \geq m_{j+1} \geq h^{a_j}$ if $a_j = a_{j+1}$.

We can deduce from (3) that the action of L_n for any n, and of W_m^a for any a and m, on a state of this sort will give a finite sum of states of the same sort, so that this definition is consistent.

Furthermore, (6) and (16) mean that the operator product of two fields which are normal ordered polynomials in L, W^a and their derivatives closes on such normal ordered combinations, so that the operator product algebra of a W-algebra is closed in this sense.

4   Simple W-Algebras

The whole subject of W-algebras started with the paper [49] of Zamolodchikov in which he presented the first non-trivial examples. He started by considering all $h^a = 1$, and found that the only algebras which resulted were direct sums of affine Lie algebras. Again, considering only $h^a = 2$ leads to direct sums of commuting Virasoro algebras.

Although considering mixtures of these two, i.e. some $h = 1$ and some $h = 2$ does lead to interesting possibilities [14], the first non-trivial example Zamolodchikov found was when he considered extending the Virasoro algebra by a single field of weight 3. The resulting algebra is known as the W_3 algebra, and since we shall illustrate most of our examples with this algebra, shall go through its derivation in detail in the next section.

4.1   The W_3 algebra

We shall consider a W-algebra with one extra field $W(z)$ of weight 3, so that

$$[L_m , W_n] = (2m - n) W_{m+n} \,, \qquad (17)$$

and so that the Hilbert space is spanned by states of the form

$$W_{-m_1} \cdots W_{-m_j} L_{-n_1} \cdots L_{-n_k} |0\rangle \,, \qquad (18)$$

where $m_i \geq m_{i+1} \geq 3$ and $n_i \geq n_{i+1} \geq 2$.

Now we can ask what the ope of $W(z)$ with $W(z')$ is, or equivalently the commutator $[W_m , W_n]$. Using (6) we have

$$W(z)\,W(z') = \frac{V(W_3 W_{-3}|0\rangle, z')}{(z - z')^6} + \frac{V(W_2 W_{-3}|0\rangle, z')}{(z - z')^5} + \frac{V(W_1 W_{-3}|0\rangle, z')}{(z - z')^4}$$

$$+ \frac{V(W_0 W_{-3}|0\rangle, z')}{(z - z')^3} + \frac{V(W_{-1} W_{-3}|0\rangle, z')}{(z - z')^2} + \frac{V(W_{-2} W_{-3}|0\rangle, z')}{z - z'}$$

$$+ \text{ regular terms} \,. \qquad (19)$$

Since the Hilbert space is spanned by the states (18), we can express the states $W_m W_{-3}|0\rangle$ which occur here as follows:

$$
\begin{aligned}
W_3\, W_{-3}|0\rangle &= \alpha_1|0\rangle\,, \\
W_2\, W_{-3}|0\rangle &= 0\,, \\
W_1\, W_{-3}|0\rangle &= \alpha_2 L_{-2}|0\rangle\,, \\
W_0\, W_{-3}|0\rangle &= (\alpha_3 L_{-3} + \alpha_4 W_{-3})\,|0\rangle\,, \\
W_{-1}\, W_{-3}|0\rangle &= (\alpha_5 L_{-4} + \alpha_6 L_{-2}L_{-2} + \alpha_7 W_{-4})\,|0\rangle\,, \\
W_{-2}\, W_{-3}|0\rangle &= (\alpha_8 L_{-5} + \alpha_9 L_{-3}L_{-2} + \alpha_{10}W_{-5} + \alpha_{11}W_{-3}L_{-2})\,|0\rangle\,.
\end{aligned}
\tag{20}
$$

We have the freedom to choose one of the 11 unknowns α_i to fix the scale of $W(z)$, and by convention we choose $\alpha_1 = c/3$.

To fix the remaining 10, we require that $[W_m, W_n]$ is antisymmetric in m and n and that the Jacobi identity

$$
[\,L_m\,,[\,W_n\,,W_p\,]\,] + [\,W_n\,,[\,W_p\,,L_m\,]\,] + [\,W_p\,,[\,L_m\,,W_n\,]\,] = 0\,,
\tag{21}
$$

holds. This results in

$$
\alpha_2 = 2,\ \alpha_3 = 1,\ \alpha_4 = \alpha_7 = \alpha_{10} = \alpha_{11} = 0\,,
$$
$$
\alpha_5 = \tfrac{3}{5} - \tfrac{6}{5}\tfrac{16}{22+5c}\,,\ \alpha_6 = \tfrac{32}{22+5c}\,,\ \alpha_8 = \tfrac{2}{5} - \tfrac{4}{5}\tfrac{16}{22+5c}\,,\ \alpha_9 = \tfrac{32}{22+5c}\,.
\tag{22}
$$

This has fixed all the unknowns in the W_3 commutators, and we still have not checked the Jacobi identity

$$
[W_m, [W_n, W_p]] + [W_n, [W_p, W_m]] + [W_p, [W_m, W_n]] = 0\,.
\tag{23}
$$

However, this holds identically, and so we have arrived at a consistent set of operator product expansions. It is conventional to define

$$
\Lambda(z) = (LL)(z) - \frac{3}{10}L''(z)\,,
\tag{24}
$$

so that the commutation relations which result are

$$
\begin{aligned}
[\,W_m\,,W_n\,] ={}& \frac{c}{360}m(m^2-1)(m^2-4)\delta_{m+n,0} + \frac{16}{22+5c}\Lambda_{m+n} \\
&+ \frac{(m-n)(2m^2 - mn + 2n^2 - 8)}{30}L_{m+n}\,.
\end{aligned}
\tag{25}
$$

Eqn (25), together with (2) and (17), defines the W_3 algebra.

Several comments are now in order:

(1) To check the Jacobi identity (21) in the manner indicated, it is necessary to work out several new commutators, for example

$$
[\,L_m\,,\Lambda_n\,] = (3m-n)\Lambda_{m+n} + \frac{(22+5c)}{30}m(m^2-1)\,L_{m+n}\,.
\tag{26}
$$

It is very easy to find this using methods analogous to those which gave (13), and is a useful exercise. To derive this commutator using the mode expansion for $\Lambda(z)$ is possible, but very quickly this method becomes impossible for normal ordered products of three or more fields.

However, in fact there are easier ways to fix 9 of the unknowns in (20), and that is to realise that they are the coefficients of states which are Virasoro descendants of the vacuum or of the Virasoro highest weight state $W_{-3}|0\rangle$; consequently, they can be found readily in terms of α_1 and α_4 using the descent equations of appendix B of [6].

(2) It is already clear that any attempt at classification of W-algebras will be hard, and that there will be many identities amongst W-algebras[1]. Consider what happens to the W_3 algebra at $c = -22/5$. The combination $16/(22+5c)$ diverges, but this is one of the structure constants of the algebra, appearing in (25). It is possible to solve this problem by rescaling W so that we then have the commutation relations

$$[L_m, \tilde{W}_m] = (2m-n)\tilde{W}_{m+n}$$

$$[\tilde{W}_m, \tilde{W}_n] = (m-n)\Lambda_{m+n}$$
$$+ \frac{22+5c}{16}\left[\frac{cm(m^2-1)(m^2-4)}{360}\delta_{m+n,0} + \frac{(m-n)(2m^2-mn+2n^2-8)}{30}L_{m+n}\right],$$

$$[L_m, \Lambda_n] = (3m-n)\Lambda_{m+n} + \frac{22+5c}{16}\left[m(m^2-1)L_{m+n}\right],$$

$$\vdots \qquad\qquad\qquad\qquad (27)$$

and if we continue we shall see that at $c = -22/5$, $\tilde{W}$ and Λ generate an ideal and so must be set to zero in any irreducible representation of the W_3 algebra, so that at $c = -22/5$ the W_3 algebra reduces to the Virasoro algebra.

However, the representation theory of the W_3 algebra naturally includes $c = -22/5$, and all expressions such as character formulae, fusion rules, etc, are regular at $c = -22/5$, so that many theorems which we might like to be true are going to be complicated by such facts.

(3) So far we have the commutation relations (2), (17) and (25) which satisfy the Jacobi identity. Is this sufficient to guarantee that the operator product is associative? In general, the answer is no. Consider the algebra with generators i, j, k and multiplication

$$i^2 = j^2 = k^2 = 1,$$
$$ij = -ji = \alpha k,$$
$$jk = -kj = \beta i, \qquad\qquad (28)$$
$$ki = -ik = \gamma j.$$

The Jacobi identity is satisfied, and in fact the commutator algebra is isomorphic to $sl(2)$, but it is an easy exercise to show that this algebra is only associative for $\alpha = \beta = \gamma = \pm 1$.

---

[1] For a fuller list of such relations, see e.g. [8,9].

However, given a commutator algebra satisfying the Jacobi identity, we can define an associative product by constructing the universal enveloping algebra and then saying that the algebra acts in this by multiplication. This is guaranteed to be an associative action.

(4) The W_3 algebra is not a Lie algebra since the commutator (25) includes an infinite sum of bilinears in the modes L_m. It is very hard to work with this algebra as a Lie algebra – it becomes necessary to introduce ever more generators such as Λ_m – but it is quite straightforward as a meromorphic conformal field theory.

In fact, if one wishes to consider a W-algebra in the same way as a Lie algebra as a space of modes with commutation relations, albeit quadratic or higher, it is quite hard to define what the space of modes is. It is clear to a physicist which expressions should make sense and which not – those which have a finite expectation value in any state – but this is rather recursive since this already uses the algebraic structure to evaluate the expectation value. We escape this dilemma by defining the W-algebra as an mcft.

5   Direct Searches for W-Algebras

Between 1990 and 1992, roughly, people looked for W-algebras, using essentially the method we have seen, that is to consider a set of weights $\{h^a\}$ and then, as we did for the W_3 algebra, consider the most general set of operator product expansions and then, by imposing the Jacobi identity (or some equivalent means), discover whether there are any consistent solutions or not. The following table gives a list of such algebras investigated so far in this way, and we then make comments on the entries in each of the columns of this table.

5.1   Allowed c Values

The allowed sets of c values fall into two classes, either an algebra is consistent for all c values, or only a finite set.

A typical example of an algebra which is only consistent for a finite set of c values is the $W(2, 5/2)$ algebra with one extra field $W(z)$ of weight 5/2 which was considered by Zamolodchikov [49]. As for the W_3 algebra, the commutator $[W_m, W_n]$ is entirely fixed by requiring that (21) vanishes, but when we insert this into (23) he found that W_{m+n+p} appears in the right hand side for all values of c except $c = -13/14$, and consequently it is only for this value of c that the Jacobi identity is satisfied.

For the purposes of this table we have allowed $c = -22/5$ as a consistent value for the W_3 algebra, although we have seen earlier that in fact it should be consistently truncated to the Virasoro algebra. This is a generic phenomenon: all the W-algebras with fields of weights greater than 2 have a finite set of c values where a structure constant is singular, and for which one must truncate the spectrum of fields. In fact, it occurs for the $WB(0, 2)$ algebra with fields of

Table 1. Constraints on the consistency of W-algebras

$\{h^a\}$	Consistent c-values	Null fields	Names	DS	Coset	Ref.
3/2	all	none	$svir$	✓	✓	49
2	all	none	$vir \oplus vir$	✓	✓	49
5/2	$-13/14$	$11/2,..$	$W(2,5/2)$			49
3	all	none	$WA_2 \equiv W_3$	✓	✓	49
7/2	$21/22, -19/6, -161/8$		$W(2,7/2)$			7
4	all	none	$WB_2 \equiv WC_2$	✓	—	12
9/2	$25/26, -7/20, -125/22,$ $-279/10, -35$		$W(2,9/2)$			7
5	$6/7, -250/11, -7, 134 \pm 60\sqrt{5}$		$W(2,5)$			7, 12, 44
11/2	$-217/16$		$W(2,11/2)$			7
6	all	none	$WG_2 \equiv W(2,6)$	✓	—	7, 30, 44
13/2	$9/34, -611/14, -111/10$		$W(2,13/2)$			7
7	$-25/2$		$W(2,7)$			7, 44
15/2	$25/28, -11/38, -39/10,$ $-473/34, -825/16, -59$		$W(2,15/2)$			7
8	$21/2, -1015/2, -224/65, -23,$ $-712/7, -3164/23,$ $350 \pm 252\sqrt{2}, -944/17$		$W(2,8)$			7, 44
9	$-1206/19, -14/11, -208/35,$ $-91/5, -71$		$W(2,9)$			22
10	$8/35, 25/26, -29, -2$		$W(2,10)$			22
11	$-36/13, -1826/23, -24$		$W(2,11)$			22
3, 3	none					7, 44
3, 4	all	none	$WA_3 \equiv WD_3$	✓	✓	7, 44
3, 5	none					7, 44
4, 5/2	all	none	$WB(0,2)$	✓	✓	22
4, 7/2	$1, -403/22$					22
4, 4	$1, -656/11$					7, 44
4, 9/2	$1, -141/2, -779/26$					22
4, 5	$1, -253/7, -1060/13$					44, 22
4, 6	all	$10, \ldots$	$Orb(svir)$			44
	all	$10, \ldots$	WD_{-1}	—	✓	44
	all	none	WB_3	✓	—	44
	all	none	WC_3	✓	—	44
6, 7/2	$561/2$					22
6, 9/2	$-304/5$					22

Table 2. Constraints on the consistency of W-algebras (cont.)

$\{h^a\}$	Consistent c-values	Null fields	Names	DS	Coset	Ref.
$3,3,3$	$-2,-30$	$6,\ldots$				43
$3,4,5$	all	none	$WA_4 \equiv W_5$	✓	✓	39
	all	$8,8,\ldots$	$W_{\{2,3\}}$	✓	✓	39
$5,5,5$	-7	$8,\ldots$				43
$7,7,7$	$-25/2$	$10,\ldots$				43
$3,4,5,6$	all	none	$WA_5 \equiv W_6$	✓	✓	39

weight 4 and 5/2 exactly at $c = -13/14$. In this case the structure constant appearing in the operator product $W^4 W^4 \sim C^4_{44} W^4$ has a singularity, and so we must re-scale this field. Doing so we in fact remove it entirely from the W-algebra and we arrive at the $W(2,5/2)$ algebra above

The algebras which are consistent for only a finite set of c values are (almost all) uninteresting as W-algebras, as they arise in this way as special cases of the other algebras when the spectrum must be truncated. For example, the $W(2,8)$ algebra with $c = 21/22, -944/17$ and $-712/7$ is a truncation of WE_8, at $c = -3164/23$ of WE_7, and all the allowable c values for the algebras $W(2,4,4)$ and $W(2,4,5)$ arise as truncations of the algebra WD_4 and WD_5 respectively [22]. However, they are certainly interesting as rational conformal field theories.

5.2   Null Fields

It may happen for an algebra that the right hand side of the Jacobi identity

$$[W^a_m, [W^b_n, W^c_p]] + \text{cyclic perms.} = X_{m+n+p}, \tag{29}$$

does not vanish identically. However, this need not be a problem if the combination X_{m+n+p} can be consistently set to zero. For example, this occurs in the $W(2,5/2)$ algebra at $c = -13/14$ where the Jacobi identity for the field W is broken by a term proportional to

$$U(z) = -28(LW') + 35(L'W) + 4W'''. \tag{30}$$

However, this field decouples from all expectation values and should be set to zero in any physical correlation function. Such a field we call a null field, and we indicate the first level h at which such a field occurs, if it is known.

While it might be that this phenomenon is solely due to the fact that the $W(2,5/2)$ algebra is only consistent for one c value, we see that this is not always the case, and that the algebras WD_{-1} and $W_{2,3}$ which are consistent for all c values are also afflicted in this way, there being fields polynomial in the W-algebra generators which must be set equal to zero for all values of c.

This causes many problems if one attempts to follow a 'classical' route to their representation theory, as the universal enveloping algebra contains a very large non-trivial ideal, much larger than one would like, which should be set to zero for all representations of physical interest.

In all the W-algebras which are defined for all c values and for which the coupling constants have been explicitly found, the coupling constants depend upon c in a typical manner: if the coefficients appearing in the singular part of the operator product expansion are

$$W^a W^b \sim \frac{c}{h^a}\delta^{ab} + \sum C^{ab}_{d_1 d_2 \ldots d_n}(W^{d_1}(W^{d_2}(\ldots W^{d_n}))) + \text{ regular terms } , \quad (31)$$

then we find that

$$C^{ab}_{d_1 d_2 \ldots d_n} \sim \begin{cases} c^{1-n} & \text{No generic null fields} \\ c^{1-n/2} & \text{Generic null fields} \end{cases} \quad (32)$$

This has implications if we try to take the classical limit, which corresponds to $c \to \infty$ (see e.g. [15, 10, 27]).

5.3   Names

There is no consistent convention for naming W–algebras; the names given in the table are some which the reader might expect to meet, but it seems that almost every group has their own favourite convention. For some algebras, the names simply indicate the field content, e.g. $W(2, 5/2)$, but as this does not uniquely specify an algebra (there are four of type $W(2, 4, 6)$) it is not perfect. Other names, such as WA_n, $WB(0, 2)$, indicate that the algebra is a 'Casimir' type algebra, of which more later.

5.4   Constructions

In the final two columns it is indicated whether this algebra is known to be found via either of the two main methods of construction W-algebras, namely the Drinfel'd Sokolov construction and the Coset construction.

Each construction has its own merits, and since the W_3 algebra is our typical example and can be found using both methods, we shall go through these two constructions in some detail in the next two sections.

5.5   Limitations of this Method

It is clear that this method can never produce a classification of W-algebras, only find a (hopefully representative) selection of algebras from which one may start to find conjectures and turn these conjectures into theorems.

However, it turns out that in fact the algebras listed in table 1 are not fully representative of W–algebras in general, for the following reason: One of the main limitations of this method is that it is technically hard to deal with more than one field of the same spin, and even harder to deal with W-algebras which include affine Lie subalgebras.

As Zamolodchikov said in his paper [49], it would surely be interesting to investigate algebras with fields of weight 1 and 2; in fact it runs out that the most general constructions of W-algebras now known will typically produce algebras with large affine Lie subalgebras, and that these were not noticed for a long time, simply because they were not looked for. As a result, the early literature is biased towards algebras with no weight 1 fields. This has now been rectified in the light of more recent work.

6   The Coset Construction

As explained in Jurgen Fuchs' lectures, we can consider an affine algebra $\hat{g}$ with subalgebra $\hat{h}$, and then the coset algebra $W(\hat{g}/\hat{h})$ is the set of all fields polynomial in the currents of $\hat{g}$ and their derivatives which commute with the currents of $\hat{h}$.

The polynomials in the currents of an untwisted affine algebra always contain a canonical Virasoro algebra $L^g(z)$, which for each semisimple $\hat{g}$ is given by the Sugawara construction

$$L^g(z) = \frac{1}{2(k + h^\vee(g))} \left(J_g^a J_g^a \right)(z) , \tag{33}$$

where k is the level of g, $h^\vee$ is the dual Coxeter number of g, and which has central charge c^g,

$$c^g = c(g, k) = \frac{k \dim(g)}{k + h^\vee} . \tag{34}$$

Similarly there is a Virasoro algebra $L^h(z)$ corresponding to the subalgebra $\hat{h}$, and the coset algebra contains a Virasoro algebra given by

$$L^{g/h} = L^g - L^h . \tag{35}$$

The currents $J_h^a(z)$ are primary fields of weight 1 with respect to both L^g and L^h, hence they commute with $L^{g/h}$, and it is easy to check that $L^{g/h}$ satisfies the Virasoro algebra operator product expansion with

$$c^{g/h} = c^g - c^h . \tag{36}$$

Furthermore, given that the vacuum representation of $\hat{g}$ is a mcft, it is easy to check that the coset algebra $W(\hat{g}/\hat{h})$ is also a mcft, with Hilbert space consisting of all states ψ such that $J_m^a \psi = 0$ for J^a in h and $m \geq 0$.

In order for this coset algebra to be a W-algebra, we have to check that $W(\hat{g}/\hat{h})$ is generated by a finite set of fields. For most cases the answer to this question is not known, but it is generally believed that this is *always* the case.

While $W(\hat{g}/\hat{h})$ may be generated by a finite set of fields, in most cases this is not freely-generated, i.e. there may be generic null fields of the sort discussed earlier which have to be consistently set to zero.

How can we check whether $W(\hat{g}/\hat{h})$ is a W-algebra? One way is by a counting argument. As explained in Fuchs' lectures, the branching functions of a coset are not necessarily single characters of irreducible representations of the coset algebra, but they are certainly sums of characters. This gives an easy proof that the coset

$$\frac{su(3)_k \oplus su(3)_1}{su(3)_{k+1}} , \tag{37}$$

contains the W_3 algebra. In this coset, g is the direct sum $su(3)_k \oplus su(3)_1$ of two affine $su(3)$ algebras at levels k and 1, with currents $J_k^a(z)$ and $J_1^a(z)$ respectively. Then the currents $J_{k+1}^a = J_k^a(z) + J_1^a(z)$ generate an $su(3)$ affine algebra at level $k+1$, which is the subalgebra h.

From the Kac-Weyl formula in Fuchs' lectures, we can calculate the branching function relating the vacuum representations of g and h for large positive integer k. So, if

$$\chi_0^g(q) = \chi_0^h(q)b(q) + \dots , \tag{38}$$

where $\chi_0^g(q)$ and $\chi_0^h(q)$ are the characters of the vacuum representations of g and h, then we find that the branching function $b(q)$ is

$$b(q) = 1 + q^2 + 2q^3 + 3q^4 + 4q^5 + \dots . \tag{39}$$

We know that the coset algebra must contain the Virasoro algebra, and so in particular the Hilbert space must contain the states

$$|0\rangle, \ L_{-2}^{g/h}|0\rangle, \ L_{-3}^{g/h}|0\rangle, \ L_{-2}^{g/h}L_{-2}^{g/h}|0\rangle, \ L_{-4}^{g/h}|0\rangle, \ L_{-3}^{g/h}L_{-2}^{g/h}|0\rangle, \ L_{-5}^{g/h}|0\rangle . \tag{40}$$

If we subtract the contributions of these states from $b(q)$, we see that there are still states at levels 3, 4 and 5 unaccounted for. If we call the state at level 3 $W_{-3}|0\rangle$, then those at levels 4 and 5 must be

$$W_{-4}|0\rangle, \ W_{-3}L_{-2}^{g/h}|0\rangle, \ W_{-5}|0\rangle , \tag{41}$$

and that exhausts $b(q)$ up to level 5. In the operator product of $W(z)$ with $W(z')$, only fields corresponding to states of weight 5 or less can arise; since we have just seen that these states are exactly those which we considered when we constructed the W_3 algebra the first time, associativity of the operator product expansions forces us to conclude that, exactly as before, the field W generates the W_3 algebra.

One must be very careful when applying this counting argument as it may be that 'generic' null vectors, or null vectors at particular c-values imply that there are relations between the states which one would believe were independent. In this case we circumvent the problem as the only null vectors which might exist amongst the states (40), (41) would be Virasoro descendants, and by taking k to be a large positive integer we are guaranteed by the representation theory of the

Virasoro algebra that this is not the case. For a fuller discussion of this problem, see e.g. [13, 10].

It is possible to find the explicit construction [2,3]. If we introduce the totally symmetric $SU(3)$ invariant tensor d_{abc},

$$d_{abc} = \mathrm{Tr}\left(T_a\{T_b, T_c\}\right),$$

then, up to an overall factor, $W(z)$ is given by

$$d_{abc}\Big(k(3+k)(3+2k)J_1^a J_1^b J_1^c - 3.4(3+k)(3+2k)J_k^a J_1^b J_1^c$$

$$+3.4.5(3+k)J_k^a J_k^b J_1^c - 3.5 J_k^a J_k^b J_k^c\Big). \tag{42}$$

The central charge of the W_3 algebra so constructed is

$$c = 2\left(1 - \frac{12}{(k+3)(k+4)}\right). \tag{43}$$

6.1   Merits of this Construction

It is known that for positive integer k, the vacuum representation of the affine algebras are unitary, and consequently the W_3 algebra has unitary representations for c-values (43), and any unitary representation of $su(3)_k \oplus su(3)_1$ for positive integer k automatically induces a unitary representation of the W_3 algebra. This is the only construction so far that is proven to lead to unitary representations of the W_3 algebra with $c < 2$, and it is firmly believed that there are no more unitary representations with $c < 2$ than those which arise this way.

Furthermore, it is believed that in this case the branching functions of the coset are actually the characters of irreducible representations of the W_3 algebra, although this is not yet proven. If this is the case, then one can now consider the classification of all the modular invariant partition functions of theories with W_3 symmetry and c given by (43).

If $\mathcal{H}$ is the full Hilbert space of a conformal field theory, including all local fields, not only those in the chiral algebra, then as Fuchs explained, it splits into the direct sum of representations of the left (z dependent) and right ($\bar{z}$ dependent) chiral algebras,

$$\mathcal{H} = \bigoplus n_{ii'} H_i \otimes \bar{H}_{i'}, \tag{44}$$

where H_i is the representation space of the left chiral algebra, and $\bar{H}_{i'}$ of the right, and $n_{ii'}$ are non-negative integers, and where the vacuum sector arises only once, i.e. $n_{00} = 1$. Then the partition function on a torus is given by

$$Z = \sum n_{ii'} \chi_i(q) \bar{\chi}_{i'}(\bar{q}), \tag{45}$$

where $\chi_i(q)$ and $\bar{\chi}_{i'}(\bar{q})$ are the characters of these spaces. If the chiral algebra is an $\widehat{su}(2)$ or $\widehat{su}(3)$ current algebra, then all possible sets $n_{ii'}$ have been classified using the fact that Z should be invariant under reparametrisations of the torus, that is if $q = \exp(-2\pi i \tau)$, then Z should be invariant under $T : \tau \to \tau + 1$ and $S : \tau \to -1/\tau$. For $\widehat{su}(2)$ this is the famous ADE classification of Cappelli et al. [16], and Gannon has found a similar result for $\widehat{su}(3)$ [36].

More recently, the classification of the modular invariant partition functions of theories with W_3 symmetry and c in (43) has been achieved by Gannon and Walton [37]. Again this is of an A-D-E type, by which is meant that there are certain infinite series of modular invariant partition functions for all k, and certain extra discrete invariants, which this time occur for $k = 4, 5, 8, 9, 20$ and 21. In the 'A' type series the chiral algebra is purely the W_3 algebra, whereas in the others it may be increased by the addition of other extra fields.

This is another case where the classification of conformal field theories using W-algebras becomes inherently difficult! These extra cases with increased symmetry naturally fall into the W_3 algebra classification, but yet their algebras are larger – similarly they will certainly also fall into the classification of modular invariant partition functions for larger algebras which have had their spectrum truncated.

7   The Drinfel'd–Sokolov Construction

The classical Drinfel'd–Sokolov construction, which is the basis of the quantum Hamiltonian reduction construction, is an old construction from the theory of classical integrable systems [20]. For the sake of brevity we shall simply state the ingredients and the method; for more details and the connection to W-algebras there are many reviews available e.g. [26].

Consider a set of currents $j^a(x)$ in a finite dimensional Lie algebra g with $x \in S^1$, whose Poisson brackets satisfy an affine Lie algebra,

$$\{j^a(x), j^b(y)\} = \delta(x - y) f_c^{ab} j^c(y) + g^{ab} \delta'(x - y) . \qquad (46)$$

Then consider the matrix J and functional X where

$$J = j^a(z) T_a , \quad X = \int_0^{2\pi} dx \, j^a(x) f_a(x) . \qquad (47)$$

Then the action of X on J is a gauge transformation:

$$\delta_X J \equiv \{J, X\} = [\hat{X}, J] + \hat{X}' , \quad \text{where} \quad \hat{X} = f_a(x) T_a , \qquad (48)$$
$$[T^a, T^b] = f_c^{ab} T^c , \quad [T^a, T_c] = -f_c^{ab} T_b .$$

We now consider various gauge-fixings and the space of functionals which are invariant under the residual gauge symmetry. For our purposes it will be sufficient to consider constructions based on an embedding

$$sl(2) \hookrightarrow g , \qquad (49)$$

and further consider only those embeddings for which g decomposes into integer spins representations of $sl(2)$. If the generators of the $sl(2)$ are I^+, I^0, I^- then we can decompose g into eigenspaces of I^0,

$$g = \oplus g_m \quad \text{where} \quad [I^0, g_m] = m \, g_m \tag{50}$$

and also split g into three parts

$$g = g_- \oplus g_0 \oplus g_+ \quad \text{where} \quad g_+ = \oplus_{m>0} g_m \; , \quad g_- = \oplus_{m<0} g_m \; , \tag{51}$$

Now we choose to fix the currents in g_- so that

$$J_+ = I_+ \; . \tag{52}$$

This leaves a residual g_- gauge symmetry generated by the Poisson brackets with the constrained currents. The polynomials in the remaining currents which are invariant under these residual gauge transformations will satisfy a classical W-algebra. We can always choose to gauge fix in the so-called 'highest weight gauge' by taking

$$J_{\text{fix}} = I_+ + \sum X^a W^a \; , \tag{53}$$

where X^a are highest weights of spin i^a for the $sl(2)$, and the W^a will have conformal weight $i^a + 1$. Let's do this explicitly in the simplest case where $g = sl(2)$.

7.1   DS Construction for $g = sl(2)$

We take $sl(2) \hookrightarrow sl(2)$ with

$$I^+ = \begin{pmatrix} 0 & 1 \\ 0 & 0 \end{pmatrix} \; , \quad I^0 = \begin{pmatrix} \frac{1}{2} & 0 \\ 0 & -\frac{1}{2} \end{pmatrix} \; , \quad I^- = \begin{pmatrix} 0 & 0 \\ 1 & 0 \end{pmatrix} \; , \tag{54}$$

and define I^a so that $\text{Tr}(I^a I_b) = \delta_b^a$. If we put

$$J = j^a I_a = \begin{pmatrix} j^0 & j^- \\ j^+ & -j^0 \end{pmatrix} \; , \quad X = \int_0^{2\pi} dx \; f_+(x) j^+(x) + f_0(x) j^0(x) + f_-(x) j^-(x) \tag{55}$$

we find that

$$\{J, X\} = \left[\hat{X}, J \right] + \hat{X}' \; , \quad \text{where} \quad \hat{X} = f_a(x) I^a = \begin{pmatrix} f_0/2 & f_+ \\ f_- & -f_0/2 \end{pmatrix} \; , \tag{56}$$

is consistent with the fundamental Poisson brackets

$$\{ j^0(x), j^\pm(y) \} = \pm j^\pm(y) \, \delta(x - y) \; ,$$
$$\{ j^+(x), j^-(y) \} = 2 \, j^0(y) \, \delta(x - y) + \delta'(x - y) \; ,$$
$$\{ j^0(x), j^0(y) \} = (1/2) \, \delta'(x - y) \; . \tag{57}$$

These Poisson brackets are indeed a classical affine $\hat{s}u(2)$ algebra, as we can check by considering the Poisson brackets of the modes

$$j_m^a = \int_0^{2\pi} \mathrm{d}x \; j^a(x) \exp(-imx) \; . \tag{58}$$

Now consider constraining J to be in the form

$$J_{\mathrm{cons}} = \begin{pmatrix} j^0 & 1 \\ j^+ & -j^0 \end{pmatrix} \; , \tag{59}$$

with the residual gauge symmetry generated by $j^-(x)$,

$$J \mapsto AJA^{-1} + A' \quad \text{where } A = \begin{pmatrix} 1 & 0 \\ a & 1 \end{pmatrix} \; . \tag{60}$$

Using this residual gauge symmetry we can always transform J into the form

$$J_{\mathrm{fix}} = \begin{pmatrix} 0 & 1 \\ l & 0 \end{pmatrix} \quad \text{where } l = j^+ + (j^0)^2 + (j^0)' \; . \tag{61}$$

Then, putting $l_m = \int_0^{2\pi} \mathrm{d}x \, l(x) \exp(-imx)$, it is an easy exercise to show that l_m the Poisson bracket algebra of the l_m is the Virasoro algebra, (up to a constant shift in l_0)

$$\{\, l_m \, , l_n \,\} = \frac{im^3}{2} \delta_{m+n,0} + i(m-n) l_{m+n} \; . \tag{62}$$

Note that by setting $j^+(x)$ to zero, we recover a classical free field construction, $l(x) = (j^0)^2 + (j^0)'$.

In the general case $sl(2) \hookrightarrow g$, as we said before, it is possible to fix J in the form

$$J_{\mathrm{fix}} = I_- + \sum X^a W^a \; , \tag{63}$$

where X_a is a highest weight of the $sl(2)$ of spin i_a and W^a is a primary field of weight $i_a + 1$. The embeddings (49) have been classified by Dynkin [18], and it is straightforward to find the spectrum of weights $\{h^a\}$ given an embedding.

For example, the embeddings of $sl(2) \hookrightarrow sl(n)$ are characterised by the partitions of n positive integers[2] and it is only the trivial partition of n which leads to no weight 1 fields in the resulting W-algebra. All other W-algebras obtained from the Drinfel'd-Sokolov construction based on $sl(n)$ will have at least a $u(1)$ subalgebra. For Drinfel'd-Sokolov constructions based on other algebras, there are more possibilities for $sl(2)$ embeddings which result in no weight 1 fields (see [15] for a full list) but still the embeddings with weight 1 fields greatly outweigh those without.

For each simple Lie algebra there is a canonical $sl(2)$ embedding with no singlets, and that is the principal embedding. In this case the spins of the representations appearing in the decomposition of the Lie algebra are exactly the

---

[2] If we want to ensure that there are only integer spin i_a representations of $sl(2)$ more constraints are necessary

exponents of the Lie algebra. For these embeddings the W-algebras are those originally found by Drinfel'd and Sokolov and are also known as the 'Casimir' W-algebras.

It is an easy exercise to show that the classical W-algebra obtained by this method from $sl(3)$ with the choice

$$
I^+ = \sqrt{2} \begin{pmatrix} 0 & 1 & 0 \\ 0 & 0 & 1 \\ 0 & 0 & 0 \end{pmatrix} , \quad I^0 = \begin{pmatrix} 1 & 0 & 0 \\ 0 & 0 & 0 \\ 0 & 0 & -1 \end{pmatrix} , \quad I^- = \sqrt{2} \begin{pmatrix} 0 & 0 & 0 \\ 1 & 0 & 0 \\ 0 & 1 & 0 \end{pmatrix} ,
$$

(64)

is the classical W_3 algebra.

7.2   Quantisation of the Drinfel'd–Sokolov Construction

The quantisation of the Drinfel'd-Sokolov construction has a somewhat che-quered history, but it is now clear that the correct way is by the BRST method, and when applied to the Drinfel'd-Sokolov construction, this now goes under the name of 'Quantum Hamiltonian Reduction'. Again, there are several nice arti-cles on this method as applied to W-algebras, e.g. [11, 28, 31], and I shall simply present an outline of the method and the results that it gives.

For notation, we shall take J^a to be currents in g, J^α to be currents in g_- which are to be constrained, and χ^α to be the value to which they are constrained. Then according to the BRST method, we introduce two fermionic fields $b^\alpha(z)$ and $c_\alpha(z)$ for each field to be constrained, with operator products

$$
b^\alpha(z)\, c_\beta(z') = \frac{\delta^\alpha_\beta}{z - z'} ,
$$

(65)

and then form the operator

$$
Q = \oint \frac{dz}{2\pi i} \left\{ (J^\alpha - \chi^\alpha)c_\alpha - \frac{1}{2} f^{\alpha\beta}_\gamma \left(b^\gamma \left(c_\alpha c_\beta \right) \right) \right\} .
$$

(66)

It is a simple exercise that $Q^2 = 0$. the W algebra is then the mcft with space of states

$$
\ker Q \, / \operatorname{Im} Q .
$$

(67)

Two important simplifications can be made:

(1) We introduce new currents $\hat{J}^a(z)$,

$$
\hat{J}^a(z) = J^a(z) + f^{a\beta}_\gamma \left(b^\gamma c_\beta \right)(z) ,
$$

(68)

since these combinations will occur in the final answer.

(2) We note that Q can be split in two as $Q = Q_0 + Q_1$,

$$
Q_0 = \oint \frac{dz}{2\pi i} \left\{ J^\alpha c_\alpha - \frac{1}{2} f^{\alpha\beta}_\gamma \left(b^\gamma \left(c_\alpha c_\beta \right) \right) \right\} , \quad Q_1 = \oint \frac{dz}{2\pi i} (-\chi^\alpha c_\alpha) .
$$

(69)

As explained in [11], the calculation of $(\ker Q/\text{Im}Q)$ can be split into separate calculations for the space of constrained currents and their associated b ghosts, and for the rest. The cohomology on the space of constrained currents and b ghosts is trivial and so we only need worry about the remaining currents, that is the c ghosts and the unconstrained currents.

De Boer and Tjin have found that this cohomology is identical as a vector space (i.e. ignoring the algebra structure) to the classical case, i.e. there is an independent field for each highest weight representation of $sl(2)$ in the decomposition of g, and furthermore these fields have the form

$$W^a = \hat{J}^a + \ldots + W[\hat{J}^a_{(0)}] \,, \tag{70}$$

where $\hat{J}^a(z)$ is the field corresponding to the highest weight X^a, and $W[\hat{J}^a_{(0)}]$ is an expression in the currents of g_0. It is further possible to show that this algebra closes as a W-algebra, although the structure constants have to be found explicitly in each case.

Some comments:

(1) One immediate consequence is that we know the relation between the levels of the affine Lie subalgebras which arise in the W-algebra. These correspond to singlets under the $sl(2)$, and so formula (70) reduces to

$$W^a = \hat{J}^a \,. \tag{71}$$

The only change in these currents from Quantum Hamiltonian reduction is the addition of ghost contributions which change the levels of each semisimple component by amounts which can be determined.

(2) Since the currents $\hat{J}^a$ still satisfy the relations of an affine Lie algebra with a shifted central term, we see that nothing is altered by restricting the expressions (70) to the term only involving the currents in g_0. If g_0 were an Abelian algebra, as happens for the principal embeddings, then this would mean that the currents in g_0 were free fields, and we would find a free-field construction of W-algebras. In fact many W-algebras were first discussed by Fateev and Luk'yanov exactly in terms of their free field constructions [23,24]. However, in the general case g_0 is not Abelian, and so we find a construction of the W-algebra in the currents of some affine Lie algebra, generalising the free field construction.

7.3   Representations of W-Algebras from Quantum Hamiltonian Reduction

Clearly we can consider the space (67) based on any representation of $\hat{g}$, and each of these cohomology space will be a representation of the relevant W-algebra. Frenkel, Kac and Wakimoto did this for various representations of $\hat{g}$, and found a consistent set of conjectures for characters of irreducible representations of W_3 minimal model representations.

The minimal models are those for which the sum in the partition function (45) has only a finite number of $n_{i,i'}$ non-zero; this is a very strong requirement, and is only possible for a discrete set of c values. Using their conjectures, Frenkel et al. found the minimal models of the W_3 algebras have

$$c = 2 \left(1 - \frac{12(p-q)^2}{pq} \right) , \qquad (72)$$

where p and q are coprime positive integers, greater than 2. The representations which occur in these models are labelled by four integers $[a, b; c, d]$ with $1 \leq a, b, c, d;\ a + b \leq p - 1;\ c + d \leq q - 1$ and subject to the identifications

$$[a, b; c, d] \equiv [b, (p - a - b); d, (q - c - d)] \equiv [(p - a - b), a; (q - c - d), d] , \quad (73)$$

so that the number of different representations is $(p - 1)(p - 2)(q - 1)(q - 2)/3$, We shall leave a precise description of these representations and their definitions until section 8.4.

For $q = p + 1$ we see that the c values (72) are the same as the series (43) obtained from the coset construction with $p = k + 3, q = k + 4$. In this case, the characters that Frenkel et al conjectured are the same as the branching functions of the coset model. Frenkel et al also computed the fusion rules of the W_3 algebra via the Verlinde formula [47], using the transformations of their characters under the modular group generated by S and T. However, although it is firmly believed to be the case, I think it is fair to say that there is still no rigorous proof that these expressions are the characters of the irreducible representations of the W_3 algebra.

8 Representations of the W_3 Algebra and Correlation Functions

8.1 Introduction

All this discussion so far has been a bit academic if the study of these symmetry algebras does not help one work out correlation functions, which are the basic physical quantities of a conformal field theory. In this section, we hope to show that indeed it is possible to use W-algebra symmetry to find correlation functions, that this is a mathematically interesting problem, and that it is substantially harder than the corresponding problem for the Virasoro algebra, i.e. for theories with pure conformal symmetry. Purely for reasons of presentation, these discussions will be limited to the W_3 algebra, but they can just as easily be applied to any algebra.

8.2   A highest weight representation of the W_3 algebra

A highest weight representation of the W_3 algebra is a space on which L_0 is diagonalisable, and for which it has a minimal eigenvalue h, say. If the representation is to be irreducible, since W_0 and L_0 commute, this space must be one-dimensional, spanned by a state $|h, w\rangle$ for which

$$
\begin{aligned}
L_m|h, w\rangle &= 0, m > 0 , & L_0|h, w\rangle &= h\,|h, w\rangle , \\
W_m|h, w\rangle &= 0, m > 0 , & W_0|h, w\rangle &= w\,|h, w\rangle ,
\end{aligned}
\tag{74}
$$

and the whole space is generated by the action of the modes W_m and L_m on this state. We shall denote an irreducible representation variously by $L_{h,w}$, $L_{[ab;cd]}$ and L_a.

Since the whole conformal field theory has two chiral algebras, we should really consider fields $\Phi_{h,w;\bar{h},\bar{w}}(z, \bar{z})$, and states which carry representations of both left and right chiral algebras, but we shall essentially ignore the $\bar{z}$ dependence and suppress the dependence on h' and w'. Accordingly, we shall loosely say that the state $|h, w\rangle$ corresponds to a field $\phi_{h,w}(z)$. Then the first things we would like to know about this field is its operator product with $L(z)$ and $W(z)$. We assume that the rules are the same as those of mcft, and write

$$
\begin{aligned}
L(z)\phi(z') &= \frac{V(L_0|h, w\rangle, z')}{(z - z')^2} + \frac{V(L_{-1}|h, w\rangle, z')}{z - z'} + O(1) \\
&= \frac{h\phi(z')}{(z - z')^2} + \frac{\phi'(z')}{z - z'} + O(1)
\end{aligned}
\tag{75}
$$

$$
\begin{aligned}
W(z)\phi(z') &= \frac{V(W_0|h, w\rangle, z')}{(z - z')^3} \frac{V(W_{-1}|h, w\rangle, z')}{(z - z')^2} + \frac{V(W_{-2}|h, w\rangle, z')}{z - z'} + O(1) \\
&= \frac{w\phi(z')}{(z - z')^3} + \frac{\hat{W}_{-1}\phi'(z')}{(z - z')^2} + \frac{\hat{W}_{-2}\phi'(z')}{z - z'} + O(1) .
\end{aligned}
\tag{76}
$$

However this is a real mess – we have had to introduce two new fields $\hat{W}_{-1}\phi(z)$ and $\hat{W}_{-2}\phi(z)$ and a priori we do not have any real understanding of their nature. It would be very nice if these fields could be interpreted geometrically, but for the moment we shall have to be happy with their algebraic properties.

Using these two operator product expansions, we can find the commutators of the modes L_m and W_m with $\phi(z)$,

$$
[\,L_m\,, \phi(z)\,] = \left[h\,z^m(m + 1) + z^{m+1}\partial/\partial z\right]\phi(z) ,
\tag{77}
$$

$$
[\,W_m\,, \phi(z)\,] = \left[w\,z^m\frac{(m+1)(m+2)}{2} + z^{m+1}(m + 2)\hat{W}_{-1} + z^{m+2}\hat{W}_{-2}\right]\phi(z) .
\tag{78}
$$

Note that, just as $[\,L_{-1}, \phi(z)\,] = \phi'(z)$, so $[\,W_{-2}, \phi(z)\,] = \hat{W}_{-2}\phi(z)$; it is very tempting to consider W_{-2} as a derivative in an extra direction, especially given that $[\,L_{-1}, W_{-2}\,] = 0$, but I know of no sensible way of incorporating such an interpretation of W_{-2} in the quantum case.

Also note that, given (77) and (78), we can find linear combinations of the L_m and W_m which have simpler commutation relations with the field $\phi(z)$:

$$[L_m - zL_{m-1}, \phi(z)] = h\, z^m\, \phi(z)\,, \tag{79}$$

$$[W_m - 2zW_{m-1} + z^2 W_{m-2}, \phi(z)] = w\, z^m\, \phi(z)\,. \tag{80}$$

One way to see this easily is to note that by multiplying (75) and (76) by $(z - z')$ and $(z - z')^2$ respectively, we find only a simple pole on the right hand side of these equations, and taking modes of both sides, we recover (79) and (80). It turns out that these new equations are really all we need to answer the next question – what is the operator product expansion of two primary fields $\phi_{h,w}(z)$ and $\phi_{h',w'}(z')$?

8.3   The Operator Product of two W_3–Primary Fields

The most natural view of the operator product of two fields $\phi_a(z)$ and $\phi_b(z')$ is as a function of z and z' taking values in representations labelled by c,

$$\phi_a(z)\phi_b(z') \sim \sum_{c,n}(z - z')^{h_c - h_a - h_b + n} V(|c;n\rangle, z')\,, \tag{81}$$

where $|c;n\rangle$ is a state in the representation c of L_0 eigenvalue $h_c + n$.

If we know a basis $\{|i\rangle\}$ for the representation c at level n, and the inner product matrix on this space, we see that we can write

$$|c;n\rangle = |i\rangle\, M_{ij}^{-1}\, \langle j|\, \phi_a(1)\, |b\rangle\,, \quad \text{where } M_{ij} = \langle i\,|\,j\rangle\,. \tag{82}$$

Two comments follow immediately:

(1) If M_{ij} has zero eigenvectors, i.e. there are null vectors in the representation c, then the operator product (81) is only defined modulo these null vectors.

(2) There is an alternative basis in which $\langle j|\phi_a(1)|b\rangle$, or equivalently $\langle b|\phi_a(1)|i\rangle$, is very easy to work out, namely one which uses the linear combinations of the modes in (79) and (80) – this is essentially the observation of Feigin and Fuchs in [29]. Let's investigate this basis in more detail.

For simplicity, let's consider trying to work out $\langle j|\phi_a(1)|b\rangle$. Using the combinations

$$l_m = L_m - 2L_{m-1} + L_{m-2}\,, \quad w_m = W_m - 3W_{m-1} + 3W_{m-2} - W_{m-3}\,, \tag{83}$$

$$h_1 = -L_{-2} + L_{-1}\,, \qquad w_1 = -W_{-3} + 2W_{-2} - W_{-1}\,, \tag{84}$$

$$h_\infty = L_{-2} - 2L_{-1} + L_0\,, \quad w_\infty = W_{-3} - 3W_{-2} + 3W_{-1} - W_0\,, \tag{85}$$

we see that we can choose a basis of the representation c consisting of states of the form

$$w_{m_1}\ldots w_{m_2} l_{n_1}\ldots l_{n_2}(w_1)^a(w_\infty)^b(h_1)^c(h_\infty)^d W_{-1}^e|c\rangle\,. \tag{86}$$

where $m_i \leq m_{i+1} < 0$, $n_i \leq n_{i+1} < 0$. The advantage of this basis is that

$$\langle b|\,\phi_a(1)\,w_m = \langle b|\,\phi_a(1)\,l_m = 0\,, \quad m < 0\,, \tag{87}$$

$$\langle b|\,\phi_a(1)\,w_1 = w_a\langle b|\,\phi_a(1)\,, \qquad \langle b|\,\phi_a(1)\,h_1 = h_a\langle b|\,\phi_a(1)\,, \tag{88}$$

$$\langle b|\,\phi_a(1)\,w_\infty = w_b\langle b|\,\phi_a(1)\,, \qquad \langle b|\,\phi_a(1)\,h_\infty = h_b\langle b|\,\phi_a(1)\,. \tag{89}$$

so that it is trivial to work out any inner product. If we take any vector i and put it in the basis (86), then it is clear that, for some coefficients α_n,

$$\langle b|\,\phi_a(1)\,|i\rangle = \sum_{c,n} \alpha_n \langle b|\,\phi_a(1)\,(W_{-1})^n|c\rangle\,. \tag{90}$$

Some comments:

(1) If $\langle b|\,\phi_a(1)\,(W_{-1})^n|c\rangle$ is not zero for any n, then the three point coupling of the representations a, b and c can a priori depend on infinitely many unknown constants, or in the language of fusion coefficients introduced by Fuchs, $N_{abc} = \infty$

(2) Conversely, if there is some level n for which $W_{-1}^n|c\rangle$ is linearly dependent on the other states at this level, we see that $N_{abc} \leq n$. As it turns out, we are very lucky, since Bajnok [4] and Furlan et al. [33] both showed that if ever the inner product matrix M_{ij} has a zero eigenvector at some level n, then necessarily this vector must be of the form $W_{-1}^n|c\rangle + \ldots$ Furthermore, the vanishings of the determinant of M_{ij} are also known explicitly, so that we know all representations L_c for which the fusion coefficients N_{abc} must be finite.

(3) Mathematically, the space of independent couplings $\langle b|\,\phi_a(1)\,(W_{-1})^n|c\rangle$ to an irreducible representation L_c appearing in (90) is the same as the quotient of the space L_c by the relations

$$w_m\,\psi = l_m\,\psi = 0\,, \quad m < 0\,, \tag{91}$$

$$(w_1 - w_a)\,\psi = 0\,, \qquad (h_1 - h_a)\,\psi = 0\,, \tag{92}$$

$$(w_\infty - w_b)\,\psi = 0\,, \qquad (h_\infty - h_b)\,\psi = 0\,. \tag{93}$$

8.4   Quasi-Rational Representations

Thus far we have implicitly considered using all of (87), (88) and (89) to help us evaluate a three point point function. For some representations L_c this will still leave an infinite number of unknowns

$$\langle b|\,\phi_a(1)\,(W_{-1})^n\,|c\rangle\,, \tag{94}$$

for any representations L_a, L_b. More interesting are the cases where (94) is zero unless h_a, h_b, w_a and w_c obey some constraints. This would be possible if singular vectors in the representation L_c allow us to simplify

$$w_1\psi \quad \text{and} \quad h_1\psi\,, \tag{95}$$

using only (91) and (93) without the use of (92). That this does happen was first shown by Bajnok et al in [5], and we shall reproduce their results here, albeit in a somewhat modified form. Representations L_c for which this happens, i.e. for which only a finite number of representations L_a can couple via (94) have also been studied by Nahm, and are also called quasi-rational representations [46].

We first introduced the parametrisation of W_3 algebra representations by four integers in the discussion of minimal models in section 7.3, but it is possible to consider the representations $[rs; tu]$ of the minimal models with c no longer a minimal value. If we parametrise c as

$$c = 50 - \frac{24}{t} - 24t , \qquad (96)$$

then the representation $[ab; cd]$ has weights given by

$$h = \frac{1}{3t} \left((at - c)^2 + (at - c)(bt - d) + (bt - d)^2 - 3(t - 1)^2 \right) ,$$
$$w = \frac{1}{27\, t^{3/2}}(at - c - bt + d)(2at + bt - 2c - d)(at + 2bt - c - 2d) . \qquad (97)$$

The minimal models are given by (96), (97) with $t = p/q$. The simplest representation $[11; 11]$ is the vacuum, with $h = w = 0$. The next simplest is $[11; 12]$, and in the next section we shall investigate the fusion of this representation.

8.5   The Fusion of the $[11; 12]$ Representation

The special property of this representation is that it has two independent singular vectors, i.e. zero eigenvectors of the inner product matrix, at levels 1 and 2, which are

$$\left(W_{-1} - \left(\frac{\sqrt{t}}{2} - \frac{5}{6\sqrt{t}} \right) \right) |11; 12\rangle , \qquad (98)$$

$$\left(L_{-1}L_{-1} + \frac{2}{3t} L_{-2} + \sqrt{t}\, W_{-2} \right) |11; 12\rangle . \qquad (99)$$

Using these two singular vectors, and their descendants, it is possible to reduce the whole space $L_{[11;12]}$ to a three dimensional space, using only (91) and (93). Since (92) still has to hold, it must be the case that h_1 and w_1 are diagonalisable on this three dimensional space, and in this way they are determined by our choice of h_b and w_b.

Explicitly, we can choose a basis of the representation $L_{[11,12]}$ modulo the constraints (87) and (89) as

$$|11; 12\rangle , \quad L_{-1}|11; 12\rangle , \quad W_{-2}|11; 12\rangle . \qquad (100)$$

In this basis, we do indeed find that h_1 and w_1 can be represented by the matrices

$$h_1 = h_b + \begin{pmatrix} \frac{-4+3t}{3t} & \frac{2(-4+3t+3th_b)}{9t^2} & \frac{-8+6t+21th_b-9t^2h_b}{27t^{5/2}} - \frac{w_b}{t} \\ 1 & 0 & \frac{-14+9t+3th_b}{9t^{3/2}} \\ 0 & \frac{1}{\sqrt{t}} & \frac{7-3t}{3t} \end{pmatrix} \qquad (101)$$

$$w_1 = w_b + \qquad\qquad\qquad\qquad\qquad\qquad\qquad\qquad (102)$$

$$\begin{pmatrix} \frac{(5-3t)(3t-4)}{27t^{3/2}} & \frac{-2(3t-4)^2+9h_bt(1-t)}{27t^{5/2}} + \frac{w_b}{t} & \frac{-2\left((3t-4)^2-h_bt(11-9t-3h_bt)\right)}{27t^3} + \frac{2\,w_b}{3\,t^{3/2}} \\ \frac{5-3t}{3\sqrt{t}} & \frac{82-81t+18t^2-9h_bt}{27t^{3/2}} & \frac{92-102t+27t^2-t(39-18t)h_b}{27t^2} + \frac{w_b}{\sqrt{t}} \\ -1 & \frac{-2+t}{t} & \frac{-68+54t-9t(t-h_b)}{27t^{3/2}} \end{pmatrix}$$

We can check explicitly that h_1 and w_1 commute, and can be diagonalised simultaneously. As a result we find that, given h_b and w_b, there are only three possible choices for h_a and w_a which are consistent with the decoupling of the singular vectors (98) and (99) from correlation functions. While the eigenvalues of the matrices (101) and (102) are not themselves very revealing, when we use the parametrisation (97) we find that the three point coupling

$$\langle 11; ab| \, \phi_{[11;cd]} \, (W_{-1})^n \, |11; 12\rangle \,, \qquad\qquad (103)$$

vanishes unless

$$(c, d) \in \{(a, b+1), (a+1, b-1), (a-1, b)\} \,. \qquad\qquad (104)$$

These are exactly the fusion rules for a tensor product of the $\bar{3}$ representation of $sl(3)$ with a general $sl(3)$ representation and agree with the fusion rules found by Frenkel et al [31] in the minimal case.

However, note that although the three point coupling (103) vanishes unless (104) holds, it does not mean that it must be non-zero in any actual conformal field theory.

Furthermore, it is not even necessarily the case that the three point coupling of three representations is determined by the coupling of the highest weight fields. For example, consider the representation $L_{[11;22]}$; in this case $L_{[11;22]}$ quotiented by (91) and (93) is 8 dimensional, and roughly speaking this representation behaves in a similar fashion to the adjoint representation of $sl(3)$. As a result, the operator product

$$\phi_{[11;22]}(z) \, \phi_{[11;22]}(z') \to L_{[11;22]} \,, \qquad\qquad (105)$$

has two independent couplings, determined by

$$\langle 11; 22| \, \phi_{[11;22]}(1) \, |11; 22\rangle \quad \text{and} \quad \langle 11; 22| \, \phi_{[11;22]}(1) \, W_{-1} \, |11; 22\rangle \,. \qquad (106)$$

It may even happen that at some c values there are extra singular vectors which further truncate this fusion, e.g. at $c = 4/5$ the first coupling is identically zero, and the second coupling is free.

8.6   Quasi-Finite Representations

As a final step we can ask ourselves if there are any representations for which we need only impose (87) for the fusion to be fully determined, in which case there will only be a finite set of $\{h_a, h_b, w_a, w_b\}$ for which the fusion is non-vanishing. The answer is yes, and these are the minimal model representations, which are the last subject in this section.

The minimal model representations L_c have the (defining) property that N_{abc} is zero unless L_a and L_b are in some finite set. For this reason they are also called quasi-finite representations. This property was proven for the minimal model Virasoro representations by Feigin and Fuchs in [29].

For the W_3 algebra, a representation is quasi-finite if the space L_c quotiented by (91) is finite dimensional, i.e. if only a finite number of the states

$$W_{-3}^a W_{-2}^b W_{-1}^c L_{-2}^d L_{-1}^e |c\rangle, \qquad (107)$$

are linearly independent modulo l_m and w_m. If this is the case, then we can directly see that the operators h_1, w_1, h_∞, and w_∞ commute modulo l_m and w_m and hence may be simultaneously diagonalised[3].

It is believed that this property holds for exactly the minimal model representations with c given by (72) and weights h, w given by (73).

8.7   Conclusions on Fusion

It is an ambitious program to prove that the minimal model representations indeed have this beautiful property. It has been completed for the Virasoro algebra by Feigin and Fuchs in [29] but for this they needed to know the whole structure of the Virasoro Verma modules.

For a long time there were not even consistent conjectures for the structure of W_3 Verma module representations, but there have recently appeared some very beautiful conjectures. These have been put forward by de Vos and van Driel in [19], and give the multiplicities of irreducible representations appearing in the composition series of Verma module representations in terms of Kazhdan-Lusztig polynomials for cosets of the affine Weyl group of the affine algebra appearing in the DS construction before Hamiltonian reduction, which in this case is $a_2^{(1)} \equiv \hat{su}(3)$.

At the very least, this definition of fusion gives a way to derive the fusion rules for any W-algebra, by suitable generalisations and can be further generalised to give 4,5,6,... point function fusion rules, and is the only known way to find differential equations for correlation functions of W-algebra primary fields, simply using the algebraic structure of a W-algebra[4]. For the first calculation performed in this way, see [5].

---

[3] It is possible that the common eigenvectors do not span the full space, but it is believed that this is not the case. For some worked examples, see [48].

[4] It may be possible to find differential equations using the methods of Furlan et al [32].

Another very promising route to studying fusion of W-algebra primary fields is that developed by Gaberdiel [34], which attacks the problem of fusion from the opposite end, that is it attempts to decompose the tensor product of two fields into representations of a single copy of the W-algebra. Of course this idea is not essentially new, as neither is the method given here, and in many ways the calculations reduce to essentially the same steps, but it has some advantages when the theory has unexpected features, as we can read in Gaberdiel and Kausch [35].

Finally, when L_c is the vacuum representation, the space of states $\langle b|$ which can couple to $\phi_a(1)\,|c\rangle$ is clearly one-dimensional – only the field conjugate to ϕ_a can have a non-zero three point function with the vacuum

$$\langle \bar{a}|\,\phi_a(1)\,|0\rangle\,,$$

so that the space of states in the vacuum representation modulo (91) counts the allowed representations in the conformal field theory. For the quasi-finite representations we have been talking about in this section, this space is finite dimensional and has been studied in greater detail by Zhu [50]. He has shown that one can also put an algebra structure on this space, (and hence it is also known as Zhu's algebra), and he has further shown that any representation of this algebra induces a representation of the full chiral algebra.

8.8   Conclusions

Clearly there are many interesting topics which have not been covered in these lectures for lack of time, and so I would like to finish off by listing some of these here.

(1) It was an assumption right from the beginning that we were only interested in W-algebras with a finite number of generating fields. However, there is a vast literature dealing with W-algebras with an infinite number of generating fields. There are at least two such algebras, which are commonly known as W_∞ and $W_{1+\infty}$ which are closely related to the W-algebras we have been studying. These are very important as they may arise as the symmetry algebra in physical models, for example in the quantum Hall effect [17]. The representation theory of these algebras has been studied in great detail, [1]. However, it is a surprising fact that each unitary representation of these algebras may be identified as a unitary representation of some standard W-algebra to which the larger algebra truncates at that value of c.

(2) Another assumption we have made is that our algebras all contain the Virasoro algebra as a subalgebra. It is equally possible to consider W-algebras which contain some superconformal algebra as a subalgebra, e.g. the $N = 1$ or $N = 2$ super Virasoro algebras. These have been investigated in a similar fashion, with searches for solutions to the Jacobi identities, coset constructions and especially with quantum Hamiltonian reductions based on affine super Lie algebras.

(3) Several times during these lectures I have mentioned that there are a large number of possible identifications between W-algebras which occur for specific c values. The most systematic investigation of these identifications have been carried out by Blumenhagen et al and by Hornfeck in [8, 9, 39, 41, 40]. During this work they also uncovered a large class of W-algebras which are not freely generated, i.e. have generic singular vectors, and which had hitherto been thought of as somehow exceptional. In fact they now appear as very regular and indeed to be in some way dual to the W-algebras we have been looking at, in that each acts as the 'unifying algebra' for the other class. It seems quite likely that the story of W-algebra truncations and identifications is not yet over.

(4) An outstanding problem is that of finding explicit formulae for the fully local correlation functions of W-primary fields. The first problem is that the presence of multiple independent three-point couplings between W-primary fields means that the evaluation of four and higher point couplings does not simply reduce to the description of a simple set of 'coupling constants', i.e. the three-point couplings between W-primary fields. One must also take into account the presence of couplings via W-descendant fields. A further problem is that the 'Dotsenko–Fateev' type integrals which should be the building blocks of these correlation functions are of new types and are not amenable to the same sorts of methods as were used for the Virasoro correlation functions.

(5) Finally I should like to stress again that many results which are proven for the Virasoro algebra are only conjectured for W-algebras, even for just the W_3 algebra. It seems to me that the results of de Vos and van Driel [19] would be the best route to understanding the structure of W-algebra representations, and from there going on to such more complicated topics as fusion etc.

Acknowledgements

I would like to thank the organisers for their invitation to present these lectures and the great hospitality they have shown here in Budapest. I would also like to thank Horst Kausch for a critical reading of the manuscript.

GMTW is supported by an EPSRC advanced fellowship.

References

[1] H. Awata, M. Fukuma, Y. Matsuo, and S. Odake, *Character and determinant formulae of quasifinite representation of the $W_{1+\infty}$ algebra*, Commun. Math. Phys. **172** (1995) 377, hep-th/9405093;
V. Kac and A. Radul, *Representation theory of the vertex algebra $W_{1+\infty}$*, hep-th/9512150.

[2] F.A. Bais, P. Bouwknegt, K. Schoutens and M. Surridge, *Extensions of the Virasoro algebra constructed from Kac-Moody algebras using higher order Casimir invariants*, Nucl. Phys. B304 (1988) 348.

[3] F.A. Bais, P. Bouwknegt, K. Schoutens and M. Surridge, *Coset construction for extended Virasoro algebras*, Nucl. Phys. B304 (1988) 371.

[4] Z. Bajnok, *Singular vectors of the WA_2 algebra*, Phys. Lett. **B329** (1994) 225, hep-th/9403032.

[5] Z. Bajnok, L. Palla and G Takács, *A_2 Toda theory in reduced WZNW framework and the representations of the W algebra*, Nucl. Phys. **B385** (1992) 329, hep-th/9206075.

[6] A.A. Belavin, A.M. Polyakov and A.B. Zamolodchikov, *Infinite conformal symmetry in two-dimensional quantum field theory*, Nucl. Phys. **B241** (1984) 333.

[7] R. Blumenhagen, M. Flohr, A. Kliem, W. Nahm, A. Recknagel, and R. Varnhagen. *W-algebras with two and three generators*, Nucl. Phys. **B361** (1991) 255.

[8] R. Blumenhagen, W. Eholzer, A. Honecker, K. Hornfeck, and R. Hübel, *Unifying W-algebras*, Phys. Lett. **B322** (1994) 51, hep-th/9404113

[9] R. Blumenhagen, W. Eholzer, A. Honecker, K. Hornfeck, and R. Hübel, *Coset realisation of unifying W-algebras*, Int. J. Mod. Phys. **A10** (1995) 2367, hep-th/9406203

[10] J. de Boer, L. Feher and A. Honecker, *A class of W algebras with infinitely generated classical limit*, Nucl. Phys. **B420** (1993) 409, hep-th/9312049.

[11] J. de Boer and T. Tjin, *The relation between quantum W algebras and Lie algebras*, Commun. Math. Phys. **160** (1994) 317, hep-th/9302006

[12] P. Bouwknegt, *Extended Conformal algebras*, Phys. Lett. **207B** (1988) 295.

[13] P. Bouwknegt and K. Schoutens, *W symmetry in conformal field theory*, Phys. Rept. **223** (1993) 183.

[14] P. Bowcock, *Representation theory of a W algebra from generalised DS reduction*, Durham preprint DTP-94-5, hep-th/9403157

[15] P. Bowcock and G.M.T. Watts, *On the classification of quantum W–algebras*, Nucl. Phys. B379 (1992) 63, hep-th/9111062.

[16] A. Cappelli, C. Itzykson and J.-B. Zuber, *The A-D-E classification of minimal and $a_1^{(1)}$ conformal field theories*, Commun. Math. Phys. **113** (1987) 1.

[17] A. Cappelli, C.A. Trugenberger and G.R. Zemba, *Stable hierarchical quantum Hall fluids as $W_{1+\infty}$ minimal models*, Nucl. Phys. **B448** (1995) 470, hep-th/9502021; *$W_{1+\infty}$ minimal models and the hierarchy of the quantum Hall effect*, Nucl. Phys. Proc. Suppl. **45A** (1996) 112.

[18] E.B. Dynkin, Transl. Am. Math. Soc. Series 2, 6 (1957) 112.

[19] P. van Driel and K. de Vos, *The Kazhdan-Lusztig conjecture for W-algebras*, Bonn preprint BONN-TH-95-14, hep-th/9508020.

[20] V.G. Drinfel'd and V.V. Sokolov, *Lie algebras and equations of Korteweg-de Vries type*, J. Sov. Math. **30** (1985) 1975.

[21] W. Eholzer, M. Flohr, A. Honecker, R. Hübel, W. Nahm and R. Varnhagen, *Representations of W-algebras with two-generators and new rational models*, Nucl. Phys. **B383** (1992) 249.

[22] W. Eholzer, A. Honecker and R. Hübel, *How complete is the classification of W-symmetries?*, Phys. Lett. **B308** (1993) 42, hep-th/9302124.

[23] V.A. Fateev and S.L. Luk'yanov, *The models of two-dimensional conformal quantum field theory with Z_n symmetry*, Int. J. Mod. Phys. **A3** (1988) 507.

[24] V.A. Fateev and S.L. Luk'yanov, *Additional symmetries and exactly-soluble models in two-dimensional conformal field theory*, Sov. Sci. Rev. **A15** (1990) 1.

[25] V.A. Fateev and A.B. Zamolodchikov, *Conformal quantum field theory models in two dimensions having Z_3 symmetry*, Nucl. Phys. **B280** [FS18] (1987) 644.

[26] L. Feher, L. O'Raifeartaigh, P. Ruelle, I.Tsutsui and A. Wipf, *On Hamiltonian reductions of the Wess-Zumino-Novikov-Witten theories*, Phys. Rep. **222** (1992) 1.

[27] L. Feher, L O'Raifeartaigh and I. Tsutsui, *The vacuum preserving Lie algebra of a classical W algebra*, Phys. Lett. **B316** (1993) 275, hep-th/9307190

[28] B. Feigin and E. Frenkel, *Quantization of the Drinfeld-Sokolov reduction*, Phys. Lett. **B246** (1990) 75.

[29] B.L. Feigin and D.B. Fuchs, *On the cohomology of some nilpotent subalgebras of Kac-Moody and the Virasoro algebras*, J. Geom. Phys. **5** (1988) 209.

[30] J.M. Figueroa-O'Farrill and S. Schrans, *The spin 6 extended conformal algebra*, Phys. Lett. **B245** (1990) 471.

[31] E.V. Frenkel, V. Kac and M. Wakimoto, *Characters and fusion rules for W algebras via quantized Drinfeld-Sokolov reductions*, Commun. Math. Phys. **147** (1992) 295.

[32] P. Furlan, A.C. Ganchev, R. Paunov and V.B. Petkova, *Solutions of the Knizhnik-Zamolodchikov equation with rational isospins and the reduction to the minimal models*, Nucl. Phys. **B394** (1993) 665, hep-th/9201080.

[33] P. Furlan, A.C. Ganchev and V.B. Petkova, *Singular vectors of W algebras via DS reduction of $A_2^{(1)}$* Nucl. Phys. **B431** (1994) 622, hep-th/9403075.

[34] M.R. Gaberdiel, *Fusion rules of chiral algebras*, Nucl. Phys. **B417** (1994) 130, hep-th/9309105; *Fusion in conformal field theory as the tensor product of the symmetry algebra*, Int. J. Mod. Phys. **A9** (1994) 4619, hep-th/9307183.

[35] M.R. Gaberdiel and H.G. Kausch, *A rational logarithmic conformal field theory*, Phys. Lett. **B386** (1996) 131, hep-th/9606050; *Indecomposable fusion products*, Nucl. Phys. **B477** (1996) 293, hep-th/9604026.

[36] T. Gannon, *The classification of affine su(3) modular invariant partition functions*, Commun. Math. Phys **161** (1994) 233, hep-th/9212060.

[37] T. Gannon and M.A. Walton, *On the classification of diagonal coset modular invariants*, Commun. Math. Phys **173** (1995) 175, hep-th/9407055

[38] P. Goddard, *Meromorphic Conformal Field Theory*, in: Infinite Dimensional Lie Algebras and Lie Groups, ed. V. G. Kac, World Scientific, 1989, CIRM-Luminy July conference on Infinite dimensional Lie algebras and Lie Groups, Marseilles 1988.

[39] K. Hornfeck, *W-algebras with set of primary fields of dimensions (3,4,5) and (3,4,5,6)*, Nucl. Phys. **B407** (1993) 237, hep-th/9212104.

[40] K. Hornfeck, *Classification of structure constants for W-algebras from highest weights*, Nucl. Phys. **B411** (1994) 307, hep-th/9307170.

[41] K. Hornfeck, *W-algebras of negative rank*, Phys. Lett. **B343** (1995) 94, hep-th/9410013.

[42] Y.-Z. Huang and J. Lepowsky, *On the D-module and formal variable approaches to vertex algebras*, in: 'Topics in Geometry: In Memory of Joseph D'Atri', ed. S. Gindikin, Progress in Nonlinear Differential Equations, Vol. 20, Birkhauser, Boston, (1996) 175, q-alg/9603020.

[43] H.G. Kausch, *Extended conformal algebras generated by a multiplet of primary fields*, Phys. Lett. **B259** (1991) 448.

[44] H.G. Kausch and G.M.T. Watts, *A study of W-algebras using Jacobi identities*, Nucl. Phys. **B354** (1991) 740.

[45] W. Nahm, *Chiral algebras of two-dimensional chiral field theories and their normal ordered products*, in 'Recent developments in conformal field theories' S. Randjbar-Daemi et al. eds, (World Scientific 1990) 81.

[46] W. Nahm, *Quasi-rational fusion products*, Int. J. Mod. Phys. **B8** (1994) 3693, hep-th/9402039.

[47] E. Verlinde, *Fusion rules and modular transformations in 2d conformal field theory*, Nucl. Phys. **B300** [FS22] (1988) 360.

[48] G.M.T. Watts, Fusion in the W_3 algebra, Commun. Math. Phys. **171** (1995) 87, hep-th/9403163.

[49] A.B. Zamolodchikov, *Infinite additional symmetries in two-dimensional conformal quantum field theory*, Theor. Mat. Fiz. **65** (1985) 347.

[50] Y.-C. Zhu, *Vertex operator algebras, elliptic functions and modular forms*, Ph.D. thesis, Yale University, 1990;
I.B. Frenkel and Y.-C. Zhu, *Vertex operator algebras associated to representations of affine and Virasoro algebras*, Duke Math. J. **66** (1992), 123.

Exact S-Matrices

Patrick Dorey

Department of Mathematical Sciences,
University of Durham, Durham DH1 3LE,
England

Abstract. The aim of these notes is to provide an elementary introduction to some of the basic elements of exact S-matrix theory. This is a large subject, and only the foothills will be explored here. A particular omission is any serious discussion of the Yang-Baxter equation; instead, I will focus on questions of analytic structure, and the bootstrap equations. Even then, what I have to say will only be a sketch of the simpler aspects. The hope is to give a hint of the many unexpected features of scattering theories in 1+1 dimensions.

1 Introduction – What's so Special About 1+1?

To get things started, I want to describe a particularly simple calculation that can be done in probably the simplest nontrivial quantum field theory imaginable, namely $\lambda\phi^4$ theory in a universe with only one spatial dimension.

The Lagrangian to consider is

$$\mathcal{L} = \frac{1}{2}(\partial\phi)^2 - \frac{1}{2}m^2\phi^2 - \frac{\lambda}{4!}\phi^4 \,,$$

resulting in the Feynman rules

$$\rule{3cm}{0.4pt} \quad = \quad \frac{i}{p^2 - m^2 + i\epsilon}$$

$$\times \quad = \quad -i\lambda$$

The task is to calculate the connected $2 \to 4$ production amplitude, at tree level. Actually, to keep track of the diagrams it is a little easier to look at the $3 \to 3$ process, leaving implicit the understanding that one of the *out* momenta will be crossed to *in* at the end. I'll label the three *in* particles as a, b, c, and the three *out* particles as d, e, f, and opt to cross c from *in* to *out* later. It also helps to adopt light-cone coordinates from the outset, using

$$(p, \bar{p}) = (p^0 + p^1, p^0 - p^1)$$

and then solving the mass-shell condition $p\bar{p} = m^2$ by writing the *in* and *out* momenta as

$$p_a = (ma, ma^{-1}) \quad , \quad p_b = (mb, mb^{-1})$$

and so on, with $a, b, \ldots$ real numbers, positive for particles travelling forwards in time. In terms of these variables, the crossing from $3 \to 3$ to $2 \to 4$ amounts to a continuation from c to $-c$. For the $3 \to 3$ amplitude there are just two classes of diagram:

$$(A) \qquad\qquad\qquad\qquad (B)$$

The internal momentum in (A) is $p = m(a{+}b{-}d, a^{-1}{+}b^{-1}{-}d^{-1})$, and so its propagator contributes

$$\frac{i}{p^2 - m^2} = \frac{i}{m^2} \frac{1}{(a{+}b{-}d)(a^{-1}{+}b^{-1}{-}d^{-1}) - 1}$$
$$= \frac{i}{m^2} \frac{-abd}{(a{+}b)(a{-}d)(b{-}d)}$$

to the total scattering amplitude. Given the agreement above that one of the *out* momenta is actually in-going, this propagator is never on-shell, and so forgetting about the $i\epsilon$ does not cause any error. The same remark applies to diagram (B), for which

$$\frac{i}{p^2 - m^2} = \frac{i}{m^2} \frac{1}{(a{+}b{+}c)(a^{-1}{+}b^{-1}{+}c^{-1}) - 1}$$
$$= \frac{i}{m^2} \frac{abc}{(a{+}b)(a{+}c)(b{+}c)} \, .$$

Adding these together, with a brief pause to check that the diagrams have been counted correctly, yields the full result at tree level:

$$\langle out|in \rangle_{\text{tree}} = -\frac{i\lambda^2}{m^2} A_{\text{legs}} H(a, b, c, d, e, f)$$

where A_{legs} contains all the factors living on external legs and so on that will be the same for all diagrams, and

$$H(a, b, c, d, e, f) = \sum_{\substack{cycl\{abc\} \\ cycl\{def\}}} \frac{-abd}{(a + b)(a - d)(b - d)} - \frac{abc}{(a + b)(b + c)(c + a)} \, ,$$

with the sum running over all cyclic permutations of $\{a, b, c\}$ and $\{d, e, f\}$.

Now I need the following fact:

$$\text{If} \quad a+b+c = d+e+f \quad \text{and} \quad \frac{1}{a}+\frac{1}{b}+\frac{1}{c} = \frac{1}{d}+\frac{1}{e}+\frac{1}{f}$$

$$\text{then} \quad H(a,b,c,d,e,f) \equiv -1 .$$

The two conditions are the so-far ignored conservation of left- and right- lightcone momenta. The formula makes no mention of the signs of the arguments to H, and certainly holds with c negative. The conclusion:

• In 1+1-dimensional $\lambda\phi^4$ theory, the $2 \rightarrow 4$ amplitude is a *constant* at tree level.

It is now very tempting to cancel this amplitude completely, by adding a term

$$-\frac{1}{6!}\frac{\lambda^2}{m^2}\phi^6$$

to the original Lagrangian. In 1+1 dimensions this does not spoil renormalisability, and gives a theory in which the $2 \rightarrow 4$ amplitude *vanishes* at tree level. With $\beta^2 = \lambda/m^2$, the new Lagrangian is

$$\mathcal{L} = \frac{1}{2}(\partial\phi)^2 - \frac{m^2}{\beta^2}\left[\frac{1}{2}\beta^2\phi^2 + \frac{1}{4!}\beta^4\phi^4 + \frac{1}{6!}\beta^6\phi^6\right] .$$

This is already curious, but it is possible to go much further. Calculating the $2 \rightarrow 6$ amplitude (left as an exercise for the energetic reader) should reveal that *this* is now constant, ready to be killed off by a judiciously-chosen ϕ^8 term, and so on. At each stage a residual constant piece can be removed by a (uniquely-determined) higher-order interaction. Keep going, and infinitely-many diagrams later you should find

$$\mathcal{L} = \frac{1}{2}(\partial\phi)^2 - \frac{m^2}{\beta^2}\left[\cosh(\beta\phi) - 1\right] ,$$

the sinh-Gordon Lagrangian. Sending β to $i\beta$ converts this into the well-known sine-Gordon model, to which the discussion will return in later lectures.

The claim of uniqueness just made deserves a small caveat. I began the calculation with no ϕ^3 term in the initial Lagrangian, and a discrete $\phi \rightarrow -\phi$ symmetry which persisted throughout. But what if I had instead started with a nonzero ϕ^3 term, and tried to play the same game? This is definitely a harder problem, but the final answer can be predicted with a fair degree of confidence:

$$\mathcal{L} = \frac{1}{2}(\partial\phi)^2 - \frac{1}{2}m^2\phi^2 - \frac{1}{3!}\lambda\phi^3 - \frac{1}{4!}\frac{17\lambda^2}{5m^2}\phi^4 - \cdots$$

$$= \frac{1}{2}(\partial\phi)^2 - \frac{m^2}{5\beta^2}\left[e^{2\beta\phi} + 2e^{-\beta\phi} - 3\right] ,$$

where this time $\beta = \frac{5}{7}\lambda/m^2$. The special properties of this Lagrangian have been been noticed by various authors over the years, the earliest probably being M.Tzitzéica, in an article published in 1910.

To summarise, it appears that in 1+1 dimensions there are some interacting Lagrangians with the remarkable property that the resulting field theories have no tree-level particle production. Arafeva and Korepin showed, in 1974, that for the sinh-Gordon model this is also true at one loop. The tree-level result was a sign of interesting classical behaviour; that it persists to one loop is evidence that the quantum theory might also be rather special.

Before continuing along this line, I want to return to the $3 \to 3$ amplitude. Should we conclude that its connected part is also zero? Contrary to initial expectations, the answer to this question is a definite *no*. For the $3 \to 3$ process, it is no longer legitimate to forget about the $i\epsilon$'s. For the diagrams of type (A), the intermediate particle can now be on-shell, and when this happens the $i\epsilon$ must be retained until all contributing diagrams have been added together. This is relevant whenever the set of ingoing momenta is equal to the set of outgoing momenta, and in such situations it turns out that the final result is indeed nonzero. Thus the connected part of the $3 \to 3$ amplitude does not vanish, but it does contain an additional delta-function which enforces the equality of the initial and final sets of momenta. We have found a model for which, at least at tree level, the connected $3 \to 3$ amplitude violates two of the usual assumptions made of an analytic S-matrix:

• it is *not* found by crossing the $2 \to 4$ amplitude;

• it is *not* analytic in the residual momenta once overall momentum conservation has been imposed.

It is clear that something odd is going on, but it is not so clear quite what, and even less clear why. Evaluating yet more Feynman diagrams is unlikely to shed much light on these questions, and besides, an infinite amount of work would be needed before we could be completely sure that any of these properties will feature in the full quantum theory. A more sophisticated approach is needed. What could force these amplitudes to vanish, irrespective of the structure of the Feynman diagrams? One possibility is that conservation laws might limit the set of *out* states accessible from any given *in* state. The far-reaching consequences of this thought are the subject of the next lecture.

2   Conserved Quantities and Factorisability

After the somewhat informal introduction, the time has come to be a little more precise, at least to the extent of pausing to set up some notation.

First, I should allow for more than one particle type, so different masses m_a, m_b and so on make an appearance. A single particle of mass m_a will be on-shell when its light-cone momenta p_a, $\bar{p}_a$ satisfy $p_a \bar{p}_a = m_a^2$. It will be convenient to solve this equation not via the variable $a = p_a/m_a$ used in the last lecture, but rather via a parameter $\theta = \log a$ called the rapidity. Thus,

$$p_a = m_a e^{\theta_a} \quad , \quad \bar{p}_a = m_a e^{-\theta_a} .$$

Recall that a was a positive real number for the forward component of the mass shell; this corresponds to θ ranging over the entire real axis. The backwards component of the mass shell, found by negating a, can be parametrised by this same rapidity so long as it is shifted onto the line $\text{Im}\,\theta = \pi$. This will be relevant when discussing the crossing of amplitudes.

An n-particle asymptotic state can now be written as

$$|A_{a_1}(\theta_1)A_{a_2}(\theta_2)\ldots A_{a_n}(\theta_n)\rangle_{\substack{in\\out}}$$

where the symbol $A_{a_i}(\theta_i)$ denotes a particle of type a_i, travelling with rapidity θ_i. By smearing the momenta a little so as to produce wavepackets, each particle can be assigned an approximate position at each moment. In a massive theory, the only sort of theory I will be bothering with, all interactions are short-ranged and so the state behaves like a collection of free particles except at times when two or more wavepackets overlap. All of this can be made more precise, but not in these lectures.

An *in* state is characterised by there being no further interactions as $t \to -\infty$. This means that the fastest particle must be on the left, the slowest on the right, with all of the others ordered in between. It is convenient to represent this situation by giving the $A_{a_i}(\theta_i)$ a life outside the $|\;\rangle_{in}$ and $|\;\rangle_{out}$ ket vectors, thinking of them as noncommuting symbols with their order on the page reflecting the spatial ordering of the particles that they represent. Thus an *in* state would be written

$$A_{a_1}(\theta_1)A_{a_2}(\theta_2)\ldots A_{a_n}(\theta_n)$$

with

$$\theta_1 > \theta_2 > \ldots > \theta_n \ .$$

Similarly, an *out* state has no further interactions as $t \to +\infty$, and so each particle must be to the left of all particles travelling faster than it, and to the right of all particles travelling slower. In terms of the non-commuting symbols, one such state is

$$A_{b_1}(\theta_1)A_{b_2}(\theta_2)\ldots A_{b_n}(\theta_n)$$

now with

$$\theta_1 < \theta_2 < \ldots < \theta_n \ .$$

Products of the symbols with other orderings of the rapidities can be thought to represent states at other times when all the particles are momentarily well-separated. Asymptotic completeness translates, at least partially, into the claim that any such product can be expanded either as a sum of products in the *in*-state ordering, or as a sum of products in the *out*-state ordering.

The S-matrix provides the mapping between the *in*-state basis and the *out*-state basis. In the new notation this reads, for a two-particle *in*-state,

$$A_{a_1}(\theta_1)A_{a_2}(\theta_2) = \sum_{n=2}^{\infty} \sum_{\theta_1' < \ldots < \theta_n'} S_{a_1 a_2}^{b_1 \ldots b_n}(\theta_1, \theta_2; \theta_1' \ldots \theta_n') A_{b_1}(\theta_1') \ldots A_{b_n}(\theta_n') \ ,$$

where $\theta_1 > \theta_2$, a sum on $b_1 \ldots b_n$ is implied, and the sum on the θ'_i will generally involve a number of integrals, with the rapidities appearing additionally constrained by the overall conservation of left- and right- lightcone momenta:

$$m_{a_1} e^{\pm\theta_1} + m_{a_2} e^{\pm\theta_2} = m_{b_1} e^{\pm\theta'_1} + \ldots + m_{b_n} e^{\pm\theta'_n} .$$

The notation works because the number of dimensions of space, namely 1, matches the 'dimensionality' of a sequence of symbols in a line of mathematics; it can't be used for higher-dimensional theories. However, at this stage it makes no mention of integrability, and can be set up for any massive quantum field theory in 1+1 dimensions.

Next, to the conserved quantities. One such is energy-momentum, a spin-one operator. In lightcone components this acts on a one-particle state as

$$P|A_a(\theta)\rangle = m_a e^{\theta}|A_a(\theta)\rangle \quad , \quad \bar{P}|A_a(\theta)\rangle = m_a e^{-\theta}|A_a(\theta)\rangle .$$

Beyond this, operators can be envisaged transforming in higher representations of the 1+1 dimensional Lorentz group:

$$Q_s|A_a(\theta)\rangle = q_a^{(s)} e^{s\theta}|A_a(\theta)\rangle .$$

The integer s is called the (Lorentz) spin of Q_s. Since $Q_{|s|}$ transforms as s copies of P, and $Q_{-|s|}$ as s copies of $\bar{P}$, it makes sense to think of Q_s and Q_{-s} as rank $|s|$ objects. The simple 'left-right' splitting is special to 1+1 dimensions.

I'll only consider those operators Q_s that come as integrals of local densities, and this has the important consequence that their action on multiparticle wavepackets is additive:

$$Q_s|A_{a_1}(\theta) \ldots A_{a_n}(\theta_n)\rangle = (q_{a_1}^{(s)} e^{s\theta_1} + \ldots + q_{a_n}^{(s)} e^{s\theta_n})|A_{a_1}(\theta) \ldots A_{a_n}(\theta_n)\rangle .$$

These are called local conserved charges and they are all in involution (they commute) since, essentially by assumption, they have been simultaneously diagonalised by the basis of asymptotic multiparticle states that I have chosen. This is not inevitable: nonlocal charges, often associated with fractional-spin operators, can be very important. The papers of Lüscher (1978), Zamolodchikov (1989c) and Bernard and Leclair (1991) are good starting-points for those interested in this aspect of the subject.

Even without the more exotic possibilities, the consequences of the extra local conserved charges are profound. In fact, Coleman and Mandula showed in 1967 that for more than one spatial dimension, the existence of even just one conserved charge transforming as a tensor of second or higher rank forces the S-matrix of the model to be trivial. (For a pauper's explanation of this fact, see later in this lecture.) This is not true in 1+1 dimensions, but nevertheless the possibilities for the S-matrix are severely limited: it must be consistent with

• no particle production;

• equality of the sets of initial and final momenta;

• factorisability of the $n \to n$ S-matrix into a product of $2 \to 2$ S-matrices.

The first two of these properties sum up the behaviour which had emerged experimentally by the end of the last lecture, and the third is a bonus, rendering the task of finding the full S-matrices of a whole class of 1+1 dimensional models genuinely feasible.

I shall outline a couple of arguments for why these properties should follow from the existence of the conserved charges.

The first simply imposes the conservation of the charges directly. Consider an $n \to m$ amplitude, with ingoing particles $A_{a_1}(\theta_1), \ldots, A_{a_n}(\theta_n)$, and outgoing particles $A_{b_1}(\theta'_1), \ldots, A_{b_m}(\theta'_m)$. If a charge Q_s is conserved, then an initial eigenstate of Q_s with a given eigenvalue must evolve into a superposition of states all sharing that same eigenvalue. For the amplitude under discussion this implies that

$$ q_{a_1}^{(s)} e^{s\theta_1} + \ldots + q_{a_n}^{(s)} e^{s\theta_n} = q_{b_1}^{(s)} e^{s\theta'_1} + \ldots + q_{b_n}^{(s)} e^{s\theta'_m} \ . $$

Now if conserved charges Q_s exist for infinitely many values of s, then there will be infinitely many such equations, and for generic *in* momenta the only way to satisfy them all will be the trivial one, namely $n = m$ and, perhaps after a reordering of the *out* momenta,

$$ \theta_i = \theta'_i \quad ; \quad q_{a_i}^{(s)} = q_{b_i}^{(s)} \qquad i = 1 \ldots n \ , $$

where s runs over the spins of the non-trivial conserved charges with nonzero spin (or over all the nonzero integers, if we agree to set $q^{(s)} \equiv 0$ for those s at which a local conserved charge cannot be defined). This does not quite imply that the outgoing set of labels, $\{b_1, \ldots b_n\}$, is equal to the ingoing set $\{a_1, \ldots a_n\}$ – they just need to agree about the values of all of the nonzero spin conserved charges. Nevertheless, it is enough to establish the absence of particle production, and the equality of the initial and final sets of momenta, though factorisability is harder to see from this point of view. One caveat should also be mentioned: in many models, it turns out that there are some solutions to the conservation constraints with $n \neq m$. However these are only found for exceptional sets of ingoing momenta, which are unphysical to boot, so this fact does not change the conclusions for the S-matrix. A more severe problem comes with the realisation that this argument hasn't escaped the infinite workload mentioned at the end of the last lecture. Consider, for example, a two-particle collision. As the relative momenta of the incident particles increases, the number of particles permitted energetically in the out state grows without limit. To be absolutely sure that, no matter how fast the two particles are fired at each other, only two particles will come out, infinitely many conservation constraints are needed. This might not matter – practical considerations are always going to limit the relative momenta to which we have access – were we not ambitious enough to hope for an *exact* formula for the S-matrix. This requires an understanding of all energy scales, and so the infinite amount of work appears to be unavoidable.

This should be motivation enough for the second argument, which can be found in a 1980 article by Parke, itself building on an observation which dates back at least to Shankar and Witten (1978). The argument also establishes factorisability and imposes the Yang-Baxter equation on the two-particle S-matrix. The key is to make use of the fact that we're dealing with a local, causal quantum field theory, by considering the effect of the conserved charges on localised wavepackets.

First take a single-particle state, with position space wavefunction

$$\psi(x) \propto \int_{-\infty}^{\infty} dp e^{-a^2(p-p_1)^2} e^{ip(x-x_1)}.$$

This describes a particle with spatial momentum approximately p_1, and position approximately x_1. Act on this with an operator giving a momentum-dependent phase factor $e^{-i\phi(p)}$. The wavefunction becomes

$$\tilde{\psi}(x) \propto \int_{-\infty}^{\infty} dp e^{-a^2(p-p_1)^2} e^{ip(x-x_1)} e^{-i\phi(p)}.$$

Most of the integral comes from $p \approx p_1$, and $\phi(p)$ can be expanded in powers of $(p-p_1)$ to find $\tilde{p}_1$ and $\tilde{x}_1$, the revised values of the momentum and position:

$$\tilde{p}_1 = p_1 \quad , \quad \tilde{x}_1 = x_1 + \phi'(p_1) .$$

For a multiparticle state a product of one-particle wavefunctions will be a good approximation when the particles are well separated, and on such a state $|p_a p_b \ldots\rangle$, the action is to shift the position of particle a by $\phi'(p_a)$, that of b by $\phi'(p_b)$, and so on.

Strictly speaking, for compatibility with the earlier discussions I should now consider the actions of the operators $Q_{|s|}$ and $Q_{-|s|}$, as Parke did in his article. However the essentials of the argument will be conveyed if I instead assume the conservation of operators P_s acting on one-particle and well-separated multiparticle states as $(P_1)^s$, with P_1 the spatial part of the two-momentum operator. Acting with $e^{-i\alpha P_s}$, the phase factor is $\phi_s(p) = \alpha p^s$, so a particle with momentum p_a will have its position shifted by $s\alpha p_a^{s-1}$. The case $s = 1$, momentum itself, just translates every particle by the same amount α. But, crucially, for $s > 1$ particles with different momenta are moved by different amounts.

The argument continues as follows. First consider a $2 \to m$ process, labelled as in figure 1.

For the amplitude to be non-vanishing, the time when the first two particles collide, call it t_{12}, must precede the time t_{23} when the trajectory of particle 2, the slower incomer, intersects that of particle 3, the fastest outgoer:

$$t_{12} \leq t_{23} .$$

Why should this be so? Nothing can happen until the wavepackets of particles 1 and 2 overlap. After this, it suffices to follow the path of the rightmost particle until all have separated in order to establish the inequality. Note that this could be violated on microscopic timescales, but not macroscopically: hence the term 'macrocausality' for this sort of property.

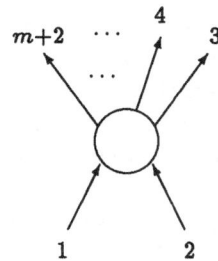

Fig. 1. A $2 \to m$ process

The constraint is rendered vastly more powerful if there is a conserved higher-spin charge P_s in the model. Since it must commute with the S-matrix, we have

$$\langle final|S|initial\rangle = \langle final|e^{i\alpha P_s} S e^{-i\alpha P_s}|initial\rangle$$

and so $e^{i\alpha P_s}$ can be used to rearrange the initial and final configurations without changing the amplitude. All that remains – and for this you should consult Parke's article – is to show that if any of the outgoing rapidities are different from θ_1 or θ_2, then shifting the configurations around in this way will give a pattern of trajectories for which $t_{12} > t_{23}$. By macrocausality the amplitude for this pattern must vanish, and then by the insensitivity of the amplitudes to shifts induced by $e^{i\alpha P_s}$, all of the other amplitudes, including the one initially under consideration, must also vanish. Hence the only possibilities for the two incoming particles are two outgoing particles with the same pair of rapidities as before the interaction, which is the result required for $n=2$.

To complete the missing step, Parke actually needed to assume the existence of *two* extra charges of higher spin. However, since a parity-conjugate pair Q_s, Q_{-s} will do, this is scarcely a problem, at least in parity-symmetric theories.

For completeness, I should mention that there is a quicker argument for this $2 \to m$ amplitude, found in Polyakov (1977), which revives the previous line of reasoning, though with a slight twist. As previously noted, if the first argument is attempted with the time in figure 1 running up the page, more and more conserved charges will be needed as m increases in order to eliminate all the undesired possibilities for the final configuration. But by T-invariance, the $2 \to m$ amplitude will only be nonvanishing if the same is true of the time-reversed $m \to 2$ amplitude. But now there are just two outgoing momenta, and these are fixed, up to a discrete ambiguity, by energy-momentum conservation. After this, *any* extra charge will suffice to eliminate the process. Economical as this argument is, it does not cover the general $m \to n$ amplitude, and factorisability and the Yang-Baxter equation are missed.

One other aside before moving on: however the higher-spin conserved charges are used to reshuffle the positions of an incident pair of particles, if their rapidities differ then their trajectories will still cross somewhere. This is special to 1+1 dimensions: with more than one spatial dimension to play with, conserved higher spin charges can be used to make trajectories miss each other completely, even on macroscopic scales. It is then but a short step to deduce that the S-matrix must be trivial – and this, in admittedly sketchy form, is an argument for the Coleman-Mandula theorem alluded to earlier.

To deal with three incoming particles, consider first how the trajectories would look were there no interactions in the model. Figure 2 shows the three distinct possibilities – which one actually occurs depends on the particular spatial positions of the incident wavepackets.

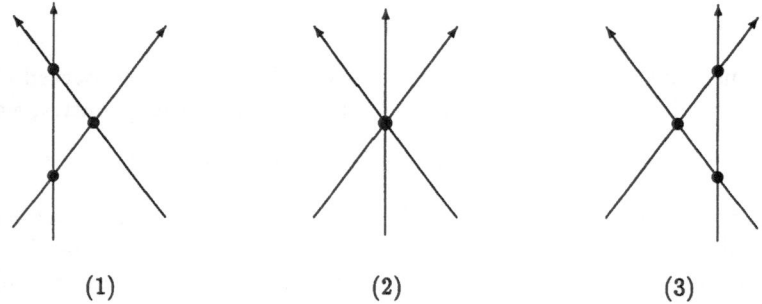

| (1) | (2) | (3) |

Fig. 2. Possibilities for a $3 \to 3$ process

In cases 1 and 3, when we switch the interaction back on again the results just established for two incident particles, together with locality, are enough to see that the pictures do not change in any essential way. Furthermore, as the interaction proceeds by a series of two-body collisions, these amplitudes must *factorise* into products of $2 \to 2$ amplitudes. Case 2 in general would give something new. However, using $e^{-i\alpha P_s}$ in the manner discussed at length above, it can be converted into one of the other cases. Hence there is never any particle production, individual momenta are conserved, and the amplitudes always factorise. In addition, the equality of amplitudes 1 and 3 gives a constraint on the two-body amplitudes, known as the Yang-Baxter equation. More on this in the next lecture, once the necessary notation has been set up.

To go beyond three incoming particles, an inductive argument can be used, showing that a set of n incident particles can always be shuffled around in such a way that the interaction occurs via a sequence of events in which at most $n-1$ particles are participating.

The ultimate conclusion is that in any local scattering theory in 1+1 dimensions with a couple of local higher-spin conserved charges (and a parity-conjugate pair $\{Q_s, Q_{-s}\}$, $s > 1$, will certainly do), there is no particle production, the final set of momenta is equal to the initial set, and the $n \to n$ S-matrix factorises into a product of $2 \to 2$ S-matrices. These are the three properties promised earlier, and now they can be established with only a finite amount of work.

Finally, I would like to mention a mild paradox that might at first sight seem troubling. If $\{p'_1 \ldots p'_n\} = \{p_1 \ldots p_n\}$ for every set of ingoing momenta, then surely $\sum (p_a)^s$ is conserved for all s, resulting in conserved charges at *all* spins, in any model for which the arguments above apply? This reasoning misses a key feature of the objects we are dealing with: for Q_s to qualify as a local conserved charge, it must be possible to write it as the integral of a local conserved density:

$$Q_s = \int_{-\infty}^{\infty} T_{s+1} dx \ .$$

There is no a priori reason why such a density should exist, even if the sums $\sum (p_a)^s$ happen to be conserved. In fact, the set of spins s at which this can be done forms a rather good fingerprint for a model, and turns out to constrain its behaviour in important ways.

3   The Two-Particle S-Matrix

Once the two-particle S-matrix is known, factorisation tells us that the entire S-matrix follows. To find the two-particle S-matrix becomes the main goal. In the algebraic notation of the last lecture, we can write

$$|A_i(\theta_1)A_j(\theta_2)\rangle_{in} = S_{ij}^{kl}(\theta_1 - \theta_2)|A_k(\theta_1)A_l(\theta_2)\rangle_{out}$$

as

$$A_i(\theta_1)A_j(\theta_2) = S_{ij}^{kl}(\theta_1 - \theta_2)A_l(\theta_2)A_k(\theta_1) \ ,$$

with $\theta_1 > \theta_2$ to ensure that *in* and *out* states are correctly represented. A sum over k and l is implied, with $k \neq i$ and $l \neq j$ being possible in those situations where some particles are not distinguished by the $Q_{s \neq 0}$ conserved charges. Lorentz boosts shift rapidities by a constant, and so S only depends on the difference $\theta_1 - \theta_2 = \theta_{12}$.

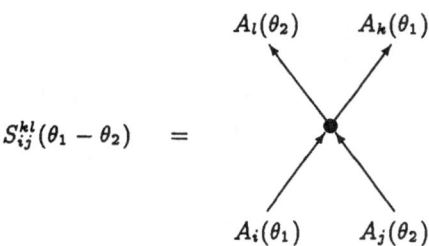

$$S_{ij}^{kl}(\theta_1 - \theta_2) \quad = $$

Fig. 3. The two-particle S-matrix

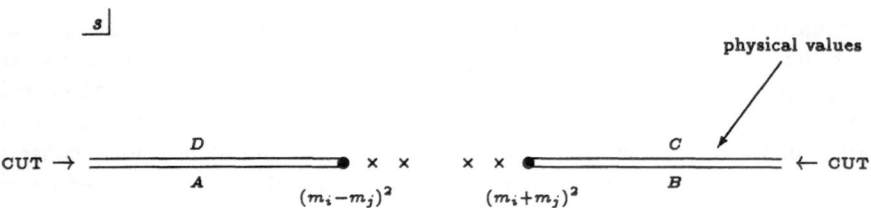

Fig. 4. The physical sheet

In a theory with r different particle types, knowledge of the r^4 functions $S_{ij}^{kl}(\theta)$ will thus give the full S-matrix. Not all of these functions are independent, and their analytic properties are heavily constrained. Such general features are the subject of this lecture.

First, the assumption of P and CT invariance requires

$$S_{ij}^{kl}(\theta) = S_{ji}^{lk}(\theta) \quad ; \quad S_{ij}^{kl}(\theta) = S_{kl}^{ij}(\theta) \ .$$

Analytic properties of the S-matrix are usually discussed in terms of the Mandelstam variables s, t and u:

$$s = (p_1 + p_2)^2 \ , \quad t = (p_1 - p_3)^2 \ , \quad u = (p_1 - p_4)^2 \ ,$$

with $s + t + u = \sum_{i=1}^{4} m_i^2$. In 1+1 dimensions only one of these is independent, and it is standard to focus on s, the square of the forward-channel momentum. In terms of the rapidity difference $\theta_{12} = \theta_1 - \theta_2$,

$$s = m_i^2 + m_j^2 + 2 m_i m_j \cosh \theta_{12} \ .$$

For a physical process, θ_{12} is real and so s is real and satisfies $s \geq (m_i + m_j)^2$. But we can consider the continuation of $S(s)$ up into the complex plane. Placing the branch cuts in the traditional way, this results in a function with the following properties:

• S is a singlevalued, meromorphic function on the complex plane with cuts on the portions of the real axis $s \leq (m_i - m_j)^2$ and $s \geq (m_i + m_j)^2$. Physical values of $S(s)$ are found for s just above the right-hand cut. This first sheet of the full Riemann surface for S is called the physical sheet.

• S is real-analytic: it takes complex-conjugate values at complex-conjugate points:

$$S_{ij}^{kl}(s^*) = \left[S_{ij}^{kl}(s) \right]^* \ .$$

In particular $S(s)$ is real if s is real and $(m_i - m_j)^2 \leq s \leq (m_i + m_j)^2$.

The situation is depicted in figure 4.

Unitarity requires that $S(s^+)S^\dagger(s^+) = 1$ whenever s^+ is a physical value for s, just above the right-hand cut: $s^+ = s + i0$, $s > (m_i + m_j)^2$. This should be understood as a matrix equation, with a sum over a complete set of asymptotic states hiding between S and $S^\dagger$. As s^+ grows, it becomes energetically possible for states with more and more particles to participate in the sum. Generally this brings the $2 \to m$ S-matrix elements into the story with $m = 3, 4, \ldots$, and gives the $2 \to 2$ S-matrix elements a series of branch points along the real axis, located at the $3, 4, \ldots$ particle thresholds. However for an integrable model these production amplitudes should all be zero, and so for all physical s^+ unitarity reads

$$S_{ij}^{kl}(s^+)\left[S_{kl}^{nm}(s^+)\right]^* = \delta_i^n \delta_l^m .$$

With the help of real analyticity this can be rewritten as

$$S_{ij}^{kl}(s^+)S_{kl}^{nm}(s^-) = \delta_i^n \delta_l^m ,$$

with $s^- = s - i0$, just below the right-hand cut. This equation shows the need for a branch cut running rightwards from the two-particle threshold $s = (m_i + m_j)^2$; if we accept that the cut actually starts at this threshold, then it is easy to see that the branch is of square-root type. The argument goes as follows. Let $S_\gamma(s)$ be the function obtained by analytic continuation of $S(s)$ once anticlockwise around the branch point. Unitarity amounts to the requirement that $S(s^+)S_\gamma(s^+) = 1$ for all physical values of s^+. When written in this way, the relation can be analytically continued to all s, so

$$S_\gamma(s) = S^{-1}(s) .$$

In particular, if s^- is a point just below the cut, then

$$S_\gamma(s^-) = S^{-1}(s^-) = S(s^+) ,$$

the last equality following from a second application of unitarity. Now $S_\gamma(s^-)$ is just the analytic continuation of $S(s^+)$ twice around $(m_i + m_j)^2$. Therefore, twice round the branch point gets us back to where we started, and the singularity is indeed a square root.

So much for the right-hand cut. The left-hand half of the figure, containing the second cut running in the opposite direction, can be understood via the fundamentally relativistic property of crossing. If one of the incoming particles, say j, is crossed to become outgoing while simultaneously one of the outgoers, say l, crosses in the opposite sense and becomes ingoing, then the amplitude for another physical two-particle scattering process results. For this new amplitude the incomers are i and $\bar{l}$, and the outgoers k and $\bar{j}$, where an overbar has been introduced to denote the (possibly trivial) operation of conjugation on particle labels. All of this amounts to looking at figure 3 from the side, with the forward-channel momentum now not s but rather $t = (p_1 - p_3)^2$. In this particular case $p_3 = p_2$, and the relation between t and s is very simple:

$$t = (p_1 - p_2)^2 = 2p_1^2 + 2p_2^2 - (p_1 + p_2)^2 = 2m_i^2 + 2m_j^2 - s .$$

Crossing symmetry states that the amplitude for this process can be obtained by analytic continuation of the previous amplitude into a region of the s plane where t becomes physical, that is $t \in \mathbb{R}$ and $t \geq (m_i + m_j)^2$. Physical amplitudes correspond to approaching this line segment from above in the t plane, and hence from below in the s plane. Thus the amplitudes are on the lower edge of the left-hand cut, marked A on figure 4. In equations:

$$S_{ij}^{kl}(s^+) = S_{i\bar{l}}^{k\bar{j}}(2m_i^2 + 2m_j^2 - s^+) \ .$$
$$\uparrow \qquad\qquad\qquad \uparrow$$
$$(\text{on } C) \qquad\qquad (\text{on } A)$$

Clearly the cross-channel branch point at $(m_i - m_j)^2$ must also be a square root, but this does *not* mean that the Riemann surface for $S(s)$ has just two sheets. Continuing through the left-hand cut can, and usually does, connect with a different sheet from that found through the right-hand cut. Stepping up and down to left and right, the typical $S(s)$, even for an integrable model, lives on an infinite cover of the physical sheet.

This looks rather complicated, but simplifies considerably if, following Zamolodchikov, attention is switched from the Mandelstam variable s to the rapidity difference θ. The transformation is

$$\theta = \cosh^{-1}\left(\frac{s - m_i^2 - m_j^2}{2m_i m_j}\right)$$
$$= \log\left[\frac{1}{2m_i m_j}\left(s - m_i^2 - m_j^2 + \sqrt{(s - (m_i+m_j)^2)(s - (m_i-m_j)^2)}\right)\right]$$

and it maps the physical sheet into the region

$$0 \leq \operatorname{Im}\theta \leq \pi$$

of the θ plane called the physical strip. Most importantly, the cuts are opened up, so that $S(\theta)$ is analytic at the images 0 and $i\pi$ of the two physical-sheet branch points, and also at the images $in\pi$ of the branch points on all of the other, unphysical, sheets. Since, by integrability, these are expected to be the only branch points, S is a meromorphic function of θ. The other sheets are mapped onto a succession of strips

$$n\pi \leq \operatorname{Im}\theta \leq (n+1)\theta \ .$$

The new image of the Riemann surface is shown in figure 5.

The previous relations can now be translated to give a list of constraints on $S(\theta)$ to be carried forward into later lectures:

- Real analyticity: $S(\theta)$ is real for θ purely imaginary;
- Unitarity: $S_{ij}^{nm}(\theta)S_{nm}^{kl}(-\theta) = \delta_i^k \delta_j^l$;
- Crossing: $S_{ij}^{kl}(\theta) = S_{i\bar{l}}^{k\bar{j}}(i\pi - \theta)$.

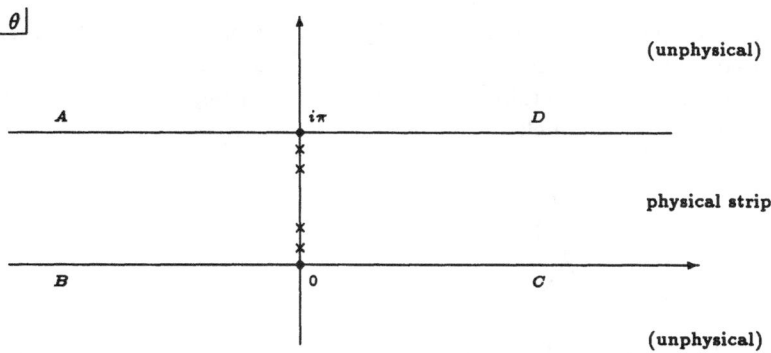

Fig. 5. The θ plane

A couple of remarks: first, both the unitarity and the crossing equations can now be analytically continued, and apply to the whole of the θ plane, not just along the line segments of physical values. Second, the unitarity constraint means that it is consistent to extend the algebraic relation

$$A_i(\theta_1)A_j(\theta_2) = S_{ij}^{kl}(\theta_1 - \theta_2)A_l(\theta_2)A_k(\theta_1)$$

to $\theta_1 < \theta_2$. Unitarity then becomes a consequence of the algebra, and the single-valued nature of products of the non-commuting symbols.

Finally to some unfinished business from the previous lecture. Shifting trajectories showed that the amplitudes (1) and (3) of figure 2 must be equal. If the two-particle S-matrix is not completely diagonal, this equality is not automatic but instead results in the following consistency condition:

$$S_{ij}^{\beta\alpha}(\theta_{12})S_{\beta k}^{n\gamma}(\theta_{13})S_{\alpha\gamma}^{ml}(\theta_{23}) = S_{jk}^{\beta\gamma}(\theta_{23})S_{i\gamma}^{\alpha l}(\theta_{13})S_{\alpha\beta}^{nm}(\theta_{12}) ,$$

where $\theta_{ab} = \theta_a - \theta_b$, and θ_1, θ_2 and θ_3 are the rapidities of particles i, j and k. This is the Yang-Baxter equation, forced by the ability of the conserved charges to shift particle trajectories around. In theories where particles appear in multiplets transforming under some symmetry group, this equation together with some minimality assumptions is often enough to conjecture the complete functional form of S. See other lectures in this school for more on this subject. The equation is equivalent to associativity for the algebra of the $A_i(\theta)$'s: moving from $A_i(\theta_1)A_j(\theta_2)A_k(\theta_3)$ to a sum of products $A_l(\theta_3)A_m(\theta_2)A_n(\theta_1)$, the result is independent of the order of the pair transpositions, if and only if the Yang-Baxter equation holds for the two-particle S-matrix elements.

4   Pole Structure and Bound States

The remaining features of figures 4 and 5 are the crosses marked between the two thresholds. The first things one might expect to find in these locations are simple poles corresponding to stable bound states, appearing either in the forward (s) or the crossed (t) channel:

This is potentially important – for example, it might signal the presence of hitherto unsuspected particles in the spectrum of the model. Most of the remaining lectures will be spent on this point. I'll start by recalling a selection of reasons why the association between simple poles in an S-matrix and bound states is natural:

• Potential scattering: in quantum mechanics, if the S-matrix for the scattering of a particle off a potential has a pole – which, as it happens, is always simple – then it is possible to use it to *construct* a wavefunction for the particle bound to the potential;

• tree-level Feynman diagrams;

• an 'axiom', justified if by nothing else by experience in 3+1 dimensions.

It turns out that life isn't so simple in 1+1 dimensions. To explain this I'll use the grandparent of all integrable field theories, the sine-Gordon model. Take the sinh-Gordon Lagrangian introduced in the first lecture, and replace β by $i\beta$. Re-zeroing the energy of the classical ground state, you will find

$$\mathcal{L} = \frac{1}{2}(\partial\phi)^2 - V(\phi)$$

with

$$V(\phi) = \frac{m^2}{\beta^2}\left[1 - \cos(\beta\phi)\right] .$$

There is extra structure here, as compared to the sinh-Gordon model, since there are infinitely-many classical vacua, $\phi(x) = 2n\pi/\beta$, $n \in \mathbb{Z}$. There is a conserved, spin-zero topological charge Q_0:

$$Q_0 = \frac{\beta}{2\pi}\int_{-\infty}^{\infty} \partial_x \phi \, dx$$

which is non-zero for configurations which interpolate between different vacua. (Formally the charge can also be defined, and is conserved, for the sinh-Gordon model – the only problem is that it is identically zero for all finite-energy configurations.)

Classically, the model has a soliton s with $Q_0 = +1$, and an antisoliton $\bar{s}$ with $Q_0 = -1$, both with mass M, say, and both interpolating between neighbouring vacua. There are no classically stable solutions with $|Q_0| > 1$. Solitons repel solitons, antisolitons repel antisolitons, but solitons and antisolitons attract. Therefore the classical theory additionally sees a continuous family of so-called breather solutions, which are $s\bar{s}$ bound states. Although not static, they are periodic in time and in most respects behave just like further particle states. Their 'masses' range from 0 (tightly-bound) to $2M$ (almost unbound).

In the quantum theory, the breather spectrum becomes discrete, just as would be expected from quantum mechanics. If s and $\bar{s}$ have mass M, then the breather masses are

$$M_k = 2M \sin \frac{\pi k}{h} , \qquad k = 1, 2, \ldots < \frac{8\pi}{\beta^2} - 1$$

where

$$h = \frac{16\pi}{\beta^2} \left(1 - \frac{\beta^2}{8\pi} \right) .$$

This was found by Dashen, Hasslacher and Neveu in 1975 via a semiclassical quantisation of the two-soliton solution, and is thought to be exact. Notice that as $\beta \to 0$, corresponding to the classical limit, the continuous breather spectrum is recovered.

The S-matrix elements of the solitons provide an illustration of the notational technology set up earlier. The model turns out to possess higher-spin conserved charges, and so all of the previous discussions apply. However at generic values of β^2 none of them breaks the $\phi \to -\phi$ symmetry of the original Lagrangian, and so none can be used to distinguish the soliton from the antisoliton. That leaves Q_0, which makes a fine job of distinguishing a single soliton from a single antisoliton but, as we shall now see, is not quite powerful enough when acting on two-particle states to rule out nondiagonal scattering.

Consider a general two-particle in-state $|A(\theta_1)_{s,\bar{s}} A(\theta_2)_{s,\bar{s}}\rangle_{in}$, each particle either a soliton or an antisoliton. The higher-spin charges can be used in the ways explained earlier to show that any out-state into which this state evolves must again contain two particles with rapidities θ_1 and θ_2, each either a soliton or an antisoliton. Thus before recourse is made to the spin-zero charge, a four-dimensional space of out-states is available. The topological charge Q_0 acts in this space as follows:

$$Q_0 \begin{pmatrix} |A_s(\theta_1)A_s(\theta_2)\rangle \\ |A_s(\theta_1)A_{\bar{s}}(\theta_2)\rangle \\ |A_{\bar{s}}(\theta_1)A_s(\theta_2)\rangle \\ |A_{\bar{s}}(\theta_1)A_{\bar{s}}(\theta_2)\rangle \end{pmatrix} = \begin{pmatrix} 2 & & & \\ & 0 & & \\ & & 0 & \\ & & & -2 \end{pmatrix} \begin{pmatrix} |A_s(\theta_1)A_s(\theta_2)\rangle \\ |A_s(\theta_1)A_{\bar{s}}(\theta_2)\rangle \\ |A_{\bar{s}}(\theta_1)A_s(\theta_2)\rangle \\ |A_{\bar{s}}(\theta_1)A_{\bar{s}}(\theta_2)\rangle \end{pmatrix} .$$

The soliton-soliton and antisoliton-antisoliton states are picked out uniquely, and therefore must scatter diagonally. The same cannot be said for the remaining two states, and it is through this loophole that nondiagonal scattering enters the story.

Taking charge conjugation symmetry into account, there are just three independent amplitudes to be determined. With $\theta = \theta_1 - \theta_2$, these can be written as:

$$
\begin{pmatrix}
|A_s(\theta_1)A_s(\theta_2)\rangle_{in} \\
|A_s(\theta_1)A_{\bar{s}}(\theta_2)\rangle_{in} \\
|A_{\bar{s}}(\theta_1)A_s(\theta_2)\rangle_{in} \\
|A_{\bar{s}}(\theta_1)A_{\bar{s}}(\theta_2)\rangle_{in}
\end{pmatrix}
=
\begin{pmatrix}
S(\theta) & & & \\
 & S_T(\theta) & S_R(\theta) & \\
 & S_R(\theta) & S_T(\theta) & \\
 & & & S(\theta)
\end{pmatrix}
\begin{pmatrix}
|A_s(\theta_1)A_s(\theta_2)\rangle_{out} \\
|A_s(\theta_1)A_{\bar{s}}(\theta_2)\rangle_{out} \\
|A_{\bar{s}}(\theta_1)A_s(\theta_2)\rangle_{out} \\
|A_{\bar{s}}(\theta_1)A_{\bar{s}}(\theta_2)\rangle_{out}
\end{pmatrix}.
$$

The same information can be given pictorially:

and also using the noncommuting symbols:

$$
A_s(\theta_1)A_s(\theta_2) = S(\theta)A_s(\theta_2)A_s(\theta_1)
$$
$$
A_s(\theta_1)A_{\bar{s}}(\theta_2) = S_T(\theta)A_{\bar{s}}(\theta_2)A_s(\theta_1) + S_R(\theta)A_s(\theta_2)A_{\bar{s}}(\theta_1) .
$$

Unitarity and crossing constrain these amplitudes. As a simple exercise, it is worthwhile to check that unitarity amounts to

$$
S(\theta)S(-\theta) = 1
$$
$$
S_T(\theta)S_T(-\theta) + S_R(\theta)S_R(-\theta) = 1
$$
$$
S_T(\theta)S_R(-\theta) + S_R(\theta)S_T(-\theta) = 0
$$

while crossing is

$$
S(i\pi - \theta) = S_T(\theta)
$$
$$
S_R(i\pi - \theta) = S_R(\theta)
$$

In 1977, Zamolodchikov was able to build on the earlier proposal of Korepin and Faddeev (1975) for the special points $h = 2n$, $n \in \mathbb{N}$ (at which $S_R(\theta)$ vanishes), to conjecture an exact formula for the S-matrix. Subsequent derivations made use of the Yang-Baxter equation, but in any event I only want to quote the physical pole structure here. (A pole is called 'physical' if, like the crosses on figure 5, it lies on the physical strip.) A moment's thought about the ways that the vacua fit together shows that:

- S_T can only form breathers in the forward channel;
- S can only form breathers in the crossed channel;
- S_R can form both.

This is precisely matched by Zamolodchikov's S-matrix: in terms of $B(\beta) = 2\beta^2/(8\pi-\beta^2) = 4/h$, the poles of S_T, S and S_R in the physical strip are found at the following points:

- S_T: $(1 - k\frac{B}{2})\pi i$, $k = 1, 2, \ldots$:

- S: $k\frac{B}{2}\pi i$, $k = 1, 2, \ldots$:

- S_R: $(1 - k\frac{B}{2})\pi i$, $k\frac{B}{2}\pi i$, $k = 1, 2, \ldots$:

(Beyond the physical strip, S_T, S and S_R have a proliferating set of unphysical poles, there to fix up crossing and unitarity, but this aspect will not be important below.) In the illustrations, the particles responsible for the poles have also been indicated. To check that these have been placed correctly, all that is needed is some elementary kinematics. Suppose that a soliton s and an antisoliton $\bar{s}$, of masses $M_s = M_{\bar{s}} = M$ and moving with respective rapidities θ_1 and $\theta_2 = -\theta_1$, fuse to form a (stationary) breather of mass M_b. The relative rapidity of the two particles is $\theta_{12} = 2\theta_1$, and the S-matrix will normally have a simple, forward-channel pole at exactly this point. Conservation of energy dictates that $M_b = 2M\cosh(\theta_{12}/2)$. It will be convenient to write this special value of θ_{12} as $iU_{s\bar{s}}^b$, where $U_{s\bar{s}}^b$ is called the fusing angle for the fusing $s\,\bar{s} \to b$:

By convention, an arrow pointing forwards in time marks a soliton, and an arrow pointing backwards an antisoliton; lines without arrows are breathers of some sort. Rotating the diagram by $\pm 2\pi/3$ gives pictures of $b\,s$ and $\bar{s}\,b$ scattering, and the corresponding fusing angles have also been indicated. If all of the poles in S_T are forward-channel, then the values of the fusing angles follow from the positions of these poles:

$$U_{s\bar{s}}^b = \left(1 - k\frac{B}{2}\right)\pi \quad , \qquad U_{bs}^s = U_{\bar{s}b}^{\bar{s}} = \left(\frac{1}{2} - k\frac{B}{4}\right)\pi .$$

The angles are all real, reflecting the fact that the bound states are below threshold and the relative rapidities at which they are formed purely imaginary. The masses of the corresponding bound states are therefore

$$M_b = 2M\cos\left(\frac{\pi}{2} - k\frac{B}{4}\pi\right) = 2M\sin\left(\frac{k\pi}{h}\right) ,$$

and these match the spectrum of breather masses.

For later use, the precise relationship between S_T, S and S_R is:

$$S_T(iu) = S(i\pi - iu) = \frac{\sin(\frac{2}{B}u)}{\sin(\frac{2}{B}\pi)} S_R(iu) .$$

The first equality is merely crossing symmetry, whilst the factor of $\sin(\frac{2}{B}u)$ multiplying S_R is there to exclude the crossed-channel poles in S_R from S_T, and the forward-channel poles in S_R from S.

There are also S-matrix elements involving the breathers. These can be deduced using bootstrap equations, to be described a little later, but for now the focus is elsewhere and so I'll just quote the required result, concerning the scattering of two copies of the first breather:

• S_{11} has poles at $i\frac{B}{2}\pi$ and $i\pi - i\frac{B}{2}\pi$:

This looks fine: it is easy to check that the pole at $i\frac{B}{2}\pi$ can be blamed on a copy of the second breather as a forward-channel bound state, and the other one on the same particle appearing in the crossed channel. But now consider what happens as β, and hence $B/2$, increases. Each time $B/2$ passes an inverse integer, a pole in S_T leaves the physical strip and the corresponding breather leaves the spectrum of the model. Finally, when $B/2$ passes $1/2$, the second breather drops out. The theory, now well into the quantum regime, has just the soliton, the antisoliton, and the first breather in its spectrum. And this is problematical: S_{11} still has a pair of simple poles. How can this be, if the particle previously invoked to explain them is no longer there?

The answer was found by Coleman and Thun in 1978, and requires a preliminary diversion into the subject of anomalous threshold singularities. These are most simply understood by asking how an individual Feynman diagram might become singular. If the external momenta are such that a number of internal propagators can find themselves simultaneously on-shell, then it turns out that the loop integrals give rise to a singularity in the amplitude. Apart from the somewhat trivial examples provided by tree-level diagrams, these singularities are always branch points in spacetimes of dimension higher than two; but in 1+1 dimensions, they can give rise to poles instead.

Once this is known, the problem of identifying the positions of such singularities becomes a geometrical exercise in gluing together a collection of on-shell vertices so as to make a pattern that closes. For three-point vertices, the on-shell requirement simply forces the relative Minkowski momenta to be equal to i times the fusing angles. If all couplings are below threshold, then all fusing angles are real and the resulting patterns can be drawn as figures in two Euclidean dimensions. These pictures are known as Landau, or on-shell, diagrams.

In fact, the characterisation as so far given also encompasses the more usual multiparticle thresholds, which are associated in perturbation theory with on-shell diagrams of the following type:

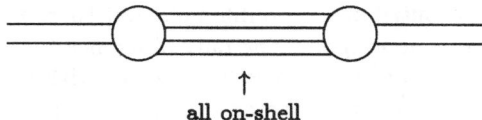

all on-shell

(Exceptionally, time is running sideways in this picture.) Here, the on-shell particles are all in the same channel, and the value of s at which the singularity is found is simply the square of the sum of the masses of the intermediate on-shell particles. To qualify as 'anomalous', something more exotic should be going on, and the position of the singularity will no longer have such a straightforward relationship with the spectrum of the model.

The moral is that when we come to analyse the pole structure of an S-matrix in 1+1 dimensions there are more things to worry about than just the tree-level processes discussed so far. Returning to the sine-Gordon model, as the point $B/2 = 1/2$ is passed, an on-shell diagram does indeed enter the game as far as the scattering of two of the first breathers is concerned:

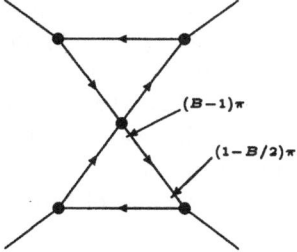

$(B-1)\pi$

$(1-B/2)\pi$

This diagram only invokes the solitons and antisolitons on the internal lines, which are present in the spectrum whatever the coupling. A couple of internal angles are marked, making it clear that the figure will only close if $B/2 \geq 1/2$. However we are not quite out of the woods yet: diagrams of this sort are expected to yield double poles when evaluated in 1+1 dimensions, and not the single poles that we are after as soon as $B/2$ passes $1/2$. (Actually at $B/2 = 1/2$, S_{11} does indeed have a double pole, but the understanding of a single extra point is scarcely major progress.) The final ingredient is to notice that for $B/2 > 1/2$, two of the internal lines must inevitably cross over. When a soliton and an antisoliton meet we should allow for reflection as well as transmission, since we have already seen that both amplitudes are generally nonzero. Thus not one but four diagrams are relevant to the amplitude near to the value of θ_{12} of interest. The full story is given by the diagrams

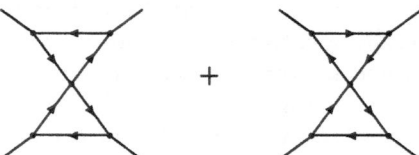

together with their overall conjugates, in which all arrows are reversed. Individually each diagram contributes a double pole, but these must be added together with the correct relative weights. The only difference between the two diagrams shown is that the central blob on the first carries with it a factor of $S_T(\theta)$, and the central blob of the second a factor of $S_R(\theta)$. At the pole position, the value of θ is fixed by the on-shell requirement to be equal to iu, with $u = (B-1)\pi$. Referring back to the earlier expression relating S_T to S_R, we have

$$S_T((B-1)\pi i) = \frac{\sin(\frac{2}{B}(B-1)\pi)}{\sin(\frac{2}{B}\pi)} S_R((B-1)\pi i)$$
$$= -S_R((B-1)\pi i)$$

Thus exactly when the individual diagrams have a double pole, $S_T + S_R = 0$, a cancellation occurs, and the field-theoritic prediction is for a simple pole, exactly as seen in the S-matrix. Coleman and Thun dubbed explanations of this sort 'prosaic', since they do not rely on properties special to integrable field theories – a non-integrable (albeit very finely-tuned) theory would be perfectly capable of exhibiting the same behaviour. Nonetheless, there is a certain miraculous quality about the result. The cancellation between S_T and S_R is very delicate: S_T describes a classically-allowed process, while S_R does not (there is no classical reflection of solitons). It is also noteworthy that the Landau diagrams expose intrinsically field-theoretical aspects of the theory, since loops are involved. Their relevance tells us that quantum mechanical intuitions about bound states and pole structure may occasionally be misleading.

Some general lessons can be drawn from all of this:

• The S-matrix can have poles between $\theta = 0$ and $\theta = i\pi$;

• these can be first order, second order, or in fact much higher order (examples up to 12th order are found in the affine Toda field theories);

• even for a first-order pole, a direct interpretation in terms of a bound state is not inevitable;

• but there is always some (prosaic^{TM}) explanation in terms of standard field theory.

5   Bootstrap Equations

If we decide that our theory does contain a bound state, then the next task is to find the S-matrix elements involving this new particle, and then to look for evidence of further bound states in these, and so on. Rather than continuing with the sine-Gordon example, which showed how complicated the story can become, I will make a tactical retreat at this point to a class of models where the behaviour is rather simpler, and the workings of the bound states can be seen more cleanly. The structure is still rewardingly rich, so this won't be too great a sacrifice.

The key concession is to assume that there are no degeneracies among the one-particle states once all of the non-zero spin conserved charges have been specified. This closes off the loophole exploited by the sine-Gordon model, and forces the scattering to be diagonal.

The S-matrix now only needs two indices:

$$S_{ij}(\theta_1 - \theta_2) \quad = \quad$$

$$A_j(\theta_2) \quad A_i(\theta_1)$$

$$A_i(\theta_1) \quad A_j(\theta_2)$$

Two of the previous constraints on the two-particle S-matrix elements can therefore be simplified:

• Unitarity: $S_{ij}(\theta)S_{ij}(-\theta) = 1$;

• Crossing: $S_{ij}(\theta) = S_{i\bar{j}}(i\pi - \theta)$.

(In contrast to its previous incarnation, there is no sum on repeated indices in the unitarity equation.) Combining these two reveals the important fact that

$$S_{ij}(\theta + 2\pi i) = S_{ij}(\theta) ,$$

so that for diagonal scattering the Riemann surface for the S-matrix really is just a double cover of the complex plane – whether you go round the left or the right branch point in figure 4, you always land up on the same unphysical sheet.

The simplification is even more drastic for the third constraint: the loss of matrix structure, already evident in the revised unitarity equation, means that the Yang-Baxter equation is trivially satisfied for any $S_{ij}(\theta)$ whatsoever.

Fortunately, a vestige of algebraic structure does remain, in the guise of the pattern of bound states. Suppose that $S_{ij}(\theta_{12})$ has a simple pole, at $\theta_{12} = iU_{ij}^k$ say, which really is due to the formation of a forward-channel bound state. Note that, in a unitary theory, forward and crossed channel poles can be distinguished by the fact that the residues are positive-real multiples of i in the forward channel, and negative-real multiples in the crossed channel. The previous picture of the scattering process can be 'expanded' near to the pole:

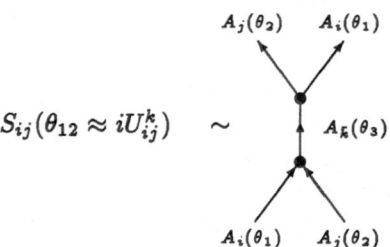

$$S_{ij}(\theta_{12} \approx iU_{ij}^k) \quad \sim$$

(The intermediate particle is labelled $\bar{k}$ for convenience, in anticipation of a convention that all indices on a three-point coupling will be ingoing.) There are a number of immediate consequences:

(1) The quantum coupling C^{ijk} is nonzero at the point where particles i, j and k are all on shell.

(2) At the rapidity difference $\theta_{12} = iU_{ij}^k$, the intermediate particle $A_{\bar{k}}(\theta_3)$ is on shell and survives for macroscopic times. On general grounds (the 'bootstrap principle', or 'nuclear democracy'), $A_{\bar{k}}$ is expected to be one of the other asymptotic one-particle states of the model.

(3) Since $s = m_k^2$ when this happens, we have

$$m_k^2 = m_i^2 + m_j^2 + 2m_i m_j \cos U_{ij}^k .$$

Of course, the U_{ij}^k are just the fusing angles already seen in the last lecture.

For a more geometrical characterisation of the fusing angles, observe that the formula just given is familiar from elementary trigonometry, and implies that U_{ij}^k is the outside angle of a 'mass triangle' of sides m_i, m_j and m_k :

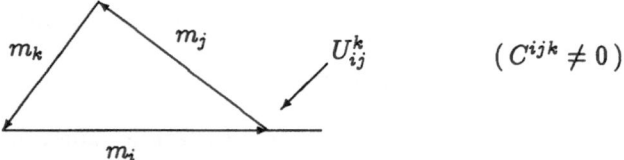

With $C^{ijk} \neq 0$, poles are also present in S_{jk} and S_{ki}. From the triangle just drawn, the three fusing angles involved satisfy

$$U_{ij}^k + U_{jk}^i + U_{ki}^j = 2\pi .$$

Concrete examples, the nonzero quantum couplings $C^{bs\bar{s}}$ in the sine-Gordon model, were mentioned in the last lecture.

Since the $\bar{k}$ is supposed to be long-lived when $\theta_{12} = iU_{jk}^k$, it should be possible evaluate a conserved charge Q_s after the fusing of i and j into $\bar{k}$, as well as before.

The action of Q_s on $|A_{\bar{k}}(\theta_3)\rangle$ and $|A_i(\theta_1)A_j(\theta_2)\rangle$ was given at the beginning of the second lecture; equating the two at the relevant rapidities gives a constraint on the numbers $q_i^{(s)}$ which characterise Q_s:

$$ C^{ijk} \neq 0 \quad \Rightarrow \quad q_{\bar{k}}^{(s)} = q_i^{(s)} e^{is\bar{U}_{ki}^j} + q_j^{(s)} e^{-is\bar{U}_{kj}^i}, $$

where $\bar{U} = \pi - U$. (To see this, switch to the frame where $\bar{k}$ is stationary. Then $\theta_1 = i\bar{U}_{ki}^j$ and $\theta_2 = -i\bar{U}_{kj}^i$.) The relations $q_{\bar{k}}^{(s)} = (-1)^{s+1} q_k^{(s)}$ and $\bar{U} + \bar{U} + \bar{U} = \pi$ can be employed to put this into a more symmetrical form:

$$ C^{ijk} \neq 0 \quad \Rightarrow \quad q_i^{(s)} + q_j^{(s)} e^{isU_{ij}^k} + q_k^{(s)} e^{is(U_{ij}^k + U_{jk}^i)} = 0. $$

Drawing this equation in the complex plane shows that it has a nice interpretation as a closure condition for a 'generalised mass triangle':

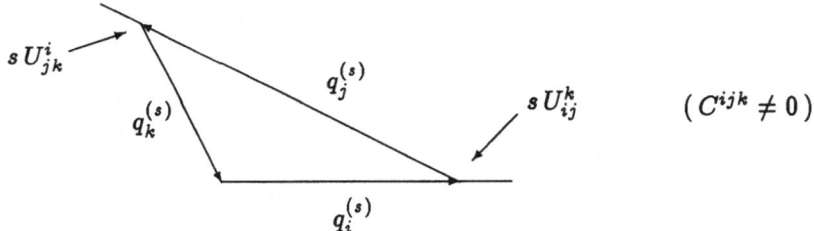

(Note that the three angles for this triangle, sU_{ij}^k, sU_{jk}^i and sU_{ki}^j, now add up to $2\pi s$ instead of just 2π.)

This set equations constitutes the conserved charge bootstrap. Given a set of masses and three-point couplings, the fusing angles can be determined by the mass triangles. The angles in the higher-spin triangles are then fixed, and at any given spin s the demand that all of the triangles at that spin should close provides an overdetermined set of conditions on the values of the $q_i^{(s)}$. Should the only solution be the trivial one, $q_i^{(s)} = 0$ for all i, then we can conclude that $Q_s \equiv 0$ and there is no conserved charge of that spin. The surprise is that there should be any choice of the initial masses and couplings such that the higher-spin triangles can be made to close for at least an infinite subset of spins. However, when this does happen, the set of spins at which the triangles do close gives access to the fingerprint of spins mentioned earlier, without the need to find the local conserved densities explicitly.

A fair amount of physical intuition has been used to arrive at these conclusions, and one particular point deserves mention. Given that the fusing angles will always be real in the applications of interest, most if not all of the states entering the discussion have unphysical momenta. I have been implicitly assuming that the states, their fusings, and the action on them of the conserved charges, continue to behave in the expected manner after the necessary analytic continuations have been made.

For the S-matrix we can use a similar argument. Consider another particle l, which might interact either before or after particles i and j fuse to form $\bar{k}$. Which depends on the impact parameter, but in an integrable model this should be irrelevant. Translating this into equations,

$$C^{ijk} \neq 0 \quad \Rightarrow \quad S_{l\bar{k}}(\theta) = S_{li}(\theta - i\bar{U}_{ki}^{j})S_{lj}(\theta + \bar{U}_{jk}^{i}),$$

and this is the S-matrix bootstrap equation. It can be given a more symmetrical appearance using crossing symmetry and unitarity, becoming:

$$C^{ijk} \neq 0 \quad \Rightarrow \quad S_{li}(\theta)S_{lj}(\theta + iU_{ij}^{k})S_{lk}(\theta + i(U_{ij}^{k}+U_{jk}^{i})) = 1.$$

Imposing these relations for each non-vanishing three-point coupling provides an overdetermined set of functional equations, and again it is rather surprising that there are any solutions. In fact the two bootstraps are rather directly related:

Exercise: with the help of a logarithmic derivative and a Fourier expansion, show that each solution to the S-matrix bootstrap contains within it a solution to the corresponding conserved charge bootstrap.

Starting from an initial guess of a single S-matrix element, we can now search for poles, infer some three-point couplings, apply the bootstrap to deduce further S-matrix elements, and then iterate away. If the process closes on a finite set of particles, then we can chalk up a success and go on to another problem; if not, then the initial guess should probably be revised. This is precisely the approach that A.B.Zamolodchikov took in his pioneering work, (1989a) and (1989b), on perturbed conformal field theories. The next lecture is devoted to a particularly interesting example of this procedure which relates to the behaviour of the $T = T_c$ Ising model, in a small magnetic field.

6   Zamolodchikov's E_8-Related S-Matrix

The critical Ising model is found at zero magnetic field, with the temperature carefully adjusted to the critical value T_c. If the continuum limit is taken at this point, the result is well-known to be described by the $c = \frac{1}{2}$ conformal field theory, a very well-understood object. In the papers (1989a) and (1989b), Zamolodchikov probed nearby points by considering the actions

$$S_{\text{pert}} = S_{\text{CFT}} + \lambda \int d^2 x \phi(x),$$

where S_{CFT} is a notional action for the $c = \frac{1}{2}$ conformal field theory, inside of which ϕ sits as one of the spinless, relevant fields. There are just two of these for the Ising model, and one of them, usually labelled σ, can be identified with the scaling limit of the local magnetisations (spins) on the lattice. Thus perturbing by σ corresponds to switching on a magnetic field. The game now is to exploit the great control that we have of the unperturbed situation to divine some information about the perturbed model. In this particular case, Zamolodchikov used an ingenious argument, based on the counting of dimensions in Virasoro representations, to establish that the perturbed model supported at the very least local conserved charges with the following spins:

$$s \; = \; 1, \; 7, \; 11, \; 13, \; 17, \; 19.$$

This tells us two things. First of all, there are certainly enough charges here to employ Parke's argument, and so the perturbed model, if massive, possesses a factorisable S-matrix (and the model must be massive, since the c-theorem tells us that the central charge of any conformal infrared limit would be less than $\frac{1}{2}$, and there is no such unitary conformal field theory). Second, we now have the first part of the fingerprint of conserved spins, and can hope to use this information to build a bridge between the ultraviolet information residing in the characterisation of the model as a perturbed conformal field theory, and the infrared information that would be revealed if we knew its S-matrix.

To commence the search for this S-matrix, suppose that the massive theory possesses a particle of mass m_1, say. In addition, assume for the time being that the model falls into the simplest class, of diagonal scattering theories. The magnetic field breaks the $\mathbb{Z}_2$ symmetry of the unperturbed model, and so there is no reason to exclude an interaction of ϕ^3 type from the effective Lagrangian of the perturbed theory. This places the model in the same general class as the second example discussed in the first lecture, and makes it natural for C^{111} to be nonzero. For this coupling the mass triangle is equilateral, and the fusing angles are therefore all equal to $2\pi/3$. The conserved charge bootstrap equation is

$$C^{111} \neq 0 \quad \Rightarrow \quad q_1^{(s)} + q_1^{(s)} e^{2\pi i s/3} + q_1^{(s)} e^{4\pi i s/3} = 0.$$

This equation has a nontrivial solution whenever s has no common divisor with 6:

$$s \; = \; 1, \; 5, \; 7, \; 11, \; 13, \; 17 \ldots \, .$$

This is too much of a good thing: the fingerprint of the theory contains rather too many spins for comfort. Whilst the unwanted charges might vanish for other reasons, it would be more satisfying if the cast of particles could be enlarged a little, so as to restrict the set of conserved spins a bit more. Besides, earlier work described in McCoy and Wu (1978) had led Zamolodchikov to suspect the presence of at least a couple of further masses in the particle spectrum. Taking things one step at a time, he first enlarged the spectrum by adding just one more particle type, with mass m_2, and supposed that both C^{112} and C^{221} were nonzero. The fusing angles are not so easily determined now, but if the ignorance is encoded in the pair of numbers $y_1 = \exp(iU_{21}^1)$ and $y_2 = \exp(iU_{12}^2)$, then two of the bootstrap equations are:

$$C^{121} \neq 0 \quad \Rightarrow \quad q_1^{(s)} + q_2^{(s)}(y_1)^s + q_1^{(s)}(y_1)^{2s} = 0 \, ;$$

$$C^{212} \neq 0 \quad \Rightarrow \quad q_2^{(s)} + q_1^{(s)}(y_2)^s + q_2^{(s)}(y_2)^{2s} = 0 \, .$$

Eliminating $q_1^{(s)}$ and $q_2^{(s)}$,

$$(y_1^s + y_1^{-s})(y_2^s + y_2^{-s}) = 1 ,$$

at least at those values of s for which there is a nontrivial conserved charge. If there are more than a couple of these, then the system is overdetermined; nevertheless, if $y_1 = \exp(4\pi i/5)$ and $y_2 = \exp(3\pi i/5)$ then there is a solution for every odd s which is not a multiple of 5. This yields the following set of fusing angles:

$$U_{12}^1 = U_{21}^1 = 4\pi/5 , \quad U_{11}^2 = 2\pi/5 ;$$

$$U_{21}^2 = U_{12}^2 = 3\pi/5 , \quad U_{22}^1 = 4\pi/5 ,$$

and the golden mass ratio

$$\frac{m_2}{m_1} = 2\cos\frac{\pi}{5} .$$

(There are other solutions such as $(-y_1, -y_2)$ or (y_2, y_1), but the choice taken is the only one which yields sensible fusing angles and $m_1 < m_2$.)

This is very promising: the multiples of 5 were exactly the values of s that had to be eliminated in order to match the sets of conserved spins. Thus encouraged, we can start to think about the S-matrix.

It has been assumed that all the particles are self-conjugate (the absence of any even spins from the fingerprint is a good hint that this assumption is correct) and so each S-matrix element S_{ij} must be individually crossing-symmetric, $S_{ij}(\theta) = S_{ij}(i\pi-\theta)$, as well as unitary. It is convenient to construct these as products of a basic 'building block' $(x)(\theta)$, where

$$(x)(\theta) = \frac{\sinh\left(\frac{\theta}{2} + \frac{i\pi x}{60}\right)}{\sinh\left(\frac{\theta}{2} - \frac{i\pi x}{60}\right)} .$$

(The 60 in the denominators has been chosen with advance knowledge of the final answer, so that all of the arguments x will turn out to be integers.) Unitarity is built into these blocks, whilst the crossing symmetry just mentioned is assured if each block (x) is always accompanied by $(30-x)$. The block (x) has a single physical-strip pole at $i\pi x/30$, and no physical-strip zeroes.

Consider first $S_{11}(\theta)$. The nonzero couplings C^{111} and C^{112} imply forward-channel poles at $iU_{11}^1 = 2\pi i/3$ and $iU_{11}^2 = 2\pi i/5$. Incorporating these and their crossed partners into a first guess for the S-matrix element gives

$$S_{11} = (10)(12)(18)(20) .$$

However this can't be the whole story. The S-matrix bootstrap equation for the ϕ^3 coupling C^{111} requires that $S_{11}(\theta-i\pi/3)S_{11}(\theta+i\pi/3)$ should be equal to $S_{11}(\theta)$. But it is easy to check that for the guess just given,

$$S_{11}(\theta-i\pi/3)S_{11}(\theta+i\pi/3) = -(2)(8)(10)(20)(22)(28) ,$$

which is not the desired answer. Stare at the equations long enough, though, and you might just spot that all will be well if the initial guess is multiplied by the factor $-(2)(28)$. Thus the minimal solution to the constraints imposed so far is

$$S_{11} = -(2)(10)(12)(18)(20)(28) \, .$$

The bootstrap equations have forced the addition of two extra poles, and the simplest option is to suppose that these are the forward- and crossed-channel signals of a further particle, with mass $m_3 = 2m_1 \cos(\pi/30)$, and a nonzero coupling C^{113}. Of course, the story is not over yet. Using the bootstrap for the fusing $1\,1 \rightarrow 2$ allows S_{12} to be obtained from the provisional S_{11}:

$$S_{12} = (6)(8)(12)(14)(16)(18)(22)(24) \, .$$

The poles from the blocks (14), (18) and (24) are correctly-placed to match forward-channel copies of particles the 3, 2 and 1 respectively, those in (6), (12) and (16) can then be blamed on the same particles in the crossed channel, but the blocks (8) and (22) are not so easily dismissed, and require the addition of yet another particle, of mass m_4 say. (Consideration of the signs of the residues shows that the forward-channel pole is at $8\pi i/30$.) Next, the bootstrap for $1\,1 \rightarrow 3$ predicts

$$S_{13} = (1)(3)(9)(11)^2(13)(17)(19)^2(21)(27)(29) \, .$$

(Exercise: check at least one of these claims.)

Apart from the double poles, which should not be too alarming after the earlier investigations of the sine-Gordon model, there is one more pair of simple poles here which cannot be explained in terms of the spectrum seen so far, and so a further mass, m_5 say, is revealed.

There is nothing to stop the mythical energetic reader from continuing with all this, and it turns out that no further backtracking is required – with just one correction to the initial guess, Zamolodchikov had arrived at a consistent conjecture for $S_{11}(\theta)$. Furthermore the final answer has a number of intriguing properties. These can be summarised in a list of what might be called 'S-matrix data':

- 8 particle types $A_1, \ldots A_8$;
- 8 masses m_i, $i = 1, \ldots 8$, which together form an eigenvector of the Cartan matrix of the Lie algebra E_8:

$$C_{ij}^{[E_8]} m_j = (2 - 2\cos\frac{\pi}{30}) m_i \; ;$$

(This allows each particle type to be attached to a spot on the E_8 Dynkin diagram – more on this later.)

- solutions to the conserved-charge bootstrap found at

$$s \;\; = \;\; 1, \, 7, \, 11, \, 13, \, 17, \, 19, \, 23, \, 29 \, \ldots$$

thus fitting the fingerprint found from perturbed conformal field theory (and also the exponents of E_8, repeated modulo 30);

• 'charges' associated with these solutions which form further eigenvectors of $C_{ij}^{[E_8]}$:

$$C_{ij}^{[E_8]} q_j^{(s)} = (2 - 2\cos\frac{\pi s}{30}) q_i^{(s)} \; ;$$

• a full two-particle S-matrix which is a collection of complicated but elementary functions, with poles at integer multiples of $i\pi/30$, all products of the elementary building blocks introduced earlier.

There is no space to record the full S-matrix here, but a complete table can be found in, for example, Braden et al. (1990).

One note of caution: elegant though it might be, it is not completely clear that this is the answer to the question originally posed, given the number of assumptions that were made along the way. Probably the most convincing reassurance comes on recalculating the central charge of the unperturbed model from the conjectured S-matrix, using a technique called the thermodynamic Bethe ansatz. Its use in this context was first advocated by Al.B.Zamolodchikov (1990), and the specific calculation for the E_8-related S-matrix can be found in Klassen and Melzer (1990).

7   Coxeter Geometry

It is a finite but lengthy task to check all of the bootstrap equations for Zamolodchikov's S-matrix, and to verify the properties listed at the end of the last lecture. However there is something not completely satisfactory about this, and a feeling that an underlying structure remains to be discovered, a structure that might help to explain quite why such an elegant solution to the bootstrap should exist at all. The purpose of this short section, something of an aside from the main development, is to show that at least some parts of this question can be answered.

One mathematical preliminary is required, a quick recap on the Weyl group of E_8. Imagine a hedgehog Φ of 240 vectors, or 'roots', sitting in eight dimensions. They all have equal length, and together they make up the root system of E_8. Each root can be written as an integer combination of the simple roots $\{\alpha_1, \ldots \alpha_8\}$:

$$\alpha \in \Phi \quad \Rightarrow \quad \alpha = \sum_{i=1}^{8} m_i \alpha_i$$

with $m_i \in \mathbb{Z}$, and the m_i either all non-negative, or all non-positive. Actually, there are 240 different eight-element subsets of Φ which could serve as the simple roots, but their geometrical properties are all identical, and can be summarised by giving the set of their mutual inner products, as encoded either in the Cartan matrix

$$C_{ij}^{[E_8]} = 2\frac{\alpha_i.\alpha_j}{\alpha_j^2} \; ,$$

or the Dynkin diagram

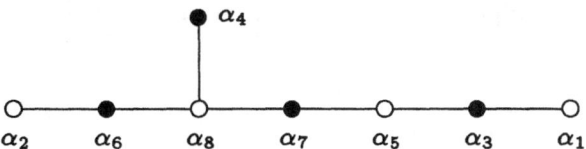

Pairs of simple roots joined by a line have inner product -1, and all other pairs are orthogonal. In particular this means that the black-coloured roots are mutually orthogonal, as are the white-coloured roots. The labelling might look random, but recall from the last lecture that the vector of masses formed an eigenvector of the Cartan matrix, so that each particle type in Zamolodchikov's S-matrix can be assigned to a point on the Dynkin diagram. The labels used here correspond to these particle labels, with $m_1 < m_2 < \ldots < m_8$.

For each $\alpha \in \Phi$, define the Weyl reflection r_α to be the reflection in the 7-dimensional hyperplane orthogonal to α :

$$r_\alpha : \quad x \mapsto x - 2\frac{\alpha . x}{\alpha^2}\alpha .$$

The products of the Weyl reflections in any order and of any length together make up W, the Weyl group of E_8. This group maps Φ to itself, and is *finite*: $|W| < \infty$. (Note that W is therefore a finite reflection group, a much simpler object than the Lie group or algebra with which it is associated.) One more fact: to generate W, it's enough to start with the set of simple reflections $\{r_{\alpha_1}, \ldots r_{\alpha_8}\}$, and I will write these as $\{r_1, \ldots r_8\}$.

Now I want to study the properties of one particular element $w \in W$. It is a Coxeter element, meaning that it is a product in some order of a set of simple reflections. Although the ordering is not crucial, the result I'm after is most transparent if I pick

$$w = r_3 r_4 r_6 r_7 r_1 r_2 r_5 r_8 .$$

This is a Steinberg ordering: reflections of one colour act first, followed by those of the other. The ordering amongst the reflections of like colour is immaterial – they all commute, since the corresponding simple roots are orthogonal. The project is to see how w^{-1} acts on Φ, and as a start we can examine the orbit of α_1 under w^{-1}. Noting that

$$r_i(\alpha_j) = \alpha_j - C_{ji}^{[E_8]}\alpha_i , \qquad \text{(no sum on } j)$$

the individual simple reflection r_i negates α_i, adds α_i to all roots α_j joined to α_i by a line on the Dynkin diagram, and leaves the others alone. With this information it doesn't take too long to compute that

$$w^{-1}(\alpha_1) = r_8 r_5 r_2 r_1 r_7 r_6 r_4 r_3(\alpha_1) = \alpha_3 + \alpha_5 .$$

(Exercise: check this!)

To continue is an easy if somewhat tedious task, acting repeatedly with w^{-1} to find $w^{-2}(\alpha_1)$, $w^{-3}(\alpha_1)$, and so on. After 30 steps, you should find yourself back at α_1 (the number 30 might just be familiar from the list at the end of the last lecture). The story for the first 14 of these steps is contained in the following table:

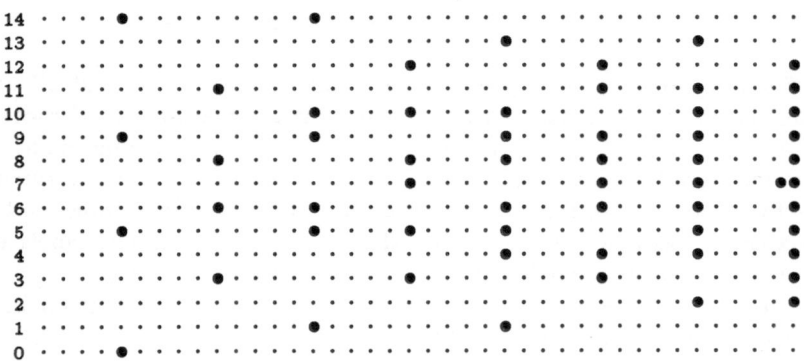

Images of α_1 under w^{-1}

The coefficient of α_i in the expansion of $w^{-p}(\alpha_1)$ is given by the number of blobs ($\bullet$) in the ith position of the pth row. For the E_8 Weyl group, $w^{15} = -1$ and so the rest of the table, rows 15 to 29, can be omitted.

All of this might seem a long way from exact S-matrices, but in fact the Weyl group computation just performed and the earlier bootstrap manipulations are in some senses one and the same calculation, just looked at from orthogonal directions. To explain this somewhat delphic remark, I will first rewrite the S-matrix elements already seen in a new and slightly more compact notation. Observe that, apart from the blocks (2) and (28) in the formula for S_{11}, every block (x) in S_{11}, S_{12} and S_{13} can be paired off with either $(x-2)$ or $(x+2)$. Noticing that $(0) = 1$ and $(30) = -1$, this pairing can be extended to the recalcitrant S_{11} as well, and in fact works for all of the other S-matrix elements too. Thus we can at least save some ink if we define a larger building block

$$\{x\} = (x-1)(x+1) \ .$$

and rewrite the S-matrix elements found previously as

$$S_{11} = \{1\}\{11\}\{19\}\{29\}$$
$$S_{12} = \{7\}\{13\}\{17\}\{23\}$$
$$S_{13} = \{2\}\{10\}\{12\}\{18\}\{20\}\{28\}$$

The next step is to introduce a pictorial representation of these formulae. Start by drawing a line segment to represent the interval from 0 to $i\pi$ on which the physical-strip poles are found. Then for each block $\{x\}$ in the S-matrix element, place a small brick ⊐ on the line segment, running from $i(x-1)\pi/30$ to

$i(x+1)\pi/30$. (Thus, the poles are located at the ends of the bricks.) The formulae just given become

$$S_{11} =$$

$$S_{12} =$$

$$S_{13} =$$

Rotate these three by 90 degrees and you should observe a neat match with the first three columns of the table on the last page, of images of α_1 under w^{-1}.

This is a glimpse of a general construction, which allows a diagonal scattering theory to be associated with every simply-laced Weyl group. Further details can be found in Dorey (1991,1992a); see also Fring and Olive (1992) and Dorey (1992b). All of these scattering theories were in fact already around in the literature: in addition to the articles by Zamolodchikov already cited, some relevant references are Köberle and Swieca (1979), Sotkov and Zhu (1989), Fateev and Zamolodchikov (1990), Christe and Mussardo (1990a,b), Braden et al. (1990), and Klassen and Melzer (1990). Why then worry about Weyl groups? This is largely a matter of taste, but it should be mentioned that the construction goes a little deeper than the curious coincidences described so far. The geometry of finite reflection groups appears to replace the rather more complicated Lie algebraic concepts that might have been a first guess as to the underlying mathematical structure. Features such as the coupling data and the pole structure can be related to simple properties of root systems, and this allows the bootstrap equations both for the conserved currents and for the S-matrices to be proved in a uniform way.

8   Affine Toda Field Theory

Zamolodchikov's E_8-related S-matrix is an example of a diagonal S-matrix with few of the subtleties that made the treatment of the sine-Gordon model so delicate. Whilst higher poles are certainly present, their orders are always just as would be predicted from an initial glance at the possible Landau diagrams. In particular, simple poles are always associated with bound states. Since the cancellations which complicated the sine-Gordon case relied on the non-diagonal nature of its S-matrix, one might suppose that diagonal scattering theories would always behave in a straightforward manner. Curiously enough, this turns out not to be true. The affine Toda field theories, the subject of this lecture, provide a number of elegant counterexamples.

The study of these models begins with a standard, albeit non-polynomial, scalar Lagrangian in 1+1 dimensions:

$$\mathcal{L} = \frac{1}{2}(\partial\phi)^2 - \frac{m^2}{\beta^2}\sum_{a=0}^{r} n_a e^{\beta\alpha_a\cdot\phi}.$$

This describes the interaction of r scalar fields, gathered together into the vector $\phi \in \mathbb{R}^r$. The set $\{\alpha_0 \ldots \alpha_r\}$ is a collection of $r+1$ further vectors in $\mathbb{R}^r$, which must be carefully picked if the model is to be integrable. It turns out that there is a classically acceptable choice for every (untwisted or twisted) affine Dynkin diagram $g^{(k)}$, thought of as encoding the mutual inner products of the α_a. By convention α_0 corresponds to the 'extra' spot on the affine diagram, and the integers n_a satisfy $n_0 = 1$ and $\sum_{a=0}^{r} n_a \alpha_a = 0$. The real constant m sets a mass scale, while β governs the strength of the interactions. When β is also real, the models generalise sinh-Gordon rather than sine-Gordon and there is no topology to worry about. In fact once we go beyond the sinh-Gordon example making β imaginary is no longer an innocent operation, since in all other cases the manifest reality of the Lagrangian is promptly lost. Despite these problems the models with β purely imaginary have received a fair amount of attention, starting with the work of Hollowood (1992). However, in this lecture I will stick to the cases where β is real.

As classical field theories, these models are all integrable, and exhibit conserved quantities at spins given by the exponents of $g^{(k)}$, repeated modulo a quantity called the kth Coxeter number, $h^{(k)}$:

$$h^{(k)} = k \sum_{a=0}^{k} n_a .$$

(For the untwisted diagrams, $k=1$ and $h^{(k)}$ is the same as the usual Coxeter number h.)

However when we turn to the quantum theory, none of the elegant classical apparatus, as described in, for example, Mikhailov et al. (1981), Wilson (1981), and Olive and Turok (1985), is immediately applicable. A more elementary approach is appropriate, studying the models with the standard perturbative tools of quantum field theory before proceeding to some exact conjectures. Arinshtein et al. were the first to try this, for the $a_n^{(1)}$ theories, in 1979. Interest in the subject was renewed following Zamolodchikov's work on perturbed conformal field theories, and the fact that in the meantime the other classically-integrable possibilities, related to the other affine Dynkin diagrams, had been uncovered. Initially, only the so-called self-dual models were understood, and elements of this story can be found in Christe and Mussardo (1990a,b) and Braden et al. (1990,1991). The other, non self-dual, cases were more tricky, since they turned out to fall into the class of less straightforward scattering theories for which simple poles do not always have simple explanations. The crucial step was made by Delius et al. (1992), and the papers by Corrigan et al. (1993) and Dorey (1993) can be consulted for the few cases not covered in their work. In the remainder of this lecture I will outline some aspects of these quantum considerations, but the discussion will perforce be very sketchy. In addition to the references just cited, the review by Corrigan (1994) is a good place to start for those interested in delving deeper into this subject. That the field is still developing is evinced by an article by Oota (1997) which appeared as these notes were being written up, indicating that the ideas discussed in the last lecture may also be relevant, if suitably q-deformed, to the non self-dual theories that had previously resisted any geometrical interpretation.

If we are to treat these models as ordinary quantum field theories, then the first step must be to find out what the multipoint couplings are. To this end, the potential term in the Lagrangian can be expanded as follows:

$$V(\phi) \equiv \frac{m^2}{\beta^2} \sum_{a=0}^{r} n_a e^{\beta \alpha_a \cdot \phi}$$

$$= \frac{m^2}{\beta^2} \sum_{a=0}^{r} n_a + \frac{1}{2}(M^2)^{ij} \phi^i \phi^j + \frac{1}{3!} C^{ijk} \phi^i \phi^j \phi^k + \dots$$

where summations on the repeated indices i, j and k running from 1 to r are implied, and the two and three index objects

$$(M^2)^{ij} = m^2 \sum_{a=0}^{r} n_a \alpha_a^i \alpha_a^j$$

and

$$C^{ijk} = m^2 \beta \sum_{a=0}^{r} n_a \alpha_a^i \alpha_a^j \alpha_a^k$$

can be thought of as the mass^2 matrix and the set of three-point couplings, at least classically. Much as for theories discussed in the first lecture, it is possible to view the C^{ijk} as containing the 'bones' of the model, with the higher couplings, hidden as ' $+ \dots$ ', there just to tidy away any residual production amplitude backgrounds that would otherwise spoil integrability. Now the general idea is the following: first diagonalise M^2 to find the classical particle masses $m_1 \dots m_r$, and then compute the C^{ijk} in the eigenbasis of M^2 to find the classical three-point couplings between the corresponding one-particle states. At this level there are already some surprises: for example, it turns out that in all of the untwisted cases, the set of masses form the eigenvector, with lowest eigenvalue, of the corresponding non-affine Cartan matrix. This was initially noticed on a case-by-case basis, before being proved in a general way by Freeman (1991). The Coxeter element, described in the last lecture, turns out to be crucial in this discussion. This work was further elaborated by Fring et al. (1991), elucidating in particular earlier observations about the three-point couplings.

However it has been obtained, once the classical data is known two things can be done: on the one hand the masses and three-point couplings can be fed into the bootstrap to make some initial conjectures as to the full quantum S-matrices, and on the other perturbation theory can be attempted in order to check these conjectures. As hinted above, the affine Toda field theories split into two classes when this programme is attempted: 'straightforward' and 'not straightforward'. To make this distinction more precise, define a duality operation on the set of all affine Dynkin diagrams by

$$\{\alpha_0 \dots \alpha_r\} \leftrightarrow \{\alpha_0^\vee \dots \alpha_r^\vee\}$$

where

$$\alpha_a^\vee \equiv \frac{2}{\alpha_a^2} \alpha_a \, .$$

(This is sometimes called Langlands duality.)

When appropriately normalised, the sets of vectors associated with the $a_n^{(1)}$, $d_n^{(1)}$, $e_n^{(1)}$ and $a_{2n}^{(2)}$ affine Dynkin diagrams are self-dual in this sense, and the corresponding affine Toda field theories are also called self-dual. These are the 'straightforward' cases: conjectures based on the classical data lead to self-consistent quantum S-matrices, and to date these have passed all perturbative checks to which they have been subjected. For example, the mass ratios, and hence the fusing angles, are preserved at one loop. As a result, the bootstrap structure is essentially blind to the value of the coupling β, which enters into the S-matrices via a function

$$B(\beta) = \frac{1}{2\pi} \frac{\beta^2}{1 + \beta^2/4\pi} \, .$$

There is a simple relationship between the S-matrices for certain perturbed conformal field theories and the S-matrices for the self-dual affine Toda models: all that has to be done is to replace building blocks of the type seen in the last lecture

$$\{x\}_{\mathrm{PCFT}} \equiv (x-1)(x+1)$$

by the slightly more elaborate blocks

$$\{x\}_{\mathrm{toda}} \equiv \frac{(x-1)(x+1)}{(x-1+B)(x+1-B)} \, .$$

Zamoldchikov's E_8-related S-matrix is related in this way to the S-matrix of the $e_8^{(1)}$ affine Toda field theory; more generally, for $g \in \{a, d, e\}$ the correspondence is between the g affine Toda field theory and a perturbation of the $g^1 \times g^1/g^2$ coset model, while $a_{2n}^{(2)}$ turns into a perturbation of the nonunitary minimal model $\mathcal{M}(2, 2n+3)$. (Note though that the factors of 60 appearing in the earlier definition of (x) should be replaced by $2h$, with h the relevant Coxeter number). For every self-dual affine Toda S-matrix, there is thus a companion 'minimal' S-matrix, sharing the same physical pole structure but lacking the coupling-constant dependent physical strip zeroes, which in the Toda theories serve to cancel the poles in the $\beta \to 0$ limit. Note also that replacing β by $4\pi/\beta$ sends B to $2-B$ and leaves the Toda blocks unchanged – a strong-weak coupling duality.

The remaining, non self-dual, models behave in a much more complicated way. The classical data is still elegant, but there is quantum trouble: conjectures based on the raw classical data are no longer self-consistent, and perturbative checks show varying mass ratios, causing the fusing angles to depend on the value of the coupling constant. These perturbative results are reinforced by the results of Kausch and Watts (1992) and Feigin and Frenkel (1993), which indicate that the correct general implementation of strong-weak coupling duality is not only to replace β by $4\pi/\beta$, but also to replace each α_a by its dual, $\alpha_a^\vee$. (Of course, for the self-dual theories this latter operation has no effect and so the earlier statement of duality remains correct for these cases.) This means that for each dual pair of classical affine Toda field theories, there should be just one quantum theory – there are 'fewer' genuinely distinct quantum theories than expected, and the different classical theories can be recovered by taking strong or weak coupling limits. The predicted dualities are:

$$b_n^{(1)} \leftrightarrow a_{2n-1}^{(2)} \qquad\qquad g_2^{(1)} \leftrightarrow d_4^{(3)}$$
$$c_n^{(1)} \leftrightarrow d_{n+1}^{(2)} \qquad\qquad f_4^{(1)} \leftrightarrow e_6^{(2)}$$

(Exercise: compare and contrast the classical conserved charge fingerprints for these models. Are they compatible with duality?)

If this picture is correct, then the mass ratios have no option but to vary: if a model is non self-dual, then (as can be checked case-by-case) its classical mass spectrum is always different from that of its dual. One spectrum is found at $\beta \to 0$, the other at $\beta \to \infty$, and the quantum theory, if it exists at all, must find some way of interpolating between the two. Given the apparent rigidity of the bootstrap equations, this looks to be rather a tall order. Nevertheless, Delius et al. (1992) decided to take the perturbatively-calculated shifts in the mass ratios seriously, and were led to a set of conjectures for most of the models in the above list (the only cases that remained were $d_4^{(3)}$, $e_6^{(2)}$ and $f_4^{(1)}$, and these were subsequently found to behave in just the same way). Whenever conjectures existed for both halves of a dual pair, they swapped over under $\beta \to 4\pi/\beta$. In fact, Delius et al. made their proposals independently of any expectations of duality; that it emerged anyway from their calculations can be seen in retrospect as strong evidence that they were on the right track. Further support came from the numerical results of Watts and Weston (1992), who examined the coupling-dependence of the single mass ratio found in the $g_2^{(1)}/d_4^{(3)}$ dual pair.

With the mass ratios depending on the coupling, it is no longer possible to use the simple building blocks $\{x\}_{\text{toda}}$ introduced earlier. A slightly more elaborate two-index block $\{x, y\}$ can be found in Dorey (1993), and is probably the most direct generalisation of the self-dual blocks. One point to note is that the natural way for the coupling to enter these blocks requires the previous definition of the function $B(\beta)$ to be slightly modified, so as to encompass the non self-dual theories as well:

$$B(\beta) = \frac{1}{2\pi} \frac{\beta^2}{h/h^\vee + \beta^2/4\pi} \, .$$

Here h is the Coxeter number of the relevant affine Dynkin diagram, and $h^\vee$ that of its (Langlands) dual.

There are, however, a couple of features of the non self-dual S-matrices which give pause for thought. Some simple poles, expected on the basis of the nonzero classical couplings and quantum mass ratios, turn out to be absent, whilst other simple poles, which are present in the quantum S-matrices, are not at locations which match any of the particle masses.

The resolution of the first problem appears to be that quantum corrections exactly cancel some of the classical three-point couplings, when evaluated on shell. This means that some quantum couplings C^{ijk} vanish even though their classical counterparts do not. The result is very delicate and has only been checked to one loop, but is probably necessary if duality is to hold, for the simple reason that the set of classical three-point couplings in a theory and its dual do not in general coincide. Those couplings which show signs of vanishing once quantum effects are taken into account are precisely those which are anyway absent at the classical level in the dual model.

As for the second problem, the mechanism is not too far removed from that operating in the sine-Gordon model. However, as mentioned at the beginning of this lecture, the fact that the scattering is diagonal means that we can no longer hope to generate simple poles through cancellations between competing Landau diagrams. Fortunately there is a compensating feature of the affine Toda S-matrices which allows the basic idea to be saved: they all exhibit zeroes as well as poles on the physical strip. These were already visible in the self-dual blocks $\{x\}_{\text{toda}}$ defined earlier, and are equally present in the more general blocks $\{x, y\}$. In the self-dual cases the zeroes are merely spectators, but in the non self-dual theories they come to play a much more central role in the pole analysis. A detailed discussion can be found in Corrigan et al. (1993), and it turns out that for every 'anomalous' simple pole in the non self-dual affine Toda S-matrices, Landau diagrams can be drawn in which some internal lines cross. Just as for sine-Gordon, the S-matrix elements for these internal crossings must be factored into the calculation before the overall order of any pole can be predicted. This time, these factors vanish individually as the diagrams are put on shell, and thereby manage to demote ostensibly higher poles into the simple poles that are required in order to match the quantum S-matrices.

Further reading, and acknowledgements

As promised, these lectures have only skimmed the surface of a large subject. In addition to the references mentioned in the main text, the review articles by Zamolodchikov and Zamolodchikov (1979), Zamolodchikov (1980) and Mussardo (1992) are recommended, as is the discussion of the sine-Gordon model given by Goebel (1986). (The opening section the first lecture was in fact inspired by a remark in this article.)

I would like to thank Zalán Horváth and Laci Palla for all their efforts in organising such a pleasant school, and also Olivier Babelon, Jean-Bernard Zuber and the other organisers of the the 'Integrable systems' semester held at the Institute Henri Poincaré, Paris, for giving me a second opportunity to talk about some of this material. I am grateful to Ed Corrigan and Gérard Watts for helpful comments, and to the UK EPSRC for an advanced fellowship.

References

Aref'eva, I.Ya. and Korepin, V.E. (1974): Scattering in two-dimensional model with Lagrangian $L = (1/\gamma) \left[(1/2)(\partial_\mu u)^2 + m^2(\cos u - 1) \right]$, JETP Lett. **20**, 312–314

Arinshtein, A.E., Fateyev, V.A. and Zamolodchikov, A.B. (1979): Quantum S-matrix of the (1+1)-dimensional Todd chain, Phys. Lett. **B87**, 389–392

Bernard, D. and Leclair, A. (1991): Quantum group symmetries and nonlocal currents in 2-D QFT, Comm. Math. Phys. **142**, 99–138

Braden, H.W., Corrigan. E., Dorey, P.E. and Sasaki, R. (1990): Affine Toda field theory and exact S-matrices, Nucl. Phys. **B338**, 689–746

Braden, H.W., Corrigan. E., Dorey, P.E. and Sasaki, R. (1991): Multiple poles and other features of affine Toda field theory, Nucl. Phys. **B356**, 469–498

Christe, P. and Mussardo, G. (1990a): Integrable systems away from criticality: the Toda field theory and S-matrix of the tricritical Ising model, Nucl. Phys. **B330**, 465–487

Christe, P. and Mussardo, G. (1990b): Elastic S-matrices in (1+1) dimensions and Toda field theories, Int. J. Mod. Phys. **A5**, 4581–4627

Coleman, S. and Mandula, J. (1967): All possible symmetries of the S-matrix, Phys. Rev. **159**, 1251–1256

Coleman, S. and Thun, H.J. (1978): On the prosaic origin of the double poles in the sine-Gordon S-matrix, Comm. Math. Phys. **61**, 31–39

Corrigan, E. (1994): Recent developments in affine Toda quantum field theory, Invited lectures at the CRM-CAP Summer School *Particles and Fields 94*, Banff, Alberta, Canada; hep-th/9412213

Corrigan, E., Dorey, P.E. and Sasaki, R. (1993): On a generalised bootstrap principle, Nucl. Phys. **B408**, 579–599

Dashen, R.F., Hasslacher, B. and Neveu, A. (1975): The particle spectrum in model field theories from semiclassical functional integral techniques, Phys. Rev. **D11**, 3424–3450

Delius, G.W., Grisaru, M.T. and Zanon, D. (1992): Exact S-matrices for nonsimply-laced affine Toda theories, Nucl. Phys. **B382**, 365–406

Dorey, P. (1991): Root systems and purely elastic S-matrices, Nucl. Phys. **B358**, 654–676

Dorey, P.E. (1992a): Root systems and purely elastic S-matrices (II), Nucl. Phys. **B374**, 741–761

Dorey, P. (1992b): Hidden geometrical structures in integrable models, in the proceedings of the NATO ARW *Integrable Quantum Field Theories*, Como, Italy (Plenum 1993); hepth/9212143

Dorey, P. (1993): A remark on the coupling dependence in affine Toda field theories, Phys. Lett. **B312**, 291–298

Fateev, V.A. and Zamolodchikov, A.B. (1990): Conformal field theory and purely elastic S-matrices, Int. J. Mod. Phys. **A5**, 1025–1048

Feigin, B.L. and Frenkel, E.V. (1993): Integrals of motion and quantum groups, Lectures given at the CIME Summer School on Integrable Systems and Quantum Groups, Montecatini Terme, Italy; hepth/9310022

Freeman, M.D. (1991): On the mass spectrum of affine Toda field theory, Phys. Lett. **B261**, 57–61

Fring, A., Liao, H.C. and Olive, D.I. (1991): The mass spectrum and coupling in affine Toda theories, Phys. Lett. **B266**, 82–86

Fring, A. and Olive, D.I. (1992): The fusing rule and the scattering matrix of affine Toda theory, Nucl. Phys. **B379**, 429–447

Goebel, C.J. (1986): On the sine-Gordon S-matrix, Prog. Theor. Phys. Supp. **86**, 261–273

Hollowood, T. (1992): Solitons in affine Toda field theories, Nucl. Phys. **B384**, 523–540

Kausch, H.G. and Watts, G.M.T. (1992): Duality in quantum Toda theory and W algebras, Nucl. Phys. **B386**, 166–192

Klassen, T.R. and Melzer, E. (1990): Purely elastic scattering theories and their ultraviolet limits, Nucl. Phys. **B338**, 485–528

Köberle, R. and Swieca, J.A. (1979): Factorizable $Z(N)$ models, Phys. Lett. **B86**, 209–210

Korepin, V.E. and Faddeev, L.D. (1975): Quantization of solitons, Teor. Mat. Fiz. **25**, 147–163

Lüscher, M. (1978): Quantum non-local charges and absence of particle production in the two-dimensional non-linear σ-model, Nucl. Phys. **B135**, 1–19

McCoy, B.M. and Wu, T.T. (1978): Two-dimensional Ising field theory in a magnetic field: Breakup of the cut in the two-point function, Phys. Rev. **D18**, 1259–1267

Mikhailov, A.V., Olshanetsky, M.A. and Perelomov, A.M. (1981): Two-dimensional generalised Toda lattice, Comm. Math. Phys. **79**, 473–488

Mussardo, G. (1992): Off-critical statistical models: factorized scattering theories and bootstrap program, Phys. Rep. **218**, 215–379

Olive, D.I. and Turok, N. (1985): Local conserved densities and zero-curvature conditions for Toda lattice field theories, Nucl. Phys. **B257**, 277–301

Oota, T. (1997): q-deformed Coxeter element in non-simply-laced affine Toda field theories, Yukawa Institute preprint YITP-97-33, hepth/9706054

Parke, S. (1980): Absence of particle production and factorization of the S matrix in (1+1)-dimensional models, Nucl. Phys. **B174**, 166–182

Polyakov, A.M. (1977): Hidden symmetry of the two-dimensional chiral fields, Phys. Lett. **B72**, 224–226

Shankar, R. and Witten, E. (1978): S matrix of the supersymmetric nonlinear σ model, Phys. Rev. **D17**, 2134–2143

Sotkov, G. and Zhu, C.-J. (1989): Bootstrap fusions and tricritical Potts model away from criticality, Phys. Lett. **B229**, 391–397

Tzitzéica, M. (1910): Sur une nouvelle classe de surfaces, Comptes Rendus Acad. Sci. **150**, 955

Watts, G.M.T. and Weston, R.A. (1992): $G_2^{(1)}$ affine Toda field theory. A numerical test of exact S-matrix results, Phys. Lett. **B289**, 61–66

Wilson, G. (1981): The modified Lax and two-dimensional Toda lattice equations associated with simple Lie algebras, Ergod. Th. & Dynam. Sys. **1**, 361–380

Zamolodchikov, A.B. (1977): Exact two-particle S-matrix of quantum sine-Gordon solitons, Comm. Math. Phys. **55**, 183–186

Zamolodchikov, A.B. (1980): Factorised S matrices and lattice statistical systems, Sov. Sci. Rev., Physics, **2**, 1–40

Zamolodchikov, A.B. (1989a): Integrals of motion and S-matrix of the (scaled) $T{=}T_c$ Ising model with magnetic field, Int. J. Mod. Phys. **A4**, 4235–4248

Zamolodchikov, A.B. (1989b): Integrable field theory from conformal field theory, Advanced Studies in Pure Mathematics **19**, 641–674

Zamolodchikov, A.B. (1989c): Fractional-spin integrals of motion in perturbed conformal field theory, in *Fields, strings and quantum gravity*, proceedings of the CCAST symposium workshop, Beijing 1989 (Gordon and Breach)

Zamolodchikov, Al.B. (1990): Thermodynamic Bethe ansatz in relativistic models. Scaling 3-state Potts and Lee-Yang models, Nucl. Phys. **B342**, 695–720

Zamolodchikov, A.B. and Zamolodchikov, Al.B. (1979): Factorised S-matrices in two dimensions as the exact solutions of certain relativistic quantum field theory models, Academic Press120, 253–291

Introduction to Simple Integrable Models of Quantum Field Theory

Harald Grosse[1], Edwin Langmann[2] and Ernst Raschhofer[1]

[1] Institut für Theoretische Physik, Universität Wien, Austria
[2] Theoretical Physics, Royal Institute of Technology, Stockholm, Sweden

Abstract. We summarize recent developments in the relationship between exactly solvable models of statistical physics and quantum field theory. After a short introduction to critical phenomena, we review the framework for quantum spin models and sketch the solution of simple models. We describe the Luttinger model and its gauged version, the Luttinger-Schwinger model. We use Boson-Fermion correspondence to solve these models and show how one can calculate Green functions. Next, we focus on algebraic aspects of quantum spin models. The structure of quantum integrability leads to the Yang-Baxter equation and eventually gives us a link to the modern developments of quantum geometry.

1 Introduction

This school was devoted to various aspects of low-dimensional models of statistical physics and quantum field theory. These parts of physics have developed almost explosively in recent years resulting in strong interrelations in a vast number of different subjects like classical integrable models, quantum spin models, vertex models, Yang-Baxter relation, braids and knots, quantum groups, Virasoro algebra representations, conformal field theory, Kac-Moody algebra representations, quantum Hall effect, anyons, fractional statistics, Chern-Simons model, topological field theory, strings, non-commutative geometry, Bose-Fermi correspondence, quantum wires, Luttinger liquid, ..., to name only some of them. Since it is impossible to cover all developments we restrict our attention mostly to algebraic aspects.

On the one hand we shall concentrate on Quantum Spin Models (Sect. (2, 5)), which are quantum integrable in the sense of the Yang–Baxter relation (Sect. 6) [1], and on the other hand we present the Luttinger model in Sect. 3 as an example of a solvable interacting field theoretical model. We can gauge this model and obtain the Luttinger-Schwinger model in Sect. 4. We solve the model, calculate the correlation functions and control the limits to relativistic invariant models explicitly.

Finally, in Sect. 7, we indicate a few of the many algebraic relationships which follow from integrability. They lead to the subject of quantum geometry.

The subject started when Lenz asked his student Ising in 1925 to study the thermodynamics of a one–dimensional spin system. The teacher became disappointed, since no nontrivial phase transition was found. Heisenberg wrote down, in 1928, a more general quantum spin model, which was supposed to describe ferromagnetism. Already in 1938 Bethe found an efficient method to obtain eigenfunctions of this Hamiltonian. The big step forward was done by Onsager in 1944, when he calculated the free energy of the two–dimensional Ising model. Already in 1950 Tomonaga formulated and solved a field theoretic model with four fermion interaction. The statistical physics model with nonlocal interaction is solvable too, as it has been shown by Luttinger in 1963. The general algebraic structure behind integrable models has been worked out first in the classical case. Gardner, Green, Kruskal and Miura invented the inverse scattering transform method in 1967. The quantum systems were treated by Yang (1967) and Baxter (1972). A number of groups worked out the second quantization of the inverse scattering method, and all recent developments resulted.

Our aim is to study models describing critical phenomena [2]: We consider as simple example the Ising model:

Assign to cubic lattice points $j = (j_1, \ldots, j_d) \rightarrow s_j \in \{1, -1\}$ a classical spin variable s_j. The set of $\{s_j\}$ is called a spin field configuration. Define the energy of such configurations by

$$H_N(\{s_j\}) = - \sum_{\langle ij \rangle}^{N} s_i s_j - h \sum_{j}^{N} s_j, \qquad (1.1)$$

where $\langle ij \rangle$ denotes nearest neighbours lattice points and h a magnetic field. At $T = 0$ we have two ground state configurations. For $d \geq 2$ and T small, two phases show up. The discrete up–down symmetry is spontaneously broken. The mean free energy becomes non–analytic in the thermodynamic limit: The free energy is defined through the partition function

$$Z_N(h, T) = \sum_{\text{conf.}} e^{-\beta H_N(\{s_j\})} = e^{-\beta F_N}, \qquad (1.2)$$

and expectation values of observables are given by

$$\langle A \rangle_N = \frac{1}{Z_N} \sum_{\text{conf.}} e^{-\beta H_N} A. \qquad (1.3)$$

Well–known examples of (1.3) are magnetization M_N and susceptibility χ_T^N

$$M_N = \left\langle \sum_{i=1}^{N} s_i \right\rangle_N = \frac{1}{\beta} \frac{\partial}{\partial h} \ln Z_N, \qquad \chi_T^N = \frac{\partial M_N}{\partial T}, \qquad (1.4)$$

as well as the inner energy U_N and the specific heat C_h^N:

$$U_N = \langle H_N \rangle_N, \qquad C_h^N = \frac{\partial U_N}{\partial T}. \qquad (1.5)$$

The connection to thermodynamics is given by $F_N = U_N - T \cdot S_N$, where S_N denotes the entropy.

The thermodynamic limit of these quantities

$$c_h = \lim_{N \to \infty} \frac{C_h^N}{N}, \qquad m = \lim_{N \to \infty} \frac{M_N}{N}, \qquad \text{etc.} \qquad (1.6)$$

shows a typical power law behaviour in the reduced temperature $\tau = (T - T_c)/T_c$ if we approach the critical point T_c:

$$c_h \simeq |\tau|^{-\alpha}, \qquad m_S \simeq |\tau|^\beta, \qquad \chi_T \simeq |\tau|^{-\gamma}, \qquad (1.7)$$

where we introduced the spontaneous magnetization $m_S = \lim_{h \searrow 0} m$, and defined critical exponents α, β, γ. m_S is called an order parameter, since it allows to distinguish between the low temperature ordered phases, where it is not vanishing, and the high temperature phase, where it is zero.

A more general class of models is given by

$$Z_N = \int \prod_{j=1}^{N} (ds_j \; e^{-V(s_j)}) e^{-\beta H(\{s_j\})}; \qquad (1.8)$$

they have either a discrete or a continuous symmetry.

Two more exponents (ν, η) are defined through cluster properties:

$$\langle s_i s_j \rangle^c \equiv \langle s_i s_j \rangle - \langle s_i \rangle \langle s_j \rangle \overset{|i-j| \to \infty}{\longrightarrow} \begin{cases} e^{-|i-j|/\xi(T)}, & T \neq T_c \\ \dfrac{\text{const}}{|i-j|^{d-2+\eta}}, & T = T_c \end{cases} \qquad (1.9)$$

where $\xi(T)$ denotes the correlation length, which diverges at $T = T_c$: $\xi(T) \overset{|\tau| \searrow 0}{\simeq} |\tau|^{-\nu}$.

We quote examples of critical exponents:

	Mean Field	$d = 2$ Ising	$d = 3$ Ising
α	0	0	0.11
2β	1	1/4	0.65
γ	1	7/4	1.24
η	0	1/4	0.03

There exist inequalities and relations among them. At $T = T_c$ we get long range order.

Such models describe gas–liquid transition, ferromagnetism, superconductivity, suprafluidity, structural phase transitions and many more.

Quantum spin 1/2 models can be defined by assigning σ–matrices to each lattice point $j \to (\sigma_j^x, \sigma_j^y, \sigma_j^z)$ with $\sigma_j^\alpha = 1 \otimes \ldots \otimes \sigma^\alpha \otimes 1 \otimes \ldots \otimes 1$ in $\bigotimes_{j=1}^{N} (\mathbf{C}^2)$ and the σ^α are at the j–th place. The Hamiltonian for the XYZ model is given by

$$H = -\sum_{\langle ij \rangle} (J_x \sigma_i^x \sigma_j^x + J_y \sigma_i^y \sigma_j^y + J_z \sigma_i^z \sigma_j^z) + h \sum_{j=1}^{N} \sigma_j. \qquad (1.10)$$

The partition function and expectation values are determined through a trace:

$$Z_N = \text{Tr } e^{-\beta H}, \qquad \langle A \rangle_N = \frac{1}{Z_N} \text{Tr } e^{-\beta H} A. \qquad (1.11)$$

Some general results are:

In $d = 1$ one cannot break spontaneously a discrete symmetry, as long as short range interactions are considered; this follows from an energy–entropy argument.

The Ising model in $d \geq 2$ has a phase transition at finite temperature; this follows from Peierls contour estimates.

The Mermin–Wagner–Hohenberg theorem states: A continuous symmetry cannot be broken spontaneously in $d = 1$ and $d = 2$; there exist thick Bloch walls and $m_S = 0$.

Due to a Goldstone mode, there exist nontrivial phase transitions in $d \geq 3$.

Quantum spin models may have long range order at $T = 0$ or at nonzero temperature. Three possibilities may occur: there is no gap; there exists a gap and the isolated ground state is non–degenerate or degenerate. We know examples to all three cases.

2 Quantum Spin Models

The deep connection between models of statistical physics and quantum field theory is also reflected in very simple relationships. There is, for example, the (Klein)–Jordan–Wigner transformation which allows to map the spin algebra for N–spins to the algebra of the canonical anticommutation relations. We compare

$$[S_n^\alpha, S_m^\beta] = 0, \qquad n \neq m; \qquad [S_n^\alpha, S_n^\beta] = i\varepsilon^{\alpha\beta\gamma} S_n^\gamma, \qquad (2.1)$$

which is a partly commutative algebra, with $\{c_n, c_m\} = 0$, and $\{c_n, c_n^\dagger\} = 1, \ldots,$ the CAR algebra.

For $N = 1$ it is easy to identify this mapping, since

$$S^+ |\downarrow\rangle = |\uparrow\rangle, \qquad S^+ |\uparrow\rangle = 0 \qquad \text{and} \qquad c^+ |0\rangle = |1\rangle, \qquad c^+ |1\rangle = 0, \ldots.$$

Both algebras can be realized in $\mathbf{C}^2$.

As for the general case we need a **Disorder– or Kink–operator**,

$$K_n = \exp\left(i\pi \sum_{j=1}^{n-1} S_j^+ S_j^- \right), \qquad (2.2)$$

which rotates spins at lattice places $j = 1, 2, \ldots, n-1$, around the z-axis; more precisely let $\sigma^1 |\pm\rangle = \pm |\pm\rangle$ and denote the product state by $|+, +, \ldots, +\rangle$ etc. Then

$$K_n |+, \ldots, +\rangle = |-, -, -, \ldots, -, \overset{n-\text{th place}}{+,} \ldots, +\rangle. \qquad (2.3)$$

Define next

$$c_n = K_n S_n^-, \qquad c_n^+ = K_n^+ S_n^+.$$

It is tedious, but simple, to prove that the two algebras are mapped into each other. This transformation can be used to map spin models into fermion models. For example, the XX–model, with Hamiltonian

$$H^{xx} = \sum_{n=1}^{N} (S_n^x S_{n+1}^x + S_n^y S_{n+1}^y) \tag{2.4}$$

and suitable boundary conditions, is mapped into the free fermion Hamiltonian

$$H^{xx} = \frac{1}{2} \sum_n^N (c_n^\dagger c_{n+1} + h.c.) = \int_{-\pi}^{\pi} dk \; \varepsilon_k c_k^\dagger c_k. \tag{2.5}$$

Through Fourier transformation the Hamiltonian is diagonalized and the dispersion relation $\varepsilon_k = \cos k$ results. Half–filling corresponds to filling all negative energy levels (Dirac sea). Occupation number distribution is given by the step function; one for energies below the Fermi energy and zero otherwise. Small interactions give small deviations.

The system is gapless and critical; no massive excitations exist. Correlation functions have power law fall off and no long range order, in the sense that $\lim_{|m-n|\to\infty} \langle s_n^+ s_m^- \rangle \to 0$ occurs.

For a study of the continuum model one identifies the even–site operators as the first components $c_{2n} \propto \psi_{1,n}$, and the odd–site operators as the second components $c_{2n+1} \propto \psi_{2,n}$ of a spinor field. The Hamiltonian $H_0 = \int dx :\psi^\dagger \alpha \frac{1}{i} \frac{\partial}{\partial x} \psi :$ results.

The same method can be applied to the XY-model [3]

$$H_\gamma = \sum_{n=1}^{N} \{(1+\gamma)S_n^x S_{n+1}^x + (1-\gamma)S_n^y S_{n+1}^y\}, \tag{2.6}$$

which interpolates between the isotropic ($\gamma = 0$) XX–model and the ($\gamma = 1$) Ising model. For $\gamma \to 0$, the x–component of the spin is completely ordered, while the y–component is completely disordered.

H_γ is transformed into

$$H_\gamma = \frac{1}{2} \sum_n^N \{c_n^\dagger c_{n+1} + \gamma c_n^\dagger c_{n+1}^\dagger + h.c.\} \qquad (+ \text{ boundary conditions}). \tag{2.7}$$

To diagonalize H_γ, a Bogoliubov transformation of the type

$$D_k = A_{km} c_m + B_{km} c_m^\dagger$$
$$D_k^\dagger = B_{km} c_m + A_{km} c_m^\dagger \tag{2.8}$$

may be applied. H_γ becomes

$$H_\gamma = \sum_k E_k^\gamma D_k^\dagger D_k, \qquad E_k^\gamma = \sqrt{1 - (1 - \gamma^2) \sin^2 k}. \qquad (2.9)$$

$\gamma = 0$ gives the previous result. $\gamma = 1$ yields $E_k^1 = 1$ for $|k| \leq k_F$ and $E_k^1 = 0$ for $|k| > k_F$. For all $\gamma \neq 0$ a gap exists.

In the continuum limit the Hamiltonian of the massive Dirac operator results.

We discuss next the, more complicated, XXX antiferromagnetic Heisenberg chain.

$$H = \sum_{n=1}^{N} S_n \cdot S_{n+1} \qquad (2.10)$$

with periodic boundary condition. The total spin operator $S = \sum_{n=1}^{N} S_n$, as well as the shift operator (on the ring) Π are conserved. The ferromagnetic ground state $|\Omega\rangle = |\uparrow, \uparrow, \ldots, \uparrow\rangle$ is easy to guess. A basis for all states is obtained from $|\Omega\rangle$ by acting with S_n^- operators: $S_n^-|\Omega\rangle, \ldots S_{n_1}^- \ldots S_{n_M}^-|\Omega\rangle$. It is tempting to diagonalize S and Π. Spin wave or magnon eigenstates of H are obtained of the form

$$|k\rangle = \sum_{n=1}^{N} e^{ikn} S_n^- |\Omega\rangle, \qquad k = \frac{2\pi}{N} p, \qquad p \in \mathbf{Z}. \qquad (2.11)$$

An ansatz for two-magnon scattering [4] is

$$|k_1 k_2\rangle = \sum_{n_1 < n_2} \psi_{n_1 n_2}^{k_1 k_2} S_{n_1}^- S_{n_2}^- |\Omega\rangle \qquad (2.12)$$

with

$$\psi_{n_1 n_2}^{k_1 k_2} = \{e^{i\Theta_{k_1 k_2}/2} e^{i(k_1 n_1 + k_2 n_2)} + k_1 \leftrightarrow k_2\}.$$

Nontrivial scattering of magnons occurs. $|k_1 k_2\rangle$ is an eigensolution if the phase shift solves a transcendental equation.

$$2 \cot \frac{\Theta_{k_1 k_2}}{2} = \cot \frac{k_1}{2} - \cot \frac{k_2}{2}. \qquad (2.13)$$

Repulsion occurs, and $\psi_{n_1 n_2}^{k_1 k_1} = 0$.

For M magnons

$$|k_1, \ldots, k_M\rangle = \sum_{n_1 < \ldots < n_M} \psi_{\underline{n}}^{\underline{k}} S_{n_1}^- \ldots S_{n_M}^- |\Omega\rangle, \qquad (2.14)$$

the ansatz

$$\psi_{\underline{n}}^{\underline{k}} = \sum_{\text{perm } \pi} \exp[i \sum_\alpha k_{\pi_\alpha} n_\alpha + \frac{i}{2} \sum_{\alpha < \beta} \Theta_{k_{\pi_\alpha} k_{\pi_\beta}}] \qquad (2.15)$$

leads to eigensolutions. The M-magnon S-matrix is factorized into a product of two–"particle" S–matrices. $\psi_{\underline{n}}^{\underline{k}}$ yields a solution if $\Theta_{k_1 k_2}$ solves again the before–mentioned transcendental equation. But, in addition, the so–called Bethe equations have to be fulfilled:

$$k_\alpha N = \sum_{\beta \neq \alpha} \Theta_{k_\alpha k_\beta} + 2\pi I_\alpha, \qquad I_\alpha \in \mathbf{Z}. \tag{2.16}$$

Exponentiating the last relation yields

$$e^{ik_\alpha N} = \prod_{\beta \neq \alpha} S_{k_\alpha k_\beta} \tag{2.17}$$

which has an interesting interpretation, connected to periodicity. One mode with quasimomentum k_α scatters on all others while returning back.

3   Luttinger Liquid

In order to solve (the simplest) model with a quartic interaction we should first briefly discuss the bosonization formulae [5]. We consider Dirac fermions with antiperiodic boundary conditions on the finite interval $\Lambda = [-\frac{L}{2}, \frac{L}{2}]$, i. e. the CAR algebra $\mathcal{A}(h)$ over the 1-particle Hilbert space $h = L^2(\Lambda, dx) \otimes \mathbb{C}^2$. The charge and current densities are defined by $\rho^\mu(x) = \bar{\psi}(x)\gamma^\mu \psi(x)$ and we can express the Hamiltonian, charge and current densities by its chiral components. They are defined by $H = H_+ + H_-$, $\rho^\pm = \frac{1}{2}[\rho(x) \pm j(x)]$. This gives:

$$H_\pm = \pm \int_\Lambda dx\, \psi_\pm^\dagger \left(-i\partial_1\right) \psi_\pm(x), \tag{3.1}$$

$$\rho^\pm(x) = \psi^\dagger(x)\frac{1}{2}(1 \pm \gamma_5)\psi(x). \tag{3.2}$$

Normal ordering is with respect to the vacuum Ω_F and

$$\hat{\psi}_+(k)\,\Omega_F = \hat{\psi}_-^\dagger(k)\,\Omega_F = 0 \qquad \forall k \in \Lambda_+,$$
$$\hat{\psi}_+^\dagger(k)\,\Omega_F = \hat{\psi}_-(k)\,\Omega_F = 0 \qquad \forall k \in \Lambda_-, \tag{3.3}$$

The chiral currents (3.2) in normal ordered form are the second quantization of the generators of local chiral phase transformations and are given by:

$$\hat{\rho}^\pm(k) = \int_{\Lambda^*} \hat{dq}\; :\psi^\dagger(k-q)\frac{1 \pm \gamma_5}{2}\psi(q): \;.$$

As it is well known, the normal ordering modifies the naive commutation relations following from the CAR. In our case

$$[\hat{\rho}^\pm(k), \hat{\rho}^\pm(p)] = \pm k\hat{\delta}(k+p) \tag{3.4}$$
$$[\hat{\rho}^\pm(k), \hat{\rho}^\mp(p)] = 0$$

with the term on the right hand side is the Kac-Moody cocycle . Moreover, the following relations between the free Hamiltonian the currents and the field operator hold:

$$[H_F, \hat{\rho}^{\pm}(k)] = \pm k \hat{\rho}^{\pm}(k),$$
$$[\psi_{\pm}(x), \hat{\rho}^{\pm}(q)] = -e^{-ikx}\psi_{\pm}(x) \tag{3.5}$$

We note, that

$$\hat{\rho}^{+}(k)\,\Omega_F = \hat{\rho}^{-}(-k)\,\Omega_F = 0 \qquad \forall k > 0 \tag{3.6}$$

wich together with (3.4) shows, that the $\hat{\rho}^{+}(k)$ (resp. ($\hat{\rho}^{-}(k)$)) give a highest (resp. lowest) weight representation of the Heisenberg (affine Kac-Moody) algebra associated with the group $U(1)$. To derive the Bose-Fermi correspondence we investigate the structure of the fermionic Fock space. There are unitary operators $R_{\pm}$ on the n-particle sector of the Fock space $\mathcal{F}(n)$ such that

$$R_{\pm}\, \hat{\psi}_{\pm}(k)R_{\pm}^{\dagger} = \hat{\psi}_{\pm}(k + \frac{2\pi}{L}) \qquad \forall k \in \Lambda \tag{3.7}$$
$$R_{\pm}\, \hat{\psi}_{\mp}(k)R_{\pm}^{\dagger} = \hat{\psi}_{\mp}(k)$$

$R_{\pm}$ are the implementers of the 'large' gauge transformations:

$$\psi_{\pm}(x) \rightarrow e^{i\frac{2\pi}{L}x}\psi(x), \quad \psi_{\mp}(x) \rightarrow \psi_{\mp}(x) \tag{3.8}$$
$$R_{\pm}\, \psi_{\pm}(x)\, R_{\pm}^{\dagger}(x) = e^{i\frac{2\pi}{L}x}\psi_{\pm}(x) \qquad \forall x \in \Lambda. \tag{3.9}$$

The following commutator relations with the chiral fermion currents and the Hamiltonian are important in the following:

$$R_{\pm}^{\dagger}\hat{\rho}^{\pm}(k)R_{\pm} = \hat{\rho}^{\pm}(k) \pm \hat{\delta}_{k,0}$$
$$R_{\pm}^{\dagger}H_{\pm}R_{\pm} = H_{\pm} \pm \frac{2\pi}{L}\left(\hat{\rho}^{\pm}(0) \pm \frac{1}{2}\right) \tag{3.10}$$

The essential point of bosonization is that the total Hilbert space $\mathcal{F}$ can be generated from Ω_F by the chiral fermion currents $\hat{\rho}^{\pm}(k)$ (3.4) and $R_{\pm}$. More precisely, for all pairs of integers $n_+, n_- \in \mathbb{Z}$ we introduce the subspaces $\mathcal{D}^{(n_+, n_-)}$ of $\mathcal{F}$ containing all linear combinations of vectors

$$\hat{\rho}^{+}(k_1)\cdots\hat{\rho}^{+}(k_{m_+})\hat{\rho}^{-}(q_1)\cdots\hat{\rho}^{-}(k_{m_-})R_+^{n_+}R_-^{-n_-}\,\Omega_F \tag{3.11}$$

where $m_{\pm} \in \mathcal{N}_0$ and $k_i, q_i \in \Lambda^*$. The basic result of the boson–fermion correspondence is the following

Lemma: The space

$$\mathcal{D} \equiv \bigoplus_{n_+, n_- \in \mathbb{Z}} D^{(n_+, n_-)}. \tag{3.12}$$

is dense in $\mathcal{F}$ [6].

Let us define boson operators $c(k)$ for $k \in \Lambda^*$ by

$$c(k) = \begin{cases} \frac{1}{\sqrt{|k|}}\hat{\rho}^{+}(k) & \text{for } k > 0 \\ \frac{1}{\sqrt{|k|}}\hat{\rho}^{-}(k) & \text{for } k < 0 \end{cases} \tag{3.13}$$

Due to the Schwinger term for the currents we see from the commutation relations (3.4) that $c(k)$ obey the CCR algebra:

$$[c(k), c^\dagger(k)] = \hat{\delta}(k - q), \tag{3.14}$$

all other commutators vanishing. From [5] we find that we can write the Hamiltonian of free Dirac fermions in three different ways:

$$
\begin{aligned}
H &= \int_{\Lambda^*} \hat{dk}\, k \; :\hat{\psi}^\dagger(k)\hat{\psi}(k): \\
&= \frac{\pi}{L}[Q_+^2 + Q_-^2] + \frac{2\pi}{L}\sum_{k>0}[\hat{\rho}^+(-k)\hat{\rho}^+(k) + \hat{\rho}^-(k)\hat{\rho}^-(-k)] \\
&= \frac{\pi}{L}[Q_+^2 + Q_-^2] + \frac{\pi}{L}\sum_{k\neq 0}|k|\,c^\dagger(k)c(k)
\end{aligned}
\tag{3.15}
$$

The boson fermion correspondence provides us with explicit formulas of the fermion operators $\psi_\pm(x)$ in terms of the operators $\hat{\rho}^\pm(k)$, $R_\pm$. We define field operators by

$$\psi_\pm(x) = \lim_{\varepsilon\to 0}\psi_\pm(x; \varepsilon), \tag{3.16}$$

where the regularized operators can be written as:

$$\psi_\pm(x; \varepsilon) = \frac{1}{\sqrt{L}}S_\pm(x) : \exp(K_\pm(x; \varepsilon)): \tag{3.17}$$

with

$$S_\pm(x) = e^{\pm i\pi x Q_\pm/L}(R_\pm)^{\mp 1}e^{\pm i\pi x Q_\pm/L} = e^{\mp \pi i x/L}(R_\pm)^{\mp 1}e^{\pm i2\pi x Q_\pm/L} \tag{3.18}$$

and

$$K_\pm(x; \varepsilon) = \mp\frac{2\pi}{L}\sum_{k\in\Lambda^*\backslash\{0\}}\frac{\hat{\rho}^\pm(-k)}{k}e^{-ikx}e^{-\varepsilon|k|} = -K_\pm(x; \varepsilon)^*. \tag{3.19}$$

The Luttinger model is now given by

$$H = H_F + H_L \tag{3.20}$$

where $H_F = H_+ + H_-$ is the free fermion Hamiltonian introduced in and the current-current interaction H_L is given by

$$H_L = \int_\Lambda dx \int_\Lambda dy\, j_\mu(x)\, V(x-y)\, j^\mu(y) \tag{3.21}$$

$$+ 4\int_\Lambda dx \int_\Lambda dy\, \rho^+(x)\, V(x-y)\, \rho^-(y), \tag{3.22}$$

where the interaction is parity invariant, $V(x) = V(-x)$. The currents j_μ, $\rho^\pm$ are defined in (3.4). Fourier transformation leads to a Hamiltonian of the form:

$$H = \int_{\Lambda^\bullet} \hat{dk}\, k \; : \hat{\psi}^\dagger(k)\gamma_5\hat{\psi}(k) : + \int_{\Lambda^\bullet} \hat{dk}\, W_k \hat{\rho}^+(k)\hat{\rho}^-(-k) \qquad (3.23)$$

The Hamiltonian (3.23) can be easily expressed by the bosonic operators $c(k)$ of (3.13):

$$H = \frac{2\pi}{L} \sum_{k \geq 0} h_k, \qquad \text{i. e.}$$

$$h_0 = \frac{1}{2}(Q_+^2 + Q_-^2 + 2W_0\, Q_+ Q_-) \qquad (3.24)$$

$$h_k = k\left[(c^\dagger(k)c(k) + c^\dagger(-k)c(-k)) + W_k(c^\dagger(k)c^\dagger(-k) + c(k)c(-k))\right].$$

The Hamiltonian (3.20), which was a sum of 2nd order and 4th order parts in fermionic fields, is now expressed in a bilinear form of $Q_\pm$ and the bosonic variables $c(k)$. A Hamiltonian which is bilinear in bosonic operators can be easily diagonalized by a Bogoliubov transformation. Since the Hamiltonian h_k (3.25) only mixes modes with momenta k, $-k$ we are led to study a system of two coupled harmonic oscillators. We define next the unitary transformation

$$\mathcal{U} = e^S, \qquad S = \sum_{k>0} S_k, \qquad (3.25)$$

with $S_k = \lambda_k[c(k)c(-k) - c^\dagger(k)c^\dagger(-k)]$, where

$$\tanh 2\lambda_k = W_k \quad \Longleftrightarrow \quad |W_k| < 1. \qquad (3.26)$$

The operator S can be shown to exist and defines an anti-selfadjoint operator if and only if the condition $\sum_{k>0}|kW_k^2| < \infty$ is fulfilled. $\mathcal{U}$ diagonalizes the Hamiltonian (3.25) of the Luttinger model and we obtain:

$$\mathcal{U}^\dagger c(\pm k)\mathcal{U} = \cosh\lambda_k c(\pm k) - \sinh\lambda_k c^\dagger(\mp k) \quad \forall k > 0. \qquad (3.27)$$

The dispersion relation ω_k and ground state energy shift η_k are given by:

$$\omega_k^2 = k^2(1 - W_k^2) \qquad \eta_k = \frac{1}{2}(|k| - \omega_k). \qquad (3.28)$$

Thus we can write the Hamiltonian (3.25) in the following diagonal form:

$$H_D = \mathcal{U}^\dagger H \mathcal{U} = \frac{2\pi}{L} h_0 + \frac{2\pi}{L}\sum_{k>0}\omega_k\left(c^\dagger(k)c(k) - c^\dagger(-k)c(-k)\right) - LE_G \qquad (3.29)$$

with the ground state energy per unit length $E_G = \frac{1}{2\pi}\int_\Lambda \hat{dk}\,\eta_k = \frac{1}{L}\sum_{k>0}k(1 - \sqrt{1 - W_k^2})$.

With the help of the algebra of vertex operators [5] it is easy to obtain the correlation functions of the model. The two point function

$$\langle \Omega | \psi_+^\dagger(x)\psi_+(0) | \Omega \rangle = \langle \Omega_F | \psi_+^\dagger(x)\psi_+(0) | \Omega_F \rangle \exp \left[\sum_{q>0} \frac{\sinh \lambda_q}{q}(e^{ixq} + e^{-ixq} - 2) \right]$$

(3.30)

is proportional to the free Greens function. The additional factor implies that

$$G(x)/G_0(x) \overset{|x| \to \infty}{\simeq} 1/|x|^\alpha$$

with $\alpha = \lambda^2/4\pi^2 v^2(0)$ and the fermi surface becomes modified.

4   The Luttinger–Schwinger Model

In [5], we studied the so called Luttinger-Schwinger model, i.e. the (1+1) dimensional model of massless Dirac fermions with a non-local 4-point interaction coupled to a U(1)-gauge field. We worked within the Hamiltonian framework on the cylinder, and constructed the field operators and observables as well-defined operators on the physical Hilbert space. The complete solution of the model was found using the boson-fermion correspondence, and the formalism for calculating all gauge invariant Green functions was provided. We discussed the role of anomalies and showed how the existence of large gauge transformations implied a fermion condensate in all physical states. The meaning of regularization and renormalization in our well-defined Hilbert space setting has been discussed. We shall illustrate these steps next and we will finally perform the limit to the Thirring–Schwinger model where the interaction becomes local.

From the action of a gauge field coupled to a fermi field we obtain a Hamiltonian of the form

$$H = \int_\Lambda dx \left(\frac{1}{2}E(x)^2 + \psi^*(x)\gamma_5 \left(-i\partial_1 + eA_1(x)\right)\psi(x) \right) +$$

$$+ 4\int_\Lambda dxdy \, \rho^+(x)V(x-y)\rho^-(y),$$

(4.1)

and the Gauss' law

$$G(x) = -\partial_1 E(x) + e\rho(x) \simeq 0.$$

(4.2)

The observables of the model are all gauge invariant operators. They leave invariant physical states. All gauge invariant objects which one can construct from $A_1(x)$ (at fixed time) are functions of

$$Y = \frac{1}{2\pi}\int_\Lambda dy A_1(y).$$

(4.3)

In fact, Y above is only invariant with respect to small gauge transformations and changes by multiples of $1/e$ under large ones. Thus the quantity which is invariant under all gauge transformations is $e^{i2\pi eY}$ which is equal to the Wilson line (holonomy)

$$W[A_1] = e^{ie \int_\Lambda dy A_1(y)} . \tag{4.4}$$

The non-trivial C(A)CR in Fourier space are

$$[\hat{A}_1(p), \hat{E}(k)] = i\hat{\delta}(k + p)$$
$$\{\hat{\psi}_\sigma(q), \hat{\psi}^*_{\sigma'}(q')\} = \delta_{\sigma\sigma'}\hat{\delta}(q - q'). \tag{4.5}$$

and the Hilbert space of the model is $\mathcal{H} = \mathcal{H}_{\text{Photon}} \otimes \mathcal{H}_{\text{Fermion}}$. For $\mathcal{H}_{\text{Photon}}$ we take the boson Fock space generated by boson field operators $b^*(k)$ obeying CCR

$$[b(k), b^*(p)] = \hat{\delta}(k - q) \quad \text{etc.} \tag{4.6}$$

and a vacuum Ω_P such that

$$b(k)\Omega_P = 0 \quad \forall k \in \Lambda^*. \tag{4.7}$$

Note that the term "vacuum" here and in the following does *not* mean that this state has anything to do with the ground state of the model; it is just one convenient state from which all other states in the Hilbert space can be generated by applying the field operators. We then set

$$\hat{A}_1(k) = \frac{1}{s}(b(k) + b^*(k)) \quad \hat{E}(k) = -\frac{is}{2}(b(k) - b^*(k)) \tag{4.8}$$

where $s^4 = \pi e^2$. For the fermions we refer to the previous section. We can now write the Gauss' law operators in Fourier space as

$$\hat{G}(k) = -ik\hat{E}(k) + e\hat{\rho}(k), \tag{4.9}$$

which implies

$$[\hat{G}(k), \hat{\rho}^\pm(p)] = \pm ke\hat{\delta}(k + p).$$

Due to the presence of the Schwinger terms, these Fermion currents no longer commute with the Gauss' law generators, hence they are not gauge invariant and no observables of the model.

To obtain Fermion currents obeying the classical relations (without Schwinger terms), we note that $[\hat{G}(k), \hat{A}_1(p)] = k\hat{\delta}(k + p)$, hence the operators

$$\bar{\rho}^\pm(k) \equiv \hat{\rho}^\pm(k) \pm e\hat{A}_1(k) \tag{4.10}$$

commute with the Gauss law generators and are thus the observables of the model corresponding to the chiral Fermion currents on the quantum level. Similarily the naive Hamiltonian is not gauge invariant on the quantum level. Gauge invariance is restored if we replace the currents with the gauge invariant currents, which is corresponds to a mass term for the photon and a modified Luttinger interaction. Thus we obtain the gauge invariant Hamiltonian of the Luttinger–Schwinger model as follows,

$$
H = \hat{H}_0 + \int_{\Lambda^{\bullet}} \hat{\mathrm{d}}k \, {}^{\times}_{\times} \left(\frac{1}{4\pi} \hat{E}(k)\hat{E}(-k) + e\hat{A}_1(k)\hat{j}(-k) + e^2 \hat{A}_1(k)\hat{A}_1(-k) \right.
$$
$$
\left. + \left[\hat{\rho}^+(k) + e\hat{A}_1(k) \right] W_k \left[\hat{\rho}^-(-k) - e\hat{A}_1(-k) \right] \right) {}^{\times}_{\times} . \quad (4.11)
$$

Kronig's identity allows us to rewrite the free Hamiltonian as

$$
H = \int_{\Lambda^{\bullet}} \hat{\mathrm{d}}k \, {}^{\times}_{\times} \left(\frac{1}{2} \left(\tilde{\rho}^+(k)\tilde{\rho}^+(-k) + \tilde{\rho}^-(k)\tilde{\rho}^-(-k) \right) \right.
$$
$$
\left. + \frac{1}{4\pi} \hat{E}(k)\hat{E}(-k) + \tilde{\rho}^+(k)W_k\tilde{\rho}^-(-k) \right) {}^{\times}_{\times} . \quad (4.12)
$$

In order to describe the solution of the model we may perform a gauge Fixing. The only gauge invariant degree of freedom of the Photon field at fixed time is the holonomy $\int_\Lambda \mathrm{d}x A_1(x)$ and one can gauge away all Fourier modes $\hat{A}_1(k)$ of the gauge field except the one for $k = 0$. Thus we can impose the gauge condition

$$
\hat{A}_1(k) = \delta_{k,0}Y, \qquad A_1(x) = \frac{2\pi}{L}Y \quad (4.13)
$$

and solve the Gauss' law $\hat{G}(k) \simeq 0$ (cf. eq. (4.9)) as

$$
\hat{E}(k) \simeq \frac{e\hat{\rho}(k)}{ik} \quad \text{for } k \neq 0. \quad (4.14)
$$

This determines all components of E except those conjugate to Y: $\hat{E}(0) = \frac{L}{2\pi} \frac{\partial}{i\partial Y}$. After that we are left with the $(k = 0)$–component of Gauss' law, viz.

$$
eQ_0 \simeq 0, \quad Q_0 = \hat{\rho}(0) = \hat{\rho}^+(0) + \hat{\rho}^-(0). \quad (4.15)
$$

Inserting this into (4.12), gives the Hamiltonian of the model on the physical Hilbert space $\mathcal{H}_{\mathrm{phys}} = \mathcal{L}^2(\mathcal{R}, \mathrm{d}Y) \otimes \mathcal{H}'_{\mathrm{Fermion}}$ (where $\mathcal{H}'_{\mathrm{Fermion}}$ is the zero charge sector of the fermionic Fock space):

$$
H = -\frac{L}{8\pi^2} \frac{\partial^2}{\partial Y^2} + \quad (4.16)
$$
$$
\frac{\pi}{L} \left(\left(\hat{\rho}^+(0) + eY \right)^2 + \left(\hat{\rho}^-(0) - eY \right)^2 + \left(\hat{\rho}^+(0) + eY \right) 2W_0 \left(\hat{\rho}^-(0) - eY \right) \right) +
$$
$$
\int_{\Lambda^{\bullet}\backslash\{0\}} {}^{\times}_{\times} \left(\frac{e^2}{4\pi k^2} \hat{\rho}(k)\hat{\rho}(-k) + \frac{1}{2} \left(\hat{\rho}^+(-k)\hat{\rho}^+(k) + \hat{\rho}^-(k)\hat{\rho}^-(-k) \right) + \hat{\rho}^+(k)W_k\hat{\rho}^-(-k) \right) {}^{\times}_{\times} .
$$

Following the treatment in Sect.3, we now can write the Hamiltonian as:

$$
H = \frac{2\pi}{L} \sum_{k \geq 0} h_k \quad (4.17)
$$

where h_k is expressed by the boson operators $c(k)$ defined in (3.13). Introducing the quantum mechanical variables P and X:

$$P = (\hat{\rho}^+(0) - \hat{\rho}^-(0) + 2eY) = Q_5 + 2eY$$

$$X = i\frac{L}{2\pi}\frac{1}{2e}\frac{\partial}{\partial Y} \tag{4.18}$$

obeying $[P, X] = -iL/2\pi$, we can write

$$h_k = \left(k + \frac{e^2}{2\pi k}\right)[c^\dagger(k)c(k) + c^\dagger(-k)c(-k)] +$$

$$+ \left(kW_k + \frac{e^2}{2\pi k}\right)[c^\dagger(k)c^\dagger(-k) + c(k)c(-k)] \quad \text{for } k > 0 \tag{4.19}$$

$$h_0 = \frac{e^2}{\pi}X^2 + \frac{1}{4}(1 - W_0)P^2 - \frac{1}{2}\sqrt{\frac{e^2}{\pi}\frac{L}{2\pi}},$$

(remember that $Q_0 = 0$). Note, that h_0 is the Hamiltonian of one harmonic oscillator. We see that the $k > 0$ part has the same structure like in the case of the Luttinger model (cf. 3.25). We find that for $k \neq 0$:

$$\tanh 2\lambda_k = \frac{2\pi k^2 W_k + e^2}{2\pi k^2 + e^2}$$

$$\omega_k^2 = k^2(1 - W_k^2) + \frac{e^2}{\pi}(1 - W_k) \tag{4.20}$$

$$\eta_k = \frac{1}{2}\left(|k| + \frac{e^2}{2\pi|k|} - \omega_k\right).$$

By the gauge fixing above we reduced the Hilbert space from $\mathcal{H}$ to $\mathcal{H}'_{\text{phys}}$ containing all states invariant under *small* gauge transformations, i.e. of the form $e^{i\alpha(x)}$ with $\alpha(L/2) = \alpha(-L/2)$. There are, however, still large gauge transformations present which are generated by $e^{i2\pi x/L}$.

The large gauge transformation $e^{i2\pi x/L}$ acts on the fields as follows

$$\psi(x) \xrightarrow{R} e^{i2\pi x/L}\psi(x) = (R_+R_-)^{-1}\psi(x)(R_+R_-),$$

$$eY \xrightarrow{R} eY - 1 \tag{4.21}$$

where $R_\pm$ are the implementers of $e^{i2\pi x/L}$ in the chiral sectors of the fermions. The large gauge transformation R obviously generates a group $\mathbb{Z}, n \to R^n$, and we denote this group as $\mathbb{Z}_R$. Our aim is to construct the states in $\mathcal{H}_{\text{phys}}$ which carry an irreducible representation of $\mathbb{Z}_R$ and especially the ground states of our model.

We start with recalling that the Fermion Fock space can be decomposed in sectors of different chiral charges $\hat{\rho}^\pm(0)$,

$$\mathcal{H}_{\text{Fermion}} = \bigoplus_{n_+, n_- \in \mathbb{Z}} \mathcal{H}^{(n_+, n_-)}$$

where

$$\mathcal{H}^{(n_+,n_-)} = \left\{ \Psi \in \mathcal{H}_{\text{Fermion}} \mid \hat{\rho}^{\pm}(0)\Psi = n_{\pm}\Psi \right\} = R_+^{n_+} R_-^{-n_-} \mathcal{H}^{(0,0)}$$

Thus,

$$\mathcal{H}_{\text{phys}} = \mathcal{L}^2(\mathbb{R}, dY) \otimes \mathcal{H}'_{\text{Fermion}} \qquad (4.22)$$

where

$$\mathcal{H}'_{\text{Fermion}} = \bigoplus_{n \in \mathbb{Z}} \mathcal{H}^{(n,-n)}, \quad \mathcal{H}^{(n,-n)} = (R_+ R_-)^n \mathcal{H}^{(0,0)} \qquad (4.23)$$

is the zero charge subspace of the Fermion Fock space and we use the Schrödinger representation for the physical degree of freedom $Y = \int_\Lambda dx A_1(x)/2\pi$ of the photon field as discussed in the last subsection. $\mathcal{H}_{\text{phys}}$ can therefore be spanned by states

$$\Psi(n) = \varphi(Y + \tfrac{n}{e})(R_+ R_-)^n \Psi, \quad \varphi \in \mathcal{L}^2(\mathbb{R}, dY), \ \Psi \in \mathcal{H}^{(0,0)} \qquad (4.24)$$

which, under a large gauge transformation (4.21), transform as

$$\Psi(n) \xrightarrow{R} \Psi(n-1). \qquad (4.25)$$

Thus the states transforming under an irreducible representation of $\mathbb{Z}_R$ are given by

$$\Psi^\theta = \sum_{n \in \mathbb{Z}} e^{i\theta n} \Psi(n) \xrightarrow{R} e^{i\theta} \Psi^\theta \qquad (4.26)$$

In our calculation of Green functions below we find it useful to use the notation

$$< \Psi_1^\theta, \Psi_2^\theta >_\theta \equiv < \Psi_1, \Psi_2 >_{\text{F}} (\varphi_1, \varphi_2)_{\mathcal{L}^2} \qquad (4.27)$$

which can be regarded as redefinition of the inner product using a simple multiplicative regularization.

We now construct the ground states of our model. As expected, the quantum mechanical variables P, X describing the zero mode h_0 of the Hamiltonian have a simple representation on the θ-states (4.26),

$$P\Psi^\theta = \sum_{n \in \mathbb{Z}} e^{i\theta n} 2e(Y + \tfrac{n}{e})\varphi(Y + \tfrac{n}{e})(R_+ R_-)^n \Psi,$$

$$X\Psi^\theta = \sum_{n \in \mathbb{Z}} e^{i\theta n} \frac{i}{2e} \frac{L}{2\pi} \frac{\partial}{\partial Y} \varphi(Y + \tfrac{n}{e})(R_+ R_-)^n \Psi. \qquad (4.28)$$

Thus the ground states of h_0 annihilated by $C(0)$ are of the form (4.24) with

$$\varphi_0(Y) = \left(\frac{\pi}{4e^2\alpha}\right)^{\frac{1}{4}} \exp\left(-\alpha(2eY)^2\right) \qquad (4.29)$$

where $\alpha = \frac{1}{L}\sqrt{\frac{\pi^3}{2e^2}(1 - W_0)}$, and the other eigenstates are the harmonic oscillator eigenfunctions $\varphi_n \propto C^*(0)^n \varphi_0$. From $C(k) = \mathcal{U}c(k)\mathcal{U}^*$ and $c(k)\Omega_{\text{F}} = 0$ it is clear that the ground state of all $h_{k>0}$ is $U\Omega_{\mathrm{F}}$. We conclude that the ground states of our model obeying $H\Psi_0^\theta = LE_0\Psi_0^\theta$ are given by

$$\Psi_0^\theta = \sum_{n\in\mathbb{Z}} e^{i\theta n}\varphi_0(Y + \tfrac{n}{e})(R_+R_-)^n U\Omega_{\mathrm{F}}. \tag{4.30}$$

We construct next gauge invariant Green functions. The observables of our model are operators on $\mathcal{H}_{\mathbf{phys}}$ where $\int_\Lambda \mathrm{d}x A_1(x)$ is represented by $2\pi Y$. The fully gauge invariant field operators are operators χ which are represented in the present gauge fixed setting by

$$\chi_\sigma(x) = e^{i2\pi eY(x-r)/L}\psi_\sigma(x).$$

These operators depend on the $r \in \Lambda$ chosen. Bilinears such as meson operators are, however, independent of r and give rise to translational invariant equal time Green functions. Moreover, on the quantum level not only the Wilson line $W[A_1]$ (4.4) but actually even

$$e\int_\Lambda \mathrm{d}x A_1(x) + \tfrac{1}{2}Q_5 \equiv w[A_1] \tag{4.31}$$

is gauge invariant (note that $W[A_1] = e^{iw[A_1]}$). This operator is represented by $eY + \tfrac{1}{2}Q_5 = P/2$. The strategy to calculate Green functions of the model using bosonization techniques is the following: One first moves the operators $R_\pm$ and combines them to some power of (R_+R_-). The operators $Q_\pm$ when applied to physical states become simple $\mathcal{C}$-numbers: $Q_\pm(R_+R_-)^n = (R_+R_-)^n(\pm n + Q_\pm)$ for all integers n, and $Q_\pm\Omega_{\mathrm{F}} = 0$. For the exponentials of boson operators we use the decomposition into creation and annihilation parts outlined in A.4. The normal ordering procedure gives a product of exponentials of commutators which are ($\mathcal{C}$-number) functions. For the correlation functions of meson operators $\chi_\sigma^*(x)\chi_{\sigma'}(y)$ we obtain:

$$\langle \Psi_0^\theta, \chi_\pm^*(x)\chi_\pm(y)\Psi_0^\theta \rangle_\theta = e^{-\frac{\pi}{4L}m(x-y)^2}e^{\Delta(x-y)}g_0^\pm(x-y), \tag{4.32}$$

$$\langle \Psi_0^\theta, \chi_\pm^*(x)\chi_\mp(y)\Psi_0^\theta \rangle_\theta = e^{\mp i\theta}e^{-i\frac{2\pi}{L}(x-y)}e^{-\frac{\pi m}{4L}((x-y)+\frac{2}{m})^2}C(L)e^{D(x-y)} \tag{4.33}$$

with

$$\Delta \doteq \sum_{k>0} \frac{2\pi}{Lk}\sinh^2(\lambda_k)[e^{ikx} + e^{-ikx} - 2],$$

$$D(x) = -\sum_{k>0} \frac{\pi}{Lk}\sinh(2\lambda_k)[e^{ikx} + e^{-ikx} - 2], \tag{4.34}$$

$$C(L) = \frac{1}{L}\exp[\sum_{k>0} \frac{2\pi}{kL}(\sinh(2\lambda_k) - 2\sinh^2(\lambda_k))].$$

where $g_0^\pm(x) = \frac{1}{L}\frac{e^{\mp i\frac{\pi}{L}x}}{1 - e^{\pm i\frac{2\pi}{L}(x\pm i\epsilon)}}$ is the 2-point function of free fermions, and the Schwinger mass is renormalized to $m^2 = \frac{e^2}{\pi(1-W_0)}$.

We next discuss the multiplicative regularization and the Thirring-Schwinger model limits. We recall that the Thirring model is formally obtained from the Luttinger model in the limits

$$L \to \infty, \quad V(x) \to g\delta(x) \tag{4.35}$$

i.e. when the interaction becomes local and space becomes infinite. A better understanding can be obtained by explicitly performing the limit (4.35) in the present setting. The idea is to find a family of Luttinger potentials $\{V_\ell(x)\}_{\ell>0}$ becoming local for $\ell \downarrow 0$, i.e. for all $\ell > 0$ the condition for unitary equivalence of the vaccum is fulfilled and $\lim_{\ell \downarrow 0} V_\ell(x) = g\delta(x)$. Then for all $\ell > 0$ everything is well-defined on the free Hilbert space and one can work out in detail how to regularize such that the correlation functions make sense for $\ell \downarrow 0$. The 2-point function of the Thirring-Schwinger model therefore become

$$\langle \Psi_0^\theta, \tilde{\chi}_\pm^*(x)\tilde{\chi}_\pm(0)\Psi_0^\theta\rangle_\theta = e^{\Delta_{\mathrm{reg}}(x)}g_0^\pm(x), \tag{4.36}$$

$$\langle \Psi_0^\theta, \tilde{\chi}_\pm^*(x)\tilde{\chi}_\mp(0)\Psi_0^\theta\rangle_\theta = e^{\mp i\theta}C_{\mathrm{reg}}\, e^{D_{\mathrm{reg}}(x)}. \tag{4.37}$$

If we define τ_0 by $\tanh(2\tau_0) = W_0$ we can write

$$
\begin{aligned}
\Delta_{\mathrm{reg}}(x) = {}& \cosh(2\tau_0)\left[K_0(|\mu x|) + \ln\frac{|\mu x|}{2} + \gamma\right] + \\
& \frac{1}{2}e^{2\tau_0}\left[1 - \frac{\pi}{2}|\mu x| - \mathrm{Ki}_2(|\mu x|)\right] + (\cosh(2\tau_0) - 1)\ln|x|, \\
D_{\mathrm{reg}}(x) = {}& -\sinh(2\tau_0)\left[K_0(|\mu x|) + \ln\frac{|\mu x|}{2} + \gamma\right] - \\
& \frac{1}{2}e^{2\tau_0}\left[1 - \frac{\pi}{2}|\mu x| - \mathrm{Ki}_2(|\mu x|)\right], \\
\ln C_{\mathrm{reg}} = {}& \gamma + \ln\frac{1}{2\pi} + e^{-2\tau_0}\ln\frac{\mu}{2}.
\end{aligned}
\tag{4.38}
$$

We checked that all Green functions of the Thirring-Schwinger model have a well-defined limit after the wave function renormalization.

We would like to stress that this procedure can be naturally interpreted as low–energy limit of the Luttinger–Schwinger model: if one is interested only in Green functions describing correlations of far–apart fermions, the precise form of the Luttinger interaction $V(x)$ should be irrelevant and only the total interaction strength $g = \int dx\, V(x)$ should matter. Thus as far as these correlators are concerned, they should be equal to the ones of the Thirring model corresponding to this coupling g.

5   From Integrable Models to Quantum Spin Models

We are very much familiar with classical (finite dimensional) integrable models. Consider a system with $2N$–dimensional phase space and local coordinates (q_i, p_i). Let $\omega = \sum_i dq_i \wedge dp_i$ be the canonical two–form and $\{f, g\} = \omega(X_f, X_g)$

be the Poisson bracket of functions f and g. Equs. of motion are then given by $\dot{q}_i = \{q_i, H\}$, $\dot{p}_i = \{p_i, H\}$. The Liouville–Arnold theorem states:

If there exist N globally conserved quantities K_i, which are in involution, $\{K_i, K_j\} = 0$, then, there exists a map from (q_i, p_i) to new coordinates (φ_i, I_i), which are cyclic action angle variables: $\varphi_i(t) = \omega_i t + \varphi_i(0)$, $I_i(t) = I_i(0)$.

For ∞ dimensional systems, a Lax pair or a zero curvature condition implies integrability [4]. Examples are the KdV or the nonlinear Schrödinger equation, the Sine–Gordon equation and many more. For the nonlinear Schrödinger equation, for example, L is given by

$$\left[i\frac{\partial}{\partial t} + M, i\frac{\partial}{\partial x} + L\right] = 0 \qquad \text{with} \qquad L = \begin{pmatrix} \lambda & \psi^\dagger \\ \psi & -\lambda \end{pmatrix}. \tag{5.1}$$

There exists a matrix M such that the consistency implies the nonlinear equation $i\frac{\partial}{\partial t}\psi = -\psi_{xx} + (\psi^\dagger\psi)\psi$. One goes over to scattering data of the L-operator. They have a free time evolution. The inverse problem allows to go back to the original field variables at time t. We remark that the monodromy matrix $g(x) = P\exp\int_{-L}^{x} L$ obeys quadratic Poisson brackets of the type: $\{g^1, g^2\} = [r_{12}, g^1 g^2]$, where

$$g^1 = g \otimes 1, \qquad g^2 = 1 \otimes g, \qquad r_{12} = \frac{1}{\lambda}P_{12}, \qquad P = \begin{pmatrix} 1 & & & \\ & 0 & 1 & \\ & 1 & 0 & \\ & & & 1 \end{pmatrix}, \tag{5.2}$$

with P being the permutation matrix. The r-matrix fulfills the classical Yang–Baxter relation:

$$[r_{12}, r_{23}] + [r_{23}, r_{31}] + [r_{31}, r_{12}] = 0. \tag{5.3}$$

Returning to Quantum Spin Models, we solve the one–dimensional Ising model by going to a QSM at **one** point:

$$Z_N = \sum_{S_1,\ldots,S_N} e^{\beta S_1 S_2}\ldots e^{\beta S_N S_1} = \operatorname{Tr} T^N, \qquad T = \begin{pmatrix} e^\beta & e^{-\beta} \\ e^{-\beta} & e^\beta \end{pmatrix},$$

$$e^{-\beta F_N} = \lambda_1^N\left(1 + \left(\frac{\lambda_2}{\lambda_1}\right)^N\right), \qquad f = \lim_{N\to\infty}\frac{F_N}{N} = -\frac{1}{\beta}\ln\lambda_1; \tag{5.4}$$

where λ_1 denotes the larger eigenvalue. This result holds in general. (T_c is here zero.) T denotes the transfer matrix. The two–dimensional analogue connects vertex models to quantum spin models like the XXZ (or XYZ) model, to which we turn next.

There exists a class of ice–like models. Ice forms a lattice of H_2O molecules. A simplified $d = 2$ model is obtained by placing O–atoms at sites of a quadratic lattice and H–atoms between them. Assume two possible positions for the H–atoms being close to a site atom or far away. Surround each O–atom by exactly two H–atoms. This allows for six vertices to which we assign weights ω_i, $i =$

$1, \ldots, 6$. Assuming a symmetry under reflections one obtains three independent constants $\omega_1 = \omega_2 = a$, $\omega_3 = \omega_4 = b$, $\omega_5 = \omega_6 = c$. The partition function is given by

$$Z = \sum_{\text{conf } \{\omega_j\}} \prod_{n,m} \omega_j(n, m). \tag{5.5}$$

The model becomes critical if $|a^2 + b^2 - c^2| \leq 2|ab|$. Correlation functions decay then algebraically. Adding a vertex with a source and one with a sink yields the eight vertex model. For the six vertex model we define a matrix

$$L^{\alpha_1 \alpha_2}_{\beta_1 \beta'_2} = \begin{pmatrix} a & 0 & 0 & 0 \\ 0 & b & c & 0 \\ \hline 0 & c & b & 0 \\ 0 & 0 & 0 & a \end{pmatrix}_{01} \qquad \text{in } \mathbf{C}_0^2 \otimes \mathbf{C}_1^2, \tag{5.6}$$

describing one vertex. N vertices of a row are encoded by the monodromy matrix

$$M_N \equiv L_{0N} \cdot \ldots \cdot L_{02} \cdot L_{01} \qquad \text{in } \mathbf{C}_0^2 \otimes \mathbf{C}_1^2 \otimes \ldots \otimes \mathbf{C}_N^2. \tag{5.7}$$

With periodic boundary conditions, $\text{tr}_0 \, M_N = T_N$, the transfer matrix, describing the interaction from row to row, is obtained. Finally $Z = \text{Tr } T_N^M$ for an $N \times M$ lattice. Tr denotes the trace in $\bigotimes_{j=1}^N \mathbf{C}^2$.

As a suitable parametrization we choose $a = \sin(\gamma - \lambda)$, $b = \sin \lambda$, $c = \sin \gamma$ where λ is called the spectral parameter.

We introduce the XXZ–Hamiltonian

$$H_{XXZ} = \frac{1}{2} \sum_{j=1}^N (\sigma_j^x \sigma_{j+1}^x + \sigma_j^y \sigma_{j+1}^y + \Delta(\sigma_j^z \sigma_{j+1}^z + 1)) \tag{5.8}$$

and claim that

$$H_{XXZ} = \sin \gamma \frac{d}{d\lambda} \ln T \Big|_{\lambda=0} \qquad \text{with } \Delta = -\cos \gamma. \tag{5.9}$$

To show (4.9) we first note that $L_{0n}(0) = \sin \gamma P_{0n}$, where P_{0n} denotes the permutation operator for vector spaces $\mathbf{C}_0^2$ and $\mathbf{C}_n^2$. Next $(\sin \gamma)^{-N} T(0) = \text{tr}_0 \, P_{0N} \ldots P_{01} = \text{tr}_0 \, P_{12} P_{23} \ldots P_{N-1,N} P_{N0} = U$, where U denotes the shift operator. Let $(\partial L_{0n}/\partial \lambda)|_{\lambda=0} \equiv \bar{L}_{0n}$. We calculate the logarithmic derivative of the transfer matrix at $\lambda = 0$ and obtain

$$\sin \gamma \frac{d}{d\lambda} \ln T \Big|_{\lambda=0} = U^{-1} \sum_{j=1}^N \text{tr}_0 \, (P_{01} \ldots P_{0j-1} \bar{L}_{0j} P_{0j+1} \ldots P_{0N})$$

$$= \sum_{j=1}^N P_{j-1j} \bar{L}_{j-1j} = \sum_{j=1}^N \begin{pmatrix} -\cos \gamma & 0 & 0 & 0 \\ 0 & 0 & 1 & 0 \\ \hline 0 & 1 & 0 & 0 \\ 0 & 0 & 0 & -\cos \gamma \end{pmatrix}_{j-1j} \tag{5.10}$$

The indices indicate the vector space in which the matrices act.

6   Quantum Integrability — Yang–Baxter Relation

We take γ fixed and consider $L_{0n}(\lambda)$ as a function of λ. We take the tensor product of two auxiliary spaces and consider $L_{0n}(\lambda) \otimes L_{\bar{0}n}(\mu)$ acting in $\mathbf{C}_0^2 \otimes \mathbf{C}_{\bar{0}}^2 \otimes (\bigotimes_{j=1}^N \mathbf{C}^2)$. It is remarkable that there exits an operator $R \in \mathrm{End}\,(\mathbf{C}_0^2 \otimes \mathbf{C}_{\bar{0}}^2)$ such that the Yang–Baxter relation [7]

$$R_{0\bar{0}}(\lambda - \mu)L_{0n}(\lambda)L_{\bar{0}n}(\mu) = L_{0n}(\mu)L_{\bar{0}n}(\lambda)R_{0\bar{0}}(\lambda - \mu) \tag{6.1}$$

holds. The solution for the quantum R–matrix is given in our case by $R = PL$, where P denotes the permutation matrix. If we introduce $L = PR$ into (5.1), push all P operators on one side and change the notation of indices we obtain

$$R_{23}(\lambda - \mu)R_{12}(\lambda)R_{23}(\mu) = R_{12}(\mu)R_{23}(\lambda)R_{12}(\lambda - \mu) \tag{6.2}$$

in $\mathbf{C}_1^2 \otimes \mathbf{C}_2^2 \otimes \mathbf{C}_3^2$. We remind the reader that R_{23} means really $\mathbf{1}_1 \otimes R_{23}$ etc. (5.1) and (5.2) are identical to the S–matrix factorization conditions used by many authors.

There exist rational, trigonometric, elliptic and hyperelliptic solutions to (5.2). If we expand R around the unit operator: $R(\lambda) \simeq 1 + i\hbar r(\lambda)$, $r(\lambda)$ obeys the classical Yang–Baxter relation. (5.2) is closely related to the braid relation, which will be explained later.

A number of consequences follow. Since L_{0n} and $L_{\bar{0}m}$ with $n \neq m$ commute it follows from (5.1) that

$$L(\lambda - \mu)M(\lambda)M(\mu) = M(\mu)M(\lambda)L(\lambda - \mu), \tag{6.3}$$

and taking the trace in $\mathbf{C}_0^2 \otimes \mathbf{C}_{\bar{0}}^2$ we conclude that

$$[T(\lambda), T(\mu)] = 0. \tag{6.4}$$

We have therefore obtained an infinite number of conserved quantities for H_{XXZ}, a signal for integrability. The first nontrivial conserved quantity for H_{XXX} is given, for example, by

$$F = \sum_\ell \sigma_\ell \cdot (\sigma_{\ell+1} \times \sigma_{\ell+2}), \tag{6.5}$$

which can be used to explain the degeneracies in the spectrum.

Relations (5.3) contain much more information: Let us introduce $2^N \times 2^N$ matrices $A_\lambda, B_\lambda, C_\lambda, D_\lambda$ as the elements of

$$M(\lambda) = \begin{pmatrix} A_\lambda & B_\lambda \\ C_\lambda & D_\lambda \end{pmatrix}. \tag{6.6}$$

Remember, that we intend to diagonalize the transfer matrix, which is now expressed in terms of the new quantities as $T(\lambda) = A_\lambda + D_\lambda$. Equations (5.3) give 16 relations among these quantities. Let us quote one of them:

$$A_\lambda B_\mu = \frac{a_{\mu-\lambda}}{b_{\mu-\lambda}} B_\mu A_\lambda - \frac{c_{\mu-\lambda}}{b_{\mu-\lambda}} B_\lambda A_\mu, \dots. \tag{6.7}$$

The strategy is now to choose (or to guess) a kind of vacuum (or highest weight vector) state. In our case it will be the ferromagnetic state. It is automatically an eigenstate of $T(\lambda)$:

$$T(\lambda)| \uparrow \ldots \uparrow\rangle = (a_\lambda^N + b_\lambda^N)| \uparrow \ldots \uparrow\rangle. \tag{6.8}$$

Next we define vectors by acting with products of $B(\lambda)$'s on the chosen reference vector: $|\lambda_1 \ldots \lambda_M\rangle = B_{\lambda_1} \ldots B_{\lambda_M}| \uparrow \ldots \uparrow\rangle$. We require that these states become eigenstates of $T(\lambda)$. This would be the case if (5.7) contains no second term on the right hand side. The eigenvalue becomes then

$$\varepsilon(\lambda_1 \ldots \lambda_M) = a_\lambda^N \prod_{i=1}^{M} \frac{a_{\lambda_i-\lambda}}{b_{\lambda_i-\lambda}} + b_\lambda^N \prod_{i=1}^{M} \frac{a_{\lambda-\lambda_i}}{b_{\lambda-\lambda_i}}. \tag{6.9}$$

Next one requires that the additional terms vanish. They cancel, if the transcendental Bethe equations

$$e^{ip_\ell N} \equiv \left(\frac{a_{\lambda_\ell}}{b_{\lambda_\ell}}\right)^N = \prod_{j\neq\ell}^{M} \frac{a_{\lambda_\ell-\lambda_j} b_{\lambda_j-\lambda_\ell}}{b_{\lambda_\ell-\lambda_j} a_{\lambda_j-\lambda_\ell}} \equiv \prod_{j\neq\ell}^{M} e^{i\varphi_{\lambda_j-\lambda_\ell}}, \qquad \ell = 1, \ldots, M \tag{6.10}$$

are fulfilled. Solutions to (5.10) can be obtained for small as well as large N. We note that $b_{\lambda_\ell-\lambda}$ vanishes at $\lambda = \lambda_\ell$, but the transfer matrix should be regular. Vanishing of the residuum at these poles explains (5.10), which rewritten is identical to Bethe's equations:

$$N_{p_\ell} = 2\pi I_\ell + \sum_{j\neq\ell} \varphi_{\lambda_j-\lambda_\ell}, \qquad I_\ell \in \mathbf{Z}. \tag{6.11}$$

Many models can be treated using the above scheme or a generalization of it (RSOS, XYZ, Hubbard, ...).

7   Algebraic Relations

A knot is a closed line embedded into $\mathbf{R}^3$ without crossings. A link is a set of knotted knots. A complete classification of links has not been achieved. In order to define braids [8] we choose n points in $\mathbf{R}^2$ and consider mappings γ_i from $[0,1] \to \mathbf{R}^2 \times [0,1] \subset \mathbf{R}^3$, $i = 1, \ldots, n$ with the properties that $\dot{\gamma}_i(t) > 0$ for $t \in [0,1]$, $\gamma_i(t) \neq \gamma_j(s)$ for $i \neq j$ and all $s, t \in [0,1]$ and $\gamma_i(0) = x_i$, $\gamma_i(1) = x_{\pi(i)}$, where π denotes a permutation from the symmetric group of n elements S_n. We identify all such mappings which can be obtained through continuous deformations. The resulting object forms a braid of n strings $\in B_n$. Algebraically B_n consists of words generated from $\{e, b_i, b_n^{-1} | i = 1, \ldots, n-1\}$. b_j exchanges the j-th and the $(j+1)$-st string, so that one string lies above the other. b_j^{-1} puts them in the opposite way. There are three types of Reidemeister moves, $e = b_i b_i^{-1} = b_i^{-1} b_i$, $b_i b_j = b_j b_i$ for $|i-j| \geq 2$ and

$$b_i b_{i+1} b_i = b_{i+1} b_i b_{i+1}. \tag{7.1}$$

B_n is defined by factorizing with respect to these relations. (6.1) is called the braiding relation and related to (5.2). If we identify upper and lower ends of a braid we obtain a link. Many braids lead to the same link. The Markov theorem identifies invariant functions on links.

If we require that b_j's have only two eigenvalues $+1$ and $-q^2 \in \mathbf{C}$, we impose the quadratic conditions $(g_i-1)(g_i+q^2) = 0$. Together with $g_i g_j = g_j g_i, |i-j| \geq 2$ and $g_i g_{i+1} g_i = g_{i+1} g_i g_{i+1}$ the Hecke algebra is defined. One representation is obtained from the R–matrix through gauging and taking the limit $\lambda \to -i\infty$,

$$g_i = \begin{pmatrix} 1 & 0 & 0 & 0 \\ 0 & 0 & q & 0 \\ 0 & q & 1-q^2 & 0 \\ 0 & 0 & 1 & 1 \end{pmatrix}_{ii+1} \quad , \quad -e^{-i\gamma} = q, \tag{7.2}$$

acting in $\bigotimes_{j=1}^N \mathbf{C}^2$. If we transform to e_i, putting $g_i = -(1+q^2)e_i + 1$, we obtain $e_i^2 = e_i = e_i^\dagger$ and

$$e_1 e_2 e_1 - \frac{1}{(q+q^{-1})^2} e_1 = e_2 e_1 e_2 - \frac{1}{(q+q^{-1})^2} e_2. \tag{7.3}$$

If both sides of (6.3) are put to zero, the Temperley–Lieb–Jones algebra is obtained.

¿From the algebraic structure behind the Yang–Baxter relation the notion of quantum groups has been deduced. We explain first the notions of coproduct, counit and coinverse within two examples:

Let G be a group and $\mathcal{A} = C(G)$ be the abelian algebra of functions over G. Let $(m_{k\ell})$ be matrix elements of the N–dimensional defining representation. The product within G implies for $\mathcal{A}$ an algebra homomorphism to $\mathcal{A} \otimes \mathcal{A}$ called the coproduct Δ:

$$(\Delta f)(g_1, g_2) = f(g_1 \cdot g_2), \qquad \Delta(m_{k\ell}) = \sum_{j=1}^N m_{kj} \otimes m_{j\ell}. \tag{7.4}$$

In the same sense the inverse mapping and the unit element implies the coinverse and counit

$$\mathcal{A} \xrightarrow{i} \mathcal{A}, \qquad (i(f))(g) = f(g^{-1}), \qquad \varepsilon(f) = f(1_G). \tag{7.5}$$

Consider $U(\mathcal{G})$ the universal enveloping algebra of a Lie algebra $\mathcal{G}$ (for example $SL(2)$), with generators H and $X^\pm$:

$$[H, X^\pm] = \pm 2X^\pm, \qquad [X^+, X^-] = H. \tag{7.6}$$

The coproduct is now given by

$$\Delta(g) = g \otimes 1 + 1 \otimes g, \qquad \Delta(1) = 1 \otimes 1, \tag{7.7}$$

the counit by $\varepsilon(g) = 0$ for $g \neq 1$ and $\varepsilon(1) = 1$.

These two examples will next be generalized to the q–deformed quantum group $\mathcal{A}_q(sl(2))$ and to the q–deformed universal enveloping algebra $U_q(sl(2))$.

We generalize the Yang–Baxter relation to [1]

$$\tilde{R}_{12}T_1T_2 = T_2T_1\tilde{R}_{12} \tag{7.8}$$

with $T = \begin{pmatrix} a & b \\ c & d \end{pmatrix}$, $T_1 = T \otimes 1$, $T_2 = 1 \otimes T$ and take as $\tilde{R}$–matrix

$$\tilde{R} = \begin{pmatrix} q & 0 & 0 & 0 \\ 0 & 1 & q - q^{-1} & 0 \\ 0 & 0 & 1 & 0 \\ 0 & 0 & 0 & q \end{pmatrix}, \tag{7.9}$$

a, b, c and d fulfill the algebra

$$ab = qba, \quad ac = qca, \quad ad - da = (q - q^{-1})bc, \quad bc = cb, \quad bd = qdb, \quad cd = qdc, \tag{7.10}$$

which follows from (6.8). There exists a central element $\det{}_qT = ad - qbc$, being the q–deformation of the determinant. The coproduct is given like in (6.4) by

$$\Delta\begin{pmatrix} a & b \\ c & d \end{pmatrix} = \begin{pmatrix} a & b \\ c & d \end{pmatrix} \otimes \begin{pmatrix} a & b \\ c & d \end{pmatrix}, \tag{7.11}$$

and represents an algebra homomorphism.

$U_q(sl(2))$ is given by generators $X^\pm$ and H and relations:

$$[H, X^\pm] = \pm 2X^\pm, \qquad [X^+, X^-] = \frac{q^H - q^{-H}}{q - q^{-1}}. \tag{7.12}$$

The coproduct for H is still the usual one $\Delta(H) = H \otimes 1 + 1 \otimes H$, but these for $X^\pm$

$$\Delta(X^\pm) = X^\pm \otimes q^{H/2} + q^{-H/2} \otimes X^\pm \tag{7.13}$$

introduce an "asymmetry", and make the algebra non cocommutative. The counit is given by $\varepsilon(H) = \varepsilon(X^\pm) = 0$; the coinverse by $S(H) = -H$ and $S(X^\pm) = -q^{\pm 1}X^\pm$.

We note that there exists an universal R–matrix and the notion of quasitriangularity. There exists a duality between Hopf algebras relating the two examples above.

The physical interesting application concerns the symmetry of the XXZ–chain in case special boundary terms are introduced ($\sin q(\sigma_1^z - \sigma_N^z)$). The resulting spin Hamiltonian $\tilde{H}^{XXZ}$ commutes with the q–deformed total spin operator

$$[\tilde{H}^{XXZ}, \tilde{X}^\pm] = [\tilde{H}^{XXZ}, \tilde{H}] = 0 \tag{7.14}$$

where

$$\tilde{H} = \sum_{j=1}^{N} S_j^z, \qquad \tilde{X}^\pm = \sum_{j=1}^{N} q^{S_3^1 + \dots S_3^{j-1}} S_j^\pm q^{-S_3^{j+1} - \dots - S_3^N} \tag{7.15}$$

and $\widetilde{H}$, $\widetilde{X}^+$ and $\widetilde{X}^-$ obey the relations (6.12). We note, that after adding these boundary terms translation invariance on the ring is lost. A different model with a nonlocal boundary term which is quantum group invariant and has a kind of translation invariance built in has been treated and solved recently [9].

We end this chapter with a few remarks showing the relationship of the above mentioned algebraic structures to various other subjects.

For $q = 1$ the $\widetilde{R}$ matrix (6.9) becomes the unit matrix and $R = P\widetilde{R} = P$ the permutation matrix. $R = \Pi^{S=1} - \Pi^{S=0}$, where Π^S denote the projection operators to spins. A similar decomposition can be achieved for $q \neq 1$: $R = \Pi_q^{S=1} - q^2\Pi_q^{S=0}$. The space on which $\widetilde{R}$ acts can be decomposed too; this way we obtain the Manin plane:

$$\mathbf{C}\langle x, y\rangle/(\Pi_q^{S=0}(x_i \otimes x_j)) : xy = qyx$$

$$\mathbf{C}\langle \xi, \eta\rangle/(\Pi_q^{S=1}(\xi_i \otimes \xi_j)) : \xi^2 = \eta^2 = 0, \ \xi\eta = -\frac{1}{q}\eta\xi.. \tag{7.16}$$

ξ and η can be identified as differentials $\xi = dx$, $\eta = dy$ and a differential calculus on the Manin plane results.

We mentioned already that link–invariants can be obtained from solutions of the Yang–Baxter equation. Following ideas of Atiyah, Witten suggested the connection to the Chern–Simons model. Topologically invariant quantities can be obtained as expectation values of products of Wilson line parallel transport operators along closed curves C_i:

$$\langle \chi_{\rho_1}(C_1)\dots\chi_{\rho_M}(C_M)\rangle, \qquad \chi_\rho(C) = \text{Tr } P\exp[\int_C A^i T_i^\rho]. \tag{7.17}$$

ρ denotes the representation and the expectation value is taken within the model with action

$$S = \frac{k}{4\pi}\int d^3x \ \text{Tr } \left(A \wedge dA + \frac{2}{3}A \wedge A \wedge A\right). \tag{7.18}$$

Recently we have studied this model within the Hamiltonian approach and proposed a quantization by deforming the Poisson structure. The quadratic algebra of monodromies becomes a quantum group algebra [10].

Many of the above mentioned aspects will be considered in detail in the other contributions to this volume.

Acknowledgement

H. G. would like to thank Prof. Zalan Horvath and Prof. Laszlo Palla for inviting me to lecture at this School.

References

1 Grosse H.: *Introduction to Integrable Models of Statistical Physics*. Lecture Notes in Physics 469, p. 107. Springer 1996
2 Grosse H.: *Models in Statistical Physics and Quantum Field Theory*. Springer, 1988
3 Mattis D. C.: *The Many-Body Problem*. World Scientific, 1993
4 Grosse H.: *An Introduction to Integrable Models and Conformal Field Theory*. Proc. of the XXIX Winter Schladming School, Springer 1991, eds. H. Mitter and W. Schweiger
5 Grosse H., Langmann E., Raschhofer E., *Ann. Phys.* **253** 310 (1997)
6 Carey A. L., Ruijsenaars S. N. M., *Acta Appl. Mat.* **10**, 1 (1987)
7 Faddeev L.: *Quantum Completely Integrable Models in Field Theory*. Les Houches Lectures, 1982
8 Fröhlich J.: *Statistics of Fields, the Yang–Baxter Equation, and the Theory of Understand Links*. Cargése Lecture 1987
9 Grosse H., Pallua S., Prester, P., Raschhofer E., Journ. Phys. A: Math. **27** (1994) 4761
10 Alekseev A. Y., Grosse H., Schomerus V., *Commun. Math. Phys.* **172** 317 (1995), *Commun. Math. Phys.* **174** 561 (1996)

Introduction to the Coordinate-Space Bethe Ansatz and to the Treatment of Bethe Ansatz Equations

Ferenc Woynarovich

Research Institute for Solid State Physics
of the Hungarian Academy of Sciences
1525 Budapest 114, Pf 49.

Abstract. The Bethe Ansatz is a method, by which the diagonalization of certain 1D many particle Hamiltonians can be reduced to the solving of a set of nonlinear algebraic equations, the so called Bethe Ansatz equations. This method is presented through the diagonalization of the Heisenberg chain and the δ-gas of spin 1/2 Fermions.

To illustrate the method to solve the Bethe Ansatz equations the case of the isotropic Heisenberg chain is treated. The ground-state is discussed, and the so called higher level Bethe Ansatz equations describing the excitations of the system are derived. The nature and properties of the excitations are discussed based on these equations. Finally also a method to calculate the spectrum of large but finite systems is presented.

1 Introduction

The rich history of Bethe Ansatz goes back some sixty years to 1931. when Bethe published his solution ([Bethe 1931]) for the Heisenberg spin chain. The main idea of this solution was an appropriate assumption on the form of the eigenfunction, by which the problem of diagonalizing the many particle Hamiltonian could be reduced to the solution of a set of nonlinear equations. For a long time it has been thought, that the method deviced by Bethe had no wider application, but by now a large class of one (space) dimensional (1D) many body systems (including the various anisotropic Heisenberg chains ([Orbach 1958, des Cloisoux and Gaudin 1966, Johnson et al 1973]), the δ-gas ([Lieb and Liniger 1963, Yang 1967]) or the Hubbard model ([Lieb and Wu 1968]), the Gross-Neveu ([Andrei and Lowenstein 1979, Andrei and Lowenstein 1980a]) and the various impurity models ([Andrei et al 1983, Tsvelick and Wiegmann 1983])) is known as soluble by Bethe Ansatz (BA).

The method developed following Bethe, that anables one to build up an educated guess for the form of wavefunction is often called real-space or coordinate-space Bethe Ansatz to distinguish from the algebraic Bethe Ansatz, what is a method to diagonalize transfer matrices for two dimensional statistical systems. The two methods are closly related to each other as they have a common mathematical background.

In this notes we would like to present a way along which one can develope a sense how the coordinate-space Bethe Ansatz works. We do this by giving the detailed solutions of two models. First we present the solution of the sympler Heisenberg chain which actually describes particles with no internal symmetry (Sec. 2). After that (Sec. 3) the more complicated δ-gas of spin-1/2 fermions will be treated. By the BA one reduces the diagonalization of the Hamiltonian to the solwing of a set of coupled nonlinear equations, the so called BA equations (BAE). The method to treat these BAE and analyse the solution we will illustrate on the example of the isotropic Heisenberg chain (Sec. 4).

2   Bethe Ansatz for the Heisenberg Chain

2.1   Heisenberg Type Chains

Consider two centers at R_1 and R_2 with potentials $V(x - R_1)$ and $V(x - R_2)$, and suppose the normalized eigenfuncion of an electron localized to one of the centers (with the other being far enough) is $\varphi(x - R_1)$ or $\varphi(x - R_2)$. Look for the two electron state in the form

$$\varphi^{\pm}(x_1, x_2) = \frac{1}{\sqrt{2}} \left(\varphi(x_1 - R_1)\varphi(x_2 - R_2) \pm \varphi(x_1 - R_2)\varphi(x_2 - R_1) \right) \quad (1)$$

As an electron eigenfunction shuld be antisymmetric, φ^{+} shold be completed by an antisymmetric (singlet) spin function, and the spin part belonging to φ^{-} should be symmetric (triplet). In a rough approximation the energy of such a state is

$$\langle \varphi^{\pm} H \varphi^{\pm} \rangle = 2E \pm J, \quad (2)$$

where E is the energy of one electron localized to one center and the J exchange is built up of terms of the type

$$\int \varphi^{*}(x_1 - R_1)V(x_1 - R_1)\varphi(x_1 - R_2) \int \varphi^{*}(x_2 - R_2)\varphi(x_2 - R_1). \quad (3)$$

The $\pm$ sign in (2) can be expressed by the spin operators of the two electrons: as

$$2\bar{S}_1\bar{S}_2 = (\bar{S}_1 + \bar{S}_2)^2 - (\bar{S}_1)^2 - (\bar{S}_2)^2 = \begin{cases} -3/2 & \text{(singlet)} \\ +1/2 & \text{(triplet)} \end{cases}, \quad (4)$$

(2) can be written in the form

$$\langle \varphi^{\pm} H \varphi^{\pm} \rangle = 2E - J/2 - 2J\bar{S}_1\bar{S}_2. \quad (5)$$

Thus the evaluation of the energy (2) is equivalent to the evaluation of the matrix elements of the spin operators. This way (5) can be considered as a Hamiltonian acting on the spins. The isotropic Heisenberg chain

$$\hat{H} = J \sum_i \bar{S}_i\bar{S}_{i+1} = J \sum_i \left(S_i^x S_{i+1}^x + S_i^y S_{i+1}^y + S_i^z S_{i+1}^z \right) \quad (6)$$

(with i labeling the sites (centers)) is a (nearest neighbour) generalisation of the above spin exchange Hamiltonian for several centers forming a chain. This

Hamiltonian, on the other side has several generalizations which are anisotropic in the spin directions: the Hamiltonians

$$\hat{H}_{XYZ} = \sum_i \left(J_x S_i^x S_{i+1}^x + J_y S_i^y S_{i+1}^y + J_z S_i^z S_{i+1}^z \right) \quad (J_x \neq J_y \neq J_y) \,, \quad (7)$$

$$\hat{H}_{XXZ} = \sum_i \left(J_x \left(S_i^x S_{i+1}^x + S_i^y S_{i+1}^y \right) + J_z S_i^z S_{i+1}^z \right) \,, \quad (8)$$

$$\hat{H}_{XY} = \sum_i \left(J_x S_i^x S_{i+1}^x + J_y S_i^y S_{i+1}^y \right) \quad (J_x \neq J_y) \,, \quad (9)$$

$$\hat{H}_{ISING} = \sum_i J_z S_i^z S_{i+1}^z \,, \quad (10)$$

correspond to the completly anisotropic or XYZ, the anisotropic or XXZ Heisenberg chains, the XY and the Ising chains respectively. All these Hamiltonians can be diagonalized exactly. In the following we shall deal with the XXZ chain.

2.2 The XXZ Hamiltonian

In this subsection we discuss some properties of the XXZ Hamiltonian on a ring, i.e. with periodic boundary condition[1]. We write the Hamiltonian in the form (the index XXZ is dropped)

$$\hat{H} = \sum_{i=1}^{N-1} \left\{ \left(S_i^x S_{i+1}^x + S_i^y S_{i+1}^y \right) + \rho \left(S_i^z S_{i+1}^z - \frac{1}{4} \right) \right\} + $$
$$+ \left\{ \left(S_N^x S_1^x + S_N^y S_1^y \right) + \rho \left(S_N^z S_1^z - \frac{1}{4} \right) \right\} \,. \quad (11)$$

Here we introduced $J_z / J_x = \rho$, added a constant and dropped an overall factor what does not restrict generality. The periodic boundary condition is usually expressed by the equivalence $S_{N+1} \equiv S_1$. This 'spelld out' gives the last term in (11). In the following when we use the shorter form, this equivalence will be understood. Introducing the spin operators

$$\sigma_j^{\pm} = S_j^x \pm i S_j^y \,, \quad \sigma_j^z = 2 S_j^z \,, \quad (12)$$

we have

$$\hat{H} = \hat{H}_{kin} + \hat{H}_{int} \,, \quad (13)$$

$$\hat{H}_{kin} = \sum_{i=1}^{N} \frac{1}{2} \left(\sigma_i^+ \sigma_{i+1}^- + \sigma_{i+1}^+ \sigma_i^- \right) \,, \quad (14)$$

$$\hat{H}_{int} = \sum_{i=1}^{N} \frac{\rho}{4} \left(\sigma_i^z \sigma_{i+1}^z - 1 \right) \,. \quad (15)$$

---

[1] The XXY Hamiltonian can be diagonalized also with other types of conditions like twisted boundary conditions, or with free or forced ends ([Hamer et al 1987, Alkaraz et al 1987, Shastry et al 1990]).

The subscripts 'kin' and 'int' refer to the fact that the corresponding terms of the Hamiltonian act as a kinetic and an interaction energy of properly choosen particles. Let us consider the sites with spins ↑ as empty, and say that on a site with spin ↓ there is a particle. It is clear that $\hat{H}_{\mathrm{kin}}$ moves a particle one site to the left or to the right (provided the neighbouring site is empty). For two particles $\hat{H}_{\mathrm{int}}$ has the value -2ρ, if the particles are apart, but its value is $-\rho$, if the particles occupy neighbouring sites, i.e. this part of the Hamiltonian describes indeed an interaction of these particles. The Hamiltonian conserves the z component of the total spin (i.e. the number of particles)

$$\left[\hat{H}, \sum S_i^z\right] = 0 \,, \tag{16}$$

and it is invariant under the translation of the chain by a lattice spacing:

$$\left[\hat{H}, e^{i\hat{P}}\right] = 0 \tag{17}$$

(with $\hat{P}$ being the operator of the (quasi-)momentum). This means that the eigenstates of $\hat{H}$ are the eigenstates of the momentum $\hat{P}$ and the $S^z = \sum S_i^z$. For $\rho = 1$ the Hamiltonian is isotropic in the spin space, for this it commutes with $S^x = \sum S_i^x$ and $S^y = \sum S_i^y$ too, and the eigenstates are the eigenstates of S^2.

2.3   Bethe Ansatz for the Many Particle Wavefunction

The Bethe Ansatz method is actually a way of constructing the many particle wavefunction which diagonalizes the Hamiltonian. It involves three main steps. First one finds the one-particle wavefunction; then one constructs the two-particle wavefunction as an appropriate combination of the one particle ones; finally one generalizes this to any number of particles. Now we follow this way.

The One-Particle Wavefunction If there are no particles, all the spins point up, the chain is in the ferromagnetic ($S^z = N/2$) state $|F\rangle$ which has no energy. In this bare vacuum particles can be created by acting on it by the σ_n^- operators. The one-particle ($S^z = N/2 - 1$) eigenstates are of the plain-wave type:

$$\left|\varphi^{(1)}\right\rangle = \sum_{n=1}^{N} a(n)\, \sigma_n^- |F\rangle \,, \qquad a(n) = e^{ikn} \,. \tag{18}$$

The eigenvalue equation

$$\hat{H}\left|\varphi^{(1)}\right\rangle = E^{(1)}\left|\varphi^{(1)}\right\rangle \tag{19}$$

written up in component form

$$\left\langle F\left|\sigma_n^+ \hat{H}\right|\varphi^{(1)}\right\rangle = E^{(1)}\left\langle F\left|\sigma_n^+\right|\varphi^{(1)}\right\rangle \tag{20}$$

yields

$$\frac{1}{2}\left(a(n-1)+a(n+1)\right)-\rho a(n)=E^{(1)}a(n),\quad 1<n<N, \tag{21}$$

$$\frac{1}{2}\left(a(N)+a(2)\right)-\rho a(1)=E^{(1)}a(1), \tag{22}$$

$$\frac{1}{2}\left(a(N-1)+a(1)\right)-\rho a(N)=E^{(1)}a(N). \tag{23}$$

Substituting the form of $a(n)$ leads to

$$E^{(1)}(k)=(\cos k-\rho), \tag{24}$$

provided

$$e^{ikN}=1\quad\text{i.e.:}\quad k=\frac{2\pi}{N}\lambda,\quad \lambda=1,2,\ldots,N \tag{25}$$

The momentum of the state is

$$P^{(1)}(k)=k=\frac{2\pi}{N}\lambda, \tag{26}$$

as it should be for the periodic boundary condition.

Two Particle States We look for the two particle ($S^z=N/2-2$) states in the form

$$\left|\varphi^{(2)}\right\rangle=\sum_{n_1<n_2}a(n_1,n_2)\,\sigma_{n_1}^{-}\sigma_{n_2}^{-}|F\rangle \tag{27}$$

(where restricting the summation to the $n_1<n_2$ region is needed, otherwise the $a(n_1,n_2)$ amplitudes can not be defined uniquely) with

$$a(n_1,n_2)=a_{12}e^{ik_1n_1+ik_2n_2}+a_{21}e^{ik_2n_1+ik_1n_2} \tag{28}$$

$$=e^{ik_1n_1+ik_2n_2+i\psi_{12}/2}+e^{ik_2n_1+ik_1n_2-i\psi_{12}/2}. \tag{29}$$

It is clear, that the introduction of the ψ_{12} to describe the a_{ij} coefficients does not restrict generality. More over we may introduce for later convenience

$$\psi_{21}=-\psi_{12},\quad \psi(k_1,k_2)=\psi_{12},\quad (\psi(k_2,k_1)=-\psi(k_1,k_2)). \tag{30}$$

The question is if k_1, k_2 and ψ_{12} can be choosen so that (27) is an eigenstate.

The eigenvalue equation for the amplitudes

$$\left\langle F\left|\sigma_{n_2}^{+}\sigma_{n_1}^{+}\hat{H}\right|\varphi^{(2)}\right\rangle=E^{(2)}\left\langle F\left|\sigma_{n_2}^{+}\sigma_{n_1}^{+}\right|\varphi^{(2)}\right\rangle \tag{31}$$

yields

$$\frac{1}{2}\{a(n_1-1,n_2)+a(n_1+1,n_2)+a(n_1,n_2-1)+a(n_1,n_2+1)\}-$$

$$-2\rho a(n_1,n_2)=E^{(2)}a(n_1,n_2) \tag{32}$$

for the case when both particles are apart from each other and from the ends
$(1 < n_1 < n_2 - 1, \; n_2 < N)$;

$$\frac{1}{2}\{a(n_1 - 1, n_1 + 1) + a(n_1, n_1 + 2)\} - \rho a(n_1, n_1 + 1) =$$
$$E^{(2)}a(n_1, n_1 + 1) \;, \qquad (33)$$

if the particles occupy neighbouring sites $(1 < n_1 = n_2 - 1, \; n_2 < N)$; and

$$\frac{1}{2}\{a(n_2, N) + a(2, n_2) + a(1, n_2 - 1) + a(1, n_2 + 1)\} -$$
$$2\rho a(1, n_2) = E^{(2)}a(1, n_2) \qquad (34)$$

for $1 = n_1 < n_2 - 1, \; n_2 < N$. (It will not be hard to show, that the case
$1 < n_1 < n_2 - 1, \; n_2 = N$ yields no new condition.)

Substituting the (29) form of the amplitudes $a(n_1, n_2)$ into (32) gives

$$E^{(2)}(k_1, k_2) = (\cos k_1 - \rho) + (\cos k_2 - \rho) \; \left(= E^{(1)}(k_1) + E^{(1)}(k_2)\right). \qquad (35)$$

Combining this, (29) and (33) we get an equation that determines the phase ψ_{12}
as a function of k_1 and k_2. The symplest way to obtain this equation is provided
by the observation, that the equation

$$\frac{1}{2}\{a(n_1, n_1) + a(n_1 + 1, n_1 + 1)\} - \rho a(n_1, n_1 + 1) = 0 \qquad (36)$$

(in which $a(n, n)$ are the formal continuations of $a(n_1, n_2)$ to equal arguments
(although these amplitudes have no meaning)) is equivalent to (33) if (35) is
satisfied: adding (36) to (33) gives (32), what is nothing but (35). Equation (36)
reads as

$$\left(e^{i\psi_{12}/2} + e^{-i\psi_{12}/2}\right) + e^{ik_1 + ik_2} \left(e^{i\psi_{12}/2} + e^{-i\psi_{12}/2}\right) -$$
$$\rho \left(e^{ik_2 + i\psi_{12}/2} + e^{ik_1 - i\psi_{12}/2}\right) = 0 \;, \qquad (37)$$

and after a straightforward manipulation it yields

$$\cot \frac{1}{2}\psi_{12} = -\rho \frac{\cot \frac{1}{2}k_1 - \cot \frac{1}{2}k_2}{(1-\rho)\cot \frac{1}{2}k_1 \cot \frac{1}{2}k_2 - (1+\rho)} \;. \qquad (38)$$

Finally we observe, that Eq. (34) is equivalent to

$$a(n_2, N) = a(0, n_2) \;, \qquad (39)$$

i.e. to

$$e^{ik_1 n_2 + ik_2 N + i\psi_{12}/2} + e^{ik_2 n_2 + ik_1 N - i\psi_{12}/2} =$$
$$e^{ik_2 n_2 + i\psi_{12}/2} + e^{ik_1 n_2 - i\psi_{12}/2} \;. \qquad (40)$$

This is satisfied for any n_2, if

$$e^{ik_1 N} - i\psi_{12} = 1 \,,$$
$$e^{ik_2 N} + i\psi_{12} = 1 \,, \tag{41}$$

i.e.:

$$N k_1 = 2\pi\lambda_1 + \psi_{12} \,,$$
$$N k_2 = 2\pi\lambda_2 - \psi_{12} \,, \tag{42}$$

with $\lambda_{1,2}$ being integers. (We note here, that $k_1 \neq k_2$, otherwise $\psi_{12} = \pi$ and the eigenvector is identically zero.) It is also worth to note that the total momentum

$$P^{(2)}(k_1, k_2) = k_1 + k_2 = \frac{2\pi}{N}(\lambda_1 + \lambda_2) \tag{43}$$

is of the form required by the periodic boundary condition (although neither k_1 nor k_2 are of this form).

The General Case Now we generalize the wavefunction to the case of $r(< N/2)$ particles[2] ($S^z = N/2 - r$). We suppose that in the

$$\left| \varphi^{(r)} \right\rangle = \sum_{n_\alpha < n_{\alpha+1}} a(n_1, n_2, \ldots, n_r) \prod_{\alpha=1}^{r} \sigma_{n_\alpha}^{-} |F\rangle \tag{44}$$

eigenvector the $a(n_1, n_2, \ldots, n_r)$ amplitudes have the form

$$a(n_1, n_2, \ldots, n_r) = \sum_{\mathcal{P}} \exp\left\{ i \sum_{\alpha} k_{\mathcal{P}\alpha} n_\alpha + \frac{i}{2} \sum_{\alpha < \beta} \psi_{\mathcal{P}\alpha, \mathcal{P}\beta} \right\} . \tag{45}$$

Here $\mathcal{P}$ is a permutation of the $1, 2, \ldots, r$ indeces, and the summation is extended over all possible permutations. It can be checked in a straigtforward way, that the state (44) with the amplitudes (45) is indeed an eigenstate of the Hamiltonian with an eigenvalue

$$E^{(r)}(k_1, k_2, \ldots, k_r) = \sum_{\alpha}^{r}(\cos k_\alpha - \rho) \left(= \sum_{\alpha}^{r} E^{(1)}(k_\alpha) \right) , \tag{46}$$

provided

$$\cot \frac{1}{2}\psi_{\alpha\beta} = -\rho \frac{\cot \frac{1}{2}k_\alpha - \cot \frac{1}{2}k_\beta}{(1-\rho)\cot \frac{1}{2}k_\alpha \cot \frac{1}{2}k_\beta - (1+\rho)} \,, \tag{47}$$

---

[2] The $r < N/2$ requirement does not restrict generality as we can choose which spin-dirrection we consider as particles.

and

$$Nk_\alpha = 2\pi\lambda_\alpha + \sum_{\beta(\neq\alpha)} \psi_{\alpha\beta}, \quad (\alpha = 1, 2, \ldots, r), \quad (\lambda_\alpha = \text{integers}). \tag{48}$$

The momentum of the state is

$$P^{(r)}(k_1, k_2, \ldots, k_r) = \sum_\alpha^r k_\alpha = \frac{2\pi}{N}\sum_\alpha^r \lambda_\alpha. \tag{49}$$

(The energy (46) is obtained by considering those equations in

$$\left\langle F\left|\prod_{\alpha=1}^r \sigma_{n_\alpha}^+ \hat{H}\right|\varphi^{(r)}\right\rangle = E^{(r)}\left\langle F\left|\prod_{\alpha=1}^r \sigma_{n_\alpha}^+\right|\varphi^{(r)}\right\rangle, \tag{50}$$

which correspond to $1 < n_\alpha < n_{\alpha+1} - 1 < N - 1$, i.e. to configurations in which no two particles occupy neighbouring sites (this is why we have to require $r \leq N/2$: otherwise no such configurations exist). The equations (47) for the ψs are given by those components of (50), in which some of the particles occupy neighbouring sites ($n_\alpha = n_{\alpha+1} - 1$ for one α). Finally (48) is a consequence of that part of (50), which corresponds to $n_1 = 1$ or $n_r = N$)

It is worth to note, that configurations in which three or more particles are at neighbouring sites ($n_\alpha = n_{\alpha+1} - 1 = n_{\alpha+2} - 2 \ldots$) bring no new conditions, the corresponding part of (50) is stisfied if (47) and (48) are satisfied. (To see this, consider a configuration where three particles are neighbours at the sites $n, n+1$ and $n+2$, all the others are apart from each other. The corresponding part of (50) reads

$$\frac{1}{2}\{\ldots + a(\ldots, n-1, n+1, n+2, \ldots) + a(\ldots, n, n+1, n+3, \ldots) + \ldots\} -$$
$$(r-2)\rho a(\ldots, n, n+1, n+2, \ldots) = E^{(r)} a(\ldots, n, n+1, n+2, \ldots) \tag{51}$$

Due to (47)

$$\frac{1}{2}\{a(\ldots, n, n, n+2, \ldots) + a(\ldots, n+1, n+1, n+2, \ldots)\} -$$
$$\rho a(\ldots, n, n+1, n+2, \ldots) = 0, \tag{52}$$

and

$$\frac{1}{2}\{a(\ldots, n, n+1, n+1, \ldots) + a(\ldots, n, n+2, n+2, \ldots)\} -$$
$$\rho a(\ldots, n, n+1, n+2, \ldots) = 0. \tag{53}$$

Adding these two equations to (51) gives an equation, which is formally the same as those in which all the particles are apart, and so, is satisfied due to (46).) This is a nontrivial property of the Heisenberg chain and this makes possible the application of the BA[3].

---

[3] A counterexample is the Bose gas on a lattice ([Choy and Haldane 1982]). This

2.4 Spin of the BA Solutions in the Isotropic Case

The solutions found in the previous section are eigenstates of the operator S^z by construction. In the isotropic case ($\rho = 1$) the Hamiltonian (11) commutes with all three components of the total spin, for this the eigenstates are expected to be also S^2 eigenstates. Now we show, that the BA solutions are indeed S^2 eigenstates with longest possible S^z projection, i.e. $S^2 = S^z(S^z + 1)$. The proof goes as follows.

Writing S^2 in the form

$$S^2 = \sigma^- \sigma^+ + S^z(S^z + 1) , \quad \sigma^\pm = \sum_{i=1}^{N} \sigma_i^\pm , \tag{54}$$

we see, that the above statement is true, if

$$\sigma^+ \left| \varphi^{(r)} \right\rangle = 0 . \tag{55}$$

For $r = 1$ the l.h.s. of (55) is formally an $S^z = N/2$ state, i.e. it is ether zero or a multiple of $|F\rangle$. Actually

$$\left\langle F \left| \sigma^+ \right| \varphi^{(1)} \right\rangle = \sum_{m=1}^{N} a(m) = \sum_{m=1}^{N} e^{ikm} \tag{56}$$

what is zero due to (25), i.e. these states are indeed $S^2 = S^z(S^z + 1)$ states.

For $r = 2$ the l.h.s. of (55) is formally an $S^z = N/2 - 1$ state, thus we have to see, that

$$\left\langle F \left| \sigma_n^+ \sigma^+ \right| \varphi^{(2)} \right\rangle = \sum_{m=1}^{n-1} a(m, n) + \sum_{m=n+1}^{N} a(n, m) = 0 \tag{57}$$

for any $1 \leq n \leq N$. Substituting (29) we have

$$\sum_{m=1}^{n-1} a(m, \ n) + \sum_{m=n+1}^{N} a(n, m) =$$

$$= \sum_{m=1}^{n-1} e^{ik_1 m + ik_2 n + i\psi_{12}/2} + e^{ik_2 m + ik_1 n - i\psi_{12}/2}$$

$$+ \sum_{m=n+1}^{N} e^{ik_1 n + ik_2 m + i\psi_{12}/2} + e^{ik_2 n + ik_1 m - i\psi_{12}/2} , \tag{58}$$

model is very symilar with the difference, that any number of particles can occupy the same site, and the interaction is an on-site-interaction. In an analogous procedure the eigenvalue equations lead to equations analogous to (47) and (48) provided only the configurations with no more than two particles on a site are considered. The configurations with three or more particles on a site introduce extra conditions which are not satisfied even if the conditions prescribed by the 'two particles on a site' configurations are satisfied. For this modell the *formal* BA solution gives a state what is not an eigenstate of the Hamiltonian.

what in a straightforward manner yields

$$\frac{e^{ik_1} - e^{ik_1 n}}{1 - e^{ik_1}} e^{ik_2 n + i\psi_{12}/2} + e^{ik_1 n} \frac{e^{ik_2(n+1)} - e^{ik_2(N+1)}}{1 - e^{ik_2}} e^{+i\psi_{12}/2} +$$

$$\frac{e^{ik_2} - e^{ik_2 n}}{1 - e^{ik_2}} e^{ik_1 n - i\psi_{12}/2} + e^{ik_2 n} \frac{e^{ik_1(n+1)} - e^{ik_1(N+1)}}{1 - e^{ik_1}} e^{-i\psi_{12}/2} .$$

$$(59)$$

This, due to (41) reduces to

$$e^{i(k_1 + k_2)n} \left\{ \frac{e^{ik_1} - i\psi_{12}/2 - e^{+i\psi_{12}/2}}{1 - e^{ik_1}} + \frac{e^{ik_2} + i\psi_{12}/2 - e^{-i\psi_{12}/2}}{1 - e^{ik_2}} \right\} ,$$

$$(60)$$

what is, however, zero as for $\rho = 1$

$$\cot \frac{1}{2}\psi_{12} - \frac{1}{2} \left(\cot \frac{1}{2}k_1 - \cot \frac{1}{2}k_2 \right) = 0 \qquad (61)$$

(see (38)).

The generalization of this calculation to the $r > 2$ case is straightforward, and we conclude, that the BA solutions for the isotropic case are $S^2 = S^z(S^z + 1)$ states. It has to be noted, that the corresponding $S^2 > S^z(S^z + 1)$ states can be constructed easily applying the σ^- operators. (We note also that $\sigma^- |\varphi^{(r)}\rangle$ is *formally* the same, as a $|\varphi^{(r+1)}\rangle$ state, in which one of the ks is equal to zero.)

2.5   Remarks Concerning the BA

BA vs. Perturbation As an alternative to the BA, the Heisenberg chain can be treated also perturbatively. In this treatment it is convenient to consider H_{kin} as the unperturbed Hamiltonian and take H_{int} as a perturbation. The eigenstates of H_{kin} are plainwaves (the same as (44-45) in the case $\rho = 0$).

$$|k_1, k_2, \ldots, k_r\rangle , \quad k_\alpha = \frac{2\pi}{N} m_\alpha \qquad (62)$$

where m_α are integers or half-odd-integers depending on the parity of r. Due to the perturbation the eigenstates will be linear combinations of these plainwaves:

$$|k_1, k_2, \ldots, k_r\rangle \implies \sum_{\{k'\}} a(k'_1, k'_2, \ldots, k'_r) |k'_1, k'_2, \ldots, k'_r\rangle , \qquad (63)$$

where the summation is extended over all possible k' sets (which yield the same total momentum). The state is practically described by an n_k distribution of the k momenta. In this treatment the shift in energy corresponding to the interaction is due to the contribution of the different plainwaves.

In the BA treatment there is one single k set, but none of the momenta is of the free $(2\pi/N)m$ form. The total energy and momentum are the sums of the contributions of the individual particles. The energy and momentum contributions of the individual particles are formally the same as those of the free particles, and the fact, that the particles interact is reflected in the deviation of the momenta (and energies) from the free values.

Interpretation of the BA Equations First consider the two particle state (27) with (29). Removing the $n_1 < n_2$ restriction it can be written in the form

$$\left|\varphi^{(2)}\right\rangle = \sum_{n_1,n_2} \exp\left\{ ik_1n_1 + ik_2n_2 + \frac{i}{2}\psi_{12}\operatorname{sgn}(n_2 - n_1)\right\} \sigma_{n_1}^-\sigma_{n_2}^-|F\rangle \ . \qquad (64)$$

This form allowes the interpretation of $-\psi_{12}$ as the *phaseshift* associated with the scattering of particle 1 on particle 2. If so, a possible (and rather fruitfull) interpretation of the BA equations is as follows. The particles move freely as long as they are apart. The energy and the momentum of the state is the sum of the energies and momenta of the individual particles:

$$E = \sum_{\alpha}(\cos k_\alpha - \rho)\,, \qquad (65)$$

$$P = \sum_{\alpha} k_\alpha\,. \qquad (66)$$

The Hamiltonian (the dynamics and the symmetry of the system) gives the connection between the momenta and the scattering phasehifts:

$$\cot \frac{1}{2}\psi_{\alpha\beta} = -\rho\frac{\cot \frac{1}{2}k_\alpha - \cot \frac{1}{2}k_\beta}{(1 - \rho)\cot \frac{1}{2}k_\alpha \cot \frac{1}{2}k_\beta - (1 + \rho)} \ . \qquad (67)$$

Finally the equations

$$Nk_\alpha = 2\pi\lambda_\alpha + \sum_{\beta(\neq\alpha)} \psi_{\alpha\beta} \qquad (68)$$

expresses that when a particle is moved around the ring, the phase of the amplitude changes partly due to the momentum and partly due to the scattering through the other particles, and this total change should be an integer multiple of 2π. The reason why the BA works is that *the many particle scatterings factorize into products of two particle scatterings*.

A Generalization The above interpretation of the BA and the BAE offers a possibility to generalize the method ([Zamolodchikov 1990]). It is possible, that for a system we have no Hamiltonian and it may be that the concept of wavefunction is not appropriate to describe the system, newertheless, we know the energies $\epsilon(\eta)$ and momenta $p(\eta)$ of the particles as functions of the rapidities η, and we also know the phaseshifts $-\psi(\eta, \eta')$ describing the scattering of the particles. If in such a system the density of particles is small enough to neglect the effect of those proceses in which three or more particles take part simultanously (even if the scattering does not factorizes), equations of the type

$$Lp(\eta) = 2\pi\lambda_\eta + \sum_{\eta'(\neq\eta)} \psi(\eta, \eta') \qquad (69)$$

can serve as (approximate) quantization conditions for the system.

3   Bethe Ansatz for the δ-Gas of Spin 1/2 Fermions

3.1   The Hamiltonian and the Symmetry of the Solution

The 1D δ-gas is described by the Hamiltonian

$$\hat{H} = \sum_{i=1}^{N} -\frac{\partial^2}{\partial x_i^2} + 2c \sum_{i<j} \delta(x_i - x_j) \,. \qquad (70)$$

Here x_i are the coordinates of the N particles. The periodic boundary condition imposed on the system means

$$\lim_{x_i \to 0} \equiv \lim_{x_i \to N} \qquad (71)$$

for any quantity.

The $\hat{H}$ acts on the coordinates only. As $\hat{H}$ is completely symmetric in the coordinates

$$\left[\hat{H}, \hat{\mathcal{P}}\right] = 0 \qquad (72)$$

(where $\hat{\mathcal{P}}$ is the operator of any permutation of the coordinates), the eigenstates of $\hat{H}$ transform according to the ireducible representations of the permutation group: the eigenvalues and eigenfunctions satisfy the relations

$$\hat{H}\varphi_\alpha^{(\nu)}(\{x_i\}) = E^{(\nu)}\varphi_\alpha^{(\nu)}(\{x_i\}) \quad (E^{(\nu)} \neq E^{(\mu)} \text{ if } \nu \neq \mu), \qquad (73)$$

$$\hat{\mathcal{P}}\varphi_\alpha^{(\nu)}(\{x_i\}) = \sum_\beta \varphi_\beta^{(\nu)}(\{x_i\})\mathcal{T}_{\beta\alpha}^{(\nu)}(\mathcal{P}). \qquad (74)$$

Here $\{x_i\}$ is a shorthand notation for $x_1, x_2, \ldots, x_i, \ldots$, and $\mathcal{T}_{\alpha\beta}^{(\nu)}(\mathcal{P})$ is the matrix of the permutation $\mathcal{P}$ in the ireducible representation of the permutation group labelled by ν.

The nature of the particles of the described system is reflected in the symmetry of the eigenfunctions describing them. Consider a system of spin 1/2 Fermions. The complete eigenfunction f, what is also a function of the spin variables σ_i, can be written up in terms of the $\varphi_\alpha^{(\nu)}$ functions. It is obvious, that in such an expansion only functions belonging to the same $E^{(\nu)}$ can occour:

$$f(\{x_i, \sigma_i\}) = \sum_\alpha \varphi_\alpha^{(\nu)}(\{x_i\})S_\alpha(\{\sigma_i\}) \qquad (75)$$

It is not hard to see, that if

$$\hat{P} f(\{x_i, \sigma_i\}) = (-1)^P f(\{x_i, \sigma_i\}) \tag{76}$$

as required for Fermions, the $S_\alpha(\{\sigma_i\})$ functions shold transform according to the irreducible representation given by the matrices

$$(-1)^P \left(T_{\alpha\beta}^{(\nu)}(\mathcal{P}) \right)^* , \tag{77}$$

i.e.:

$$\hat{P} S_\alpha(\{\sigma_i\}) = \sum_\beta S_\beta(\{\sigma_i\})(-1)^P \left(T_{\beta\alpha}^{(\nu)}(\mathcal{P}) \right)^* , \tag{78}$$

(where $*$ denotes complex conjugation). This means, if $S_\alpha(\{\sigma_i\})$ is an $S^2 = l(l+1)$, $S^z = m$ spin eigenfunction, $\varphi_\alpha(\{x_i\})$ must transform according to

$$T_{\alpha\beta}^{(\nu)}(\mathcal{P}) = (-1)^P \left(T_{\alpha\beta}^{(l,m)}(\mathcal{P}) \right)^* \tag{79}$$

where $T^{(l,m)}$ denotes the irreducible representation according to which spin functions with l and m transform.

3.2   BA for the $\varphi_\alpha(\{x_i\})$ Functions

In the present section we build up the BA for the $\varphi_\alpha(\{x_i\})$ functions —just as in the case of the Heisenberg chain — in three steps. First we write up the one particle wavefunctions, after that we construct the two particle eigenfunctions as a combination of the one particle ones , and finally we generalize this for any particle number and any symmetry of the state. The result of this procedure is a set of simultanous eigenvalue problems, which will be solved in a later section by a second BA.

One Particle Wavefunction If there is only one particle, the eigenfunctions of the Hamiltonian (70) are free plainwaves:

$$-\frac{\partial^2}{\partial x^2}\varphi_k(x) = E(k)\varphi_k(x), \quad \varphi_k(x) = e^{ikx}, \quad E(k) = k^2. \tag{80}$$

The periodic boundary condition (71) yields for the momentum

$$P(k) = k = \frac{2\pi}{L}n, \quad (n = \text{integer}). \tag{81}$$

Two Particles For two particles the Hamiltonian (70) reads:

$$-\frac{\partial^2}{\partial x_1^2} - \frac{\partial^2}{\partial x_2^2} + 2c\delta(x_1 - x_2) \tag{82}$$

Introducing center of mass and relative coordinates ($\xi = (x_1 + x_2)/2$ resp. $\eta = x_1 - x_2$) we have

$$-\frac{1}{2}\frac{\partial^2}{\partial \xi^2} + 2\left(-\frac{\partial^2}{\partial \eta^2} + c\delta(\eta)\right) \tag{83}$$

This tells us that a $\varphi(\xi, \eta)$ eigerfunction of (83) is continuous in $\eta = 0$, but its derivative has a jump:

$$\varphi(\eta = +0) = \varphi(\eta = -0), \quad \text{but} \quad \left.\frac{\partial\varphi}{\partial\eta}\right|_{\eta=+0} - \left.\frac{\partial\varphi}{\partial\eta}\right|_{\eta=-0} = c\varphi(\eta = 0) \tag{84}$$

at any ξ. Writing back the original x_1 and x_2 koordinates we arrive at

$$\lim_{x_1 \to x_2 + 0} \varphi = \lim_{x_1 \to x_2 - 0} \varphi,$$

$$\lim_{x_1 \to x_2 + 0}\left(\frac{\partial\varphi}{\partial x_1} - \frac{\partial\varphi}{\partial x_2}\right) - \lim_{x_1 \to x_2 - 0}\left(\frac{\partial\varphi}{\partial x_1} - \frac{\partial\varphi}{\partial x_2}\right) = 2c \lim_{x_1 \to x_2} \varphi. \tag{85}$$

Now we suppose, that the eigenfunction is a suitable combination of the one-particle wavefunctions:

$$\varphi(x_1, x_2) = \begin{cases} a_{12}^{12}\,e^{ik_1x_1 + ik_2x_2} + a_{21}^{12}\,e^{ik_2x_1 + ik_1x_2} & (\text{if } x_1 < x_2), \\ a_{12}^{21}\,e^{ik_1x_2 + ik_2x_1} + a_{21}^{21}\,e^{ik_2x_2 + ik_1x_1} & (\text{if } x_2 < x_1). \end{cases} \tag{86}$$

Here the $a_{k,l}^{i,j}$ coefficients have to be choosen so, that (86) satisfys (85), and the wavenumbers must have values at which (71) also holds. If these requirements are met, it is obvious, that the eiganstate has an energy and a momentum

$$E = E(k_1) + E(k_2), \quad P = P(k_1) + P(k_2). \tag{87}$$

The continuity of φ at $x_1 = x_2$ requires

$$a_{12}^{12} + a_{21}^{12} = a_{12}^{21} + a_{21}^{21}, \tag{88}$$

and for the right discontinuity of the derivatives (85)

$$i(k_1 - k_2)\left(-a_{12}^{21} + a_{21}^{21} - a_{12}^{12} + a_{21}^{12}\right) = 2c\left(a_{12}^{12} + a_{21}^{12}\right) \tag{89}$$

must hold. These two equations are equivalent to

$$a_{21}^{12} = \frac{(k_1 - k_2)a_{12}^{21} - ic\,a_{12}^{12}}{(k_1 - k_2) + ic}, \tag{90}$$

$$a_{21}^{21} = \frac{(k_1 - k_2)a_{12}^{12} - ic\,a_{12}^{21}}{(k_1 - k_2) + ic}. \tag{91}$$

If we introduce the vectors

$$\xi_{12} = \begin{pmatrix} a_{12}^{12} \\ a_{12}^{21} \end{pmatrix} , \quad \xi_{21} = \begin{pmatrix} a_{21}^{12} \\ a_{21}^{21} \end{pmatrix} , \tag{92}$$

and the operators

$$Y^{(i,j)}(k_\alpha, k_\beta) = \frac{(k_\alpha - k_\beta)\mathcal{P}^{(ij)} - ic}{(k_\alpha - k_\beta) + ic} , \quad \{i, j\} = \{1, 2\}, \{2, 1\} ,$$
$$\{\alpha, \beta\} = \{1, 2\}, \{2, 1\} , \tag{93}$$

(where the operator $\mathcal{P}^{(12)}(= \mathcal{P}^{(21)})$ interchanges the (12) and (21) components of the ξ vectors,) the above equations take the form

$$\xi_{21} = Y^{(1,2)}(k_1, k_2)\xi_{12} , \quad \xi_{12} = Y^{(2,1)}(k_2, k_1)\xi_{21} . \tag{94}$$

It is easy to check that these two equations are compatible as

$$Y^{(1,2)}(k_1, k_2) Y^{(2,1)}(k_2, k_1) = 1 . \tag{95}$$

The periodic boundary condition (71) (with $x_i = x_1$) yields

$$a_{12}^{12} e^{ik_2 x_2} + a_{21}^{12} e^{ik_1 x_2} = a_{12}^{21} e^{ik_1 x_2 + ik_2 L} + a_{21}^{21} e^{ik_2 x_2 + ik_1 L} \tag{96}$$

This equation holds for any x_2, if

$$a_{12}^{12} = a_{21}^{21} e^{ik_1 L} , \quad a_{21}^{12} = a_{12}^{21} e^{ik_2 L} . \tag{97}$$

In a symilar way (71) with $x_i = x_2$ leads to

$$a_{12}^{21} = a_{21}^{12} e^{ik_1 L} , \quad a_{21}^{21} = a_{12}^{12} e^{ik_2 L} . \tag{98}$$

Writing (97) and (98) in vector form we have

$$\xi_{21} = e^{ik_2 L} \mathcal{P}^{(12)}\xi_{12} , \tag{99}$$
$$\xi_{12} = e^{ik_1 L} \mathcal{P}^{(12)}\xi_{21} , \tag{100}$$

what are equivalent to

$$e^{ik_2 L}\xi_{12} = \mathcal{P}^{(12)}Y^{(1,2)}(1, 2)\xi_{12} , \tag{101}$$
$$e^{ik_1 L}\xi_{12} = Y^{(1,2)}(2, 1)\mathcal{P}^{(21)}\xi_{12} . \tag{102}$$

Apparently ξ_{12} is an eigenvector of two operators. This is possible as the two operators commute: with the notation

$$X_{i,j} = \mathcal{P}^{(ij)}Y^{(i,j)}(k_i, k_j) = \frac{(k_i - k_j) - ic\mathcal{P}^{(ij)}}{(k_i - k_j) + ic} , \tag{103}$$

$$[X_{1,2}, X_{2,1}] = 0 , \tag{104}$$

as it can be checked dirrectly. (We shall see later, that the $X_{i,j}$ operators are closely related to the scattering matrices of the particles.) Finally, after diagonalizing $\mathcal{P}^{(12)}$ we arrive at the following four equations:

$$\left(e^{ik_2 L} - \frac{(k_1 - k_2) - ic}{(k_1 - k_2) + ic}\right)\frac{1}{\sqrt{2}}\left(a_{12}^{12} + a_{12}^{21}\right) = 0 , \tag{105}$$

$$\left(e^{ik_1 L} - \frac{(k_2 - k_1) - ic}{(k_2 - k_1) + ic}\right)\frac{1}{\sqrt{2}}\left(a_{12}^{12} + a_{12}^{21}\right) = 0 , \tag{106}$$

$$\left(e^{ik_2 L} - 1\right)\frac{1}{\sqrt{2}}\left(a_{12}^{12} - a_{12}^{21}\right) = 0 , \tag{107}$$

$$\left(e^{ik_1 L} - 1\right)\frac{1}{\sqrt{2}}\left(a_{12}^{12} - a_{12}^{21}\right) = 0 , \tag{108}$$

Here (105) and (107) are the components of (101), while (106) and (108) are the components of (102). This set of equations is equivalent to two sets of differernt structure:

$$e^{ik_2 L} = \frac{(k_1 - k_2) - ic}{(k_1 - k_2) + ic} ,$$

$$e^{ik_1 L} = \frac{(k_2 - k_1) - ic}{(k_2 - k_1) + ic} , \tag{109}$$

$$a_{12}^{12} = a_{12}^{21} , \quad (\Longrightarrow a_{21}^{12} = a_{21}^{21}) ,$$

and

$$e^{ik_2 L} = 1 ,$$

$$e^{ik_1 L} = 1 , \quad \left(k_{1,2} = \frac{2\pi}{L}l_{1,2}\right) , \tag{110}$$

$$a_{12}^{12} = -a_{12}^{21} , \quad (\Longrightarrow a_{21}^{12} = -a_{21}^{21}) .$$

The solution of (109) corresponds to a symmetric $\varphi(x_1, x_2)$, i.e. the corresponding $S(\sigma_1, \sigma_2)$ is antisymmetric (singlet), while (109) yields an antisymmetric $\varphi(x_1, x_2)$, corresponding to a symmetric (triplet) $S(\sigma_1, \sigma_2)$.

The N Particle Case In the case of N particle the eigenfunctions of the Hamiltonian (70) must be continuous, but the derivatives must satisfy conditions analogous to (85) when any two coordinates become equal:

$$\lim_{x_i \to x_j + 0} \varphi = \lim_{x_i \to x_j - 0} \varphi , \tag{111}$$

$$\lim_{x_i \to x_j + 0}\left(\frac{\partial\varphi}{\partial x_i} - \frac{\partial\varphi}{\partial x_j}\right) - \lim_{x_i \to x_j - 0}\left(\frac{\partial\varphi}{\partial x_i} - \frac{\partial\varphi}{\partial x_j}\right) = 2c\lim_{x_i \to x_j}\varphi .$$

We look for the eigenfunction in the form

$$\varphi(x_1, x_2, \ldots, x_N) = \sum_{\mathcal{P}} a_{\mathcal{P}}^{\mathcal{Q}} \exp\left\{ i \sum_j k_{\mathcal{P}j} x_{\mathcal{Q}j} \right\} . \tag{112}$$

Here the permutation $\mathcal{Q}$ is defined as the permutation which puts the koordinates into increasing order

$$0 \le x_{\mathcal{Q}1} \le x_{\mathcal{Q}2} \le \ldots \le x_{\mathcal{Q}N} \le L , \tag{113}$$

and the summation is extended over all possible permutations $\mathcal{P}$ of the wavenumbers. It is clear, that if the $a_{\mathcal{P}}^{\mathcal{Q}}$ coefficients and the k wavenumbers can be choosen so that (112) meets both the continuity/discontinuity requirements and the periodic boundary conditions, the energy and the momentum of the corresponding eigenstate is

$$E = \sum_j k_j^2 , \quad P = \sum_j k_j . \tag{114}$$

It is not hard to convince ourselvs, that at $x_{\mathcal{Q}i} = x_{\mathcal{Q}i+1}$ the derivative of (112) will have the proper discontinuity (111) if

$$a_{\mathcal{P}(i,i+1)\mathcal{P}}^{\mathcal{Q}} = \frac{(k_{\mathcal{P}i} - k_{\mathcal{P}i+1}) a_{\mathcal{P}}^{\mathcal{P}(i,i+1)\mathcal{Q}} - ic\, a_{\mathcal{P}}^{\mathcal{Q}}}{(k_{\mathcal{P}i} - k_{\mathcal{P}i+1}) + ic} , \tag{115}$$

(where $\mathcal{P}^{(i,j)}$ is the permutation which interchanges the elements at the positions i and j). Let us arrange the $a_{\mathcal{P}}^{\mathcal{Q}}$ in an $N! \times N!$ matrix so that the columns are labelled by the $\mathcal{P}$s and the rows by the $\mathcal{Q}$s, and denote the colums by $\xi_{\mathcal{P}}$. These columnvectors satisfy the relations

$$\xi_{\mathcal{P}(i,i+1)\mathcal{P}} = \frac{(k_{\mathcal{P}i} - k_{\mathcal{P}i+1}) \mathcal{T}(\mathcal{P}^{(i,i+1)}) - ic}{(k_{\mathcal{P}i} - k_{\mathcal{P}i+1}) + ic} \xi_{\mathcal{P}} , \tag{116}$$

where $\mathcal{T}(\mathcal{P}^{(i,i+1)})$ is the matrix of $\mathcal{P}^{(i,i+1)}$ in the regular representation:

$$\mathcal{T}_{\mathcal{Q}\mathcal{Q}'}(\mathcal{P}) = \delta_{\mathcal{Q}, \mathcal{P}\mathcal{Q}'} . \tag{117}$$

By the repeted application of (116) from any $\xi_{\mathcal{P}}$ any other $\xi_{\mathcal{P}'}$ can be consructed. As the steps of such a construction are not determined uniquely, the result will be unique only if the matrices

$$Y^{(i,j)}(k_\alpha, k_\beta) = \frac{(k_\alpha - k_\beta) \mathcal{T}(\mathcal{P}^{(i,j)}) - ic}{(k_\alpha - k_\beta) + ic} \tag{118}$$

satisfy certain relations. (In other words the $(N-1) \times N!$ equations satisfied by the $N!$ $\xi_{\mathcal{P}}$ vectors must be compatible.) To find this relations consider a permutation product

$$\mathcal{P}_2 = \mathcal{P}\mathcal{P}_1 . \tag{119}$$

The permutation $\mathcal{P}$ can be factorized into a product of permuting neghbour elements. Such a decomposition into transpositions is not unique, but any decomposition can be transformed into any other by the subsequent applications of the identities

$$\mathcal{P}^{(i,i+1)}\mathcal{P}^{(i,i+1)} = 1\,, \tag{120}$$

$$\mathcal{P}^{(i,i+1)}\mathcal{P}^{(i+1,i+2)}\mathcal{P}^{(i,i+1)} = \mathcal{P}^{(i+1,i+2)}\mathcal{P}^{(i,i+1)}\mathcal{P}^{(i+1,i+2)}\,. \tag{121}$$

It is not hard to see, that if the analogous relations

$$Y^{(i,i+1)}(k_\beta, k_\alpha)Y^{(i,i+1)}(k_\alpha, k_\beta) = 1\,, \tag{122}$$

and

$$Y^{(i,i+1)}(k_\beta, k_\gamma)\ Y^{(i+1,i+2)}(k_\alpha, k_\gamma)Y^{(i,i+1)}(k_\alpha, k_\beta) =$$
$$= Y^{(i+1,i+2)}(k_\alpha, k_\beta)Y^{(i,i+1)}(k_\alpha, k_\gamma)Y^{(i+1,i+2)}(k_\beta, k_\gamma)\,, \tag{123}$$

hold, in

$$\xi_{\mathcal{P}_2} = Y^{(\mathcal{P})}(\mathcal{P}_1)\xi_{\mathcal{P}_1} \tag{124}$$

the operator $Y^{(\mathcal{P})}(\mathcal{P}_1)$ will be unique (allthough its decomposition into a product of the Ys of (118) is not)[4]. Eq. (123) is the Yang-Baxter equations for the 1D δ-gas. It is named after Yang right for it appeard in this context first in his work (1967). It is not hard to see by dirrect substitution that (122) and (123) hold. This means, that if one ξ is given, any other can be constructed uniquely.

To see the effect of the periodic boundary condition consider the cases when $x_i \to 0$ and $x_i \to L$. Suppose, that in the first case it is the permutation Q which puts the coordinates into increasing order. It is than clear, that in the second case that Q' will do the same, which is

$$Q' = C^{(1\to N)}Q \tag{125}$$

with $C^{(i\to j)}$ being the ciclic permutation moving the element at the position i to the position j (not permuting the other elements among themselvs). Comparing the wavefunctions of the two cases we find, that

$$e^{ik_{\mathcal{P}1}L}\,a^{C^{(1\to N)}Q}_{C^{(1\to N)}\mathcal{P}} = a^{Q}_{\mathcal{P}}\,, \tag{126}$$

what can be written (after also renaming the permutations) in the form

$$e^{ik_{\mathcal{P}N}L}\xi_{\mathcal{P}} = T(C^{(1\to N)})\xi_{C^{(N\to1)}\mathcal{P}}\,. \tag{127}$$

The number of these equations is $N!$, but only N of them are independent: all those, for which $k_{\mathcal{P}N}$ are the same, (say $k_{\mathcal{P}N} = k_j$) are equivalent to

$$e^{ik_jL}\xi_{\mathcal{P}_0} = Y^{(j,j+1)}(k_{j+1}, k_j)\,Y^{(j+1,j+2)}(k_{j+2}, k_j)\ldots Y^{(N-1,N)}(k_N, k_j) \times$$
$$T(C^{(1\to N)})Y^{(1,2)}(k_1, k_j)\,Y^{(2,3)}(k_2, k_j)\ldots Y^{(j-1,j)}(k_{j-1}, k_j)\,\xi_{\mathcal{P}_0}\,, \tag{128}$$

---

[4] Note, that $Y^{(\mathcal{P})}(\mathcal{P}_1)$ depends explicitely on the $\mathcal{P}_1$ through the ks, so this operators do not form a representation of the permutation group.

where $\xi_{\mathcal{P}_0}$ is the ξ corresponding to that permutation $\mathcal{P}_0$ in which the ks are in the order $k_1, k_2, \ldots k_j \ldots k_N$. (The main steps of showing (128) are as follows. A permutation in which $k_{\mathcal{P}N} = k_j$ can be factorized as

$$\mathcal{P} = \mathcal{P}' C^{(j \to N)} \tag{129}$$

where $\mathcal{P}'$ permutes the elements at the positions $1 \ldots N - 1$. Then

$$C^{(N \to 1)} \mathcal{P} = \mathcal{P}'' C^{(j \to 1)} , \tag{130}$$

where $\mathcal{P}''$ permutes the elements at the positions $2 \ldots N$, more over

$$\mathcal{P}'' = C^{(N \to 1)} \mathcal{P}' C^{(1 \to N)} , \quad \text{or} \quad \mathcal{P}' = C^{(1 \to N)} \mathcal{P}'' C^{(N \to 1)} . \tag{131}$$

At the same time (127) can be written in the (symbolic) form

$$e^{ik_j L} \; Y^{(\mathcal{P}')}(C^{(j \to N)}) Y^{(C^{(j \to N)})}(\mathcal{P}_0) \xi_{\mathcal{P}_0} = T(C^{(1 \to N)}) Y^{(\mathcal{P}'')}(C^{(j \to 1)}) Y^{(C^{(j \to 1)})}(\mathcal{P}_0) \xi_{\mathcal{P}_0} , \tag{132}$$

where $Y^{(\mathcal{P})}(\mathcal{R})$ stands for that combination of the $Y^{(i,j)}(k_\alpha, k_\beta)$ matrices, which acting on $\xi_{\mathcal{R}}$ cretes $\xi_{\mathcal{P R}}$. This relation due to the identity

$$T(C^{(1 \to N)}) Y^{(\mathcal{P}'')}(C^{(j \to 1)}) T(C^{(N \to 1)}) = Y^{(\mathcal{P}')}(C^{(j \to N)}) \tag{133}$$

yields

$$e^{ik_j L} \xi_{\mathcal{P}_0} = \left(Y^{(C^{(j \to N)})}(\mathcal{P}_0) \right)^{-1} T(C^{(1 \to N)}) Y^{(C^{(j \to 1)})}(\mathcal{P}_0) \xi_{\mathcal{P}_0} . \tag{134}$$

This equation, however, written out explicitly is (128)).

Eqs. (128), using the matrices

$$X_{i,j} = T(\mathcal{P}^{(i,j)}) Y^{(i,j)}(k_i, k_j) = \frac{(k_i - k_j) - ic\, T(\mathcal{P}^{(i,j)})}{(k_i - k_j) + ic} \tag{135}$$

take the form

$$e^{ik_j L} \xi_{\mathcal{P}_0} = X_{j+1,j} X_{j+2,j} \ldots X_{N,j} X_{1,j} X_{2,j} \ldots X_{j-1,j} \xi_{\mathcal{P}_0} \tag{136}$$

As we see, $\xi_{\mathcal{P}_0}$ must be an eigenvector of N different matrices:

$$X_{j+1,j} X_{j+2,j} \ldots X_{N,j} X_{1,j} X_{2,j} \ldots X_{j-1,j} , \quad j = 1, 2, \ldots, N . \tag{137}$$

This is possible as the N matrices commute: this can be shown dirrectly using

$$\begin{aligned} &X_{i,j} X_{j,i} = 1 , \\ &X_{j,k} X_{i,k} X_{i,j} = X_{i,j} X_{i,k} X_{j,k} , \\ &X_{i,j} X_{k,l} = X_{k,l} X_{i,j} \quad \text{if } i, j, k, l \text{ are all unequal} . \end{aligned} \tag{138}$$

3.3   The Right Representation of the Permutation Group

The Y and X matrices are built up out of the matrices of the regular representation of the permutation gruop. This matrices can be transformed into a blockdiagonal form:

$$\mathcal{T}_{\text{BL.DIAG.}} = U \mathcal{T}_{\text{REG.}} U^{-1} \tag{139}$$

with

$$
U = \begin{pmatrix}
\vdots & & \vdots & & \vdots \\
\sqrt{\frac{d^\mu}{N!}} \mathcal{T}^\mu_{d^\mu d^\mu}(Q_1) & \cdots & \sqrt{\frac{d^\mu}{N!}} \mathcal{T}^\mu_{d^\mu d^\mu}(Q_l) & \cdots & \sqrt{\frac{d^\mu}{N!}} \mathcal{T}^\mu_{d^\mu d^\mu}(Q_{N!}) \\
\sqrt{\frac{d^\nu}{N!}} \mathcal{T}^\nu_{11}(Q_1) & \cdots & \sqrt{\frac{d^\nu}{N!}} \mathcal{T}^\nu_{11}(Q_l) & \cdots & \sqrt{\frac{d^\nu}{N!}} \mathcal{T}^\nu_{11}(Q_{N!}) \\
\sqrt{\frac{d^\nu}{N!}} \mathcal{T}^\nu_{21}(Q_1) & \cdots & \sqrt{\frac{d^\nu}{N!}} \mathcal{T}^\nu_{21}(Q_l) & \cdots & \sqrt{\frac{d^\nu}{N!}} \mathcal{T}^\nu_{21}(Q_{N!}) \\
\vdots & & \vdots & & \vdots \\
\sqrt{\frac{d^\nu}{N!}} \mathcal{T}^\nu_{d^\nu 1}(Q_1) & \cdots & \sqrt{\frac{d^\nu}{N!}} \mathcal{T}^\nu_{d^\nu 1}(Q_l) & \cdots & \sqrt{\frac{d^\nu}{N!}} \mathcal{T}^\nu_{d^\nu 1}(Q_{N!}) \\
\sqrt{\frac{d^\nu}{N!}} \mathcal{T}^\nu_{12}(Q_1) & \cdots & \sqrt{\frac{d^\nu}{N!}} \mathcal{T}^\nu_{12}(Q_l) & \cdots & \sqrt{\frac{d^\nu}{N!}} \mathcal{T}^\nu_{12}(Q_{N!}) \\
\sqrt{\frac{d^\nu}{N!}} \mathcal{T}^\nu_{22}(Q_1) & \cdots & \sqrt{\frac{d^\nu}{N!}} \mathcal{T}^\nu_{22}(Q_l) & \cdots & \sqrt{\frac{d^\nu}{N!}} \mathcal{T}^\nu_{22}(Q_{N!}) \\
\vdots & & \vdots & & \vdots \\
\sqrt{\frac{d^\nu}{N!}} \mathcal{T}^\nu_{d^\nu 2}(Q_1) & \cdots & \sqrt{\frac{d^\nu}{N!}} \mathcal{T}^\nu_{d^\nu 2}(Q_l) & \cdots & \sqrt{\frac{d^\nu}{N!}} \mathcal{T}^\nu_{d^\nu 2}(Q_{N!}) \\
\vdots & & \vdots & & \vdots \\
\sqrt{\frac{d^\nu}{N!}} \mathcal{T}^\nu_{d^\nu d^\nu}(Q_1) & \cdots & \sqrt{\frac{d^\nu}{N!}} \mathcal{T}^\nu_{d^\nu d^\nu}(Q_l) & \cdots & \sqrt{\frac{d^\nu}{N!}} \mathcal{T}^\nu_{d^\nu d^\nu}(Q_{N!}) \\
\sqrt{\frac{d^\eta}{N!}} \mathcal{T}^\eta_{11}(Q_1) & \cdots & \sqrt{\frac{d^\eta}{N!}} \mathcal{T}^\eta_{11}(Q_l) & \cdots & \sqrt{\frac{d^\eta}{N!}} \mathcal{T}^\eta_{11}(Q_{N!}) \\
\vdots & & \vdots & & \vdots
\end{pmatrix}
\tag{140}
$$

Here d^ν and $\mathcal{T}^\nu_{ij}(Q)$ are the dimension and the elements of the matrices of the irreducible representation $\mathcal{T}^\nu$, and each irreducible representation is present in the U matrix once. In the blockdiagonal form the blocks in the diagonal are the matrices of the irreducible representations of the permutation gruop. Each representation occours as many times as large its dimension is, (i.e. $\mathcal{T}^\nu$ has d^ν copies in the diagonal). We can transform also the $\xi_\mathcal{P}$ vectors by the transformation U:

$$\zeta_\mathcal{P} = U \xi_\mathcal{P} . \tag{141}$$

We look for a $\xi_{\mathcal{P}_0}$ solution of (128) (i.e. of (136)), among those vectors for which the $\zeta_{\mathcal{P}_0}$ has nonzero elements only at a given copy of a given irreducible representation $\mathcal{T}^\nu$. Than the d^ν different $\xi_{\mathcal{P}_0}$, which belong to the d^ν different copies of the same $\mathcal{T}^\nu$ will generate d^ν eigenfunctions which transform according to the irreducible representation $(\mathcal{T}^\nu)^*$.

To see this, first let us define the Y^ν (X^ν) operators so that they are the same as the original Ys (Xs) just the $N! \times N!$ matrices of the regular representation are replaced by the $d^\nu \times d^\nu$ matrices of the representation $\mathcal{T}^\nu$. Than consider Eq. (128) ((136)) with the Ys (Xs) replaced by the Y^νs (X^νs), and suppose that the (d^ν dimensional) vector $\zeta(\mathcal{P}_0)$ is a solution. It is clear (thruogh (139)), that the $N!$ dimensional vectors $\xi_{\alpha,\mathcal{P}_0}$ with the components (labelled by the permutations $\mathcal{Q}$)

$$a_{\alpha,\mathcal{P}_0}^{\mathcal{Q}} = \sum_\beta (T_{\beta,\alpha}^\nu(\mathcal{Q}))^* \zeta_\beta(\mathcal{P}_0) , \quad (\alpha = 1, 2, \ldots d^\nu) . \tag{142}$$

solve Eq. (128) ((136)). As a next step construct all the different $\xi_{\alpha,\mathcal{P}}$ vectors belonging to the different $\mathcal{P}$ permutations by the repeted application of (116) type relations. As a result one obtains the vectors $\xi_{\alpha,\mathcal{P}}$ with components

$$a_{\alpha,\mathcal{P}}^{\mathcal{Q}} = \sum_\beta (T_{\beta,\alpha}^\nu(\mathcal{Q}))^* \zeta_\beta(\mathcal{P}) , \quad (\alpha = 1, 2, \ldots d^\nu) . \tag{143}$$

Here the $\zeta(\mathcal{P})$ are those vectors which are obtained from $\zeta(\mathcal{P}_0)$ by applying the same combination of Y^νs, which combination of the Ys connects $\xi_\mathcal{P}$ to $\xi_{\mathcal{P}_0}$. Obviously all of the d^ν functions

$$\varphi_\alpha(x_1, x_2, \ldots, x_N) = \sum_\mathcal{P} a_{\alpha,\mathcal{P}}^{\mathcal{Q}} \exp\left\{ i \sum_j k_{\mathcal{P}j} x_{\mathcal{Q}j} \right\} \quad (\alpha = 1, 2, \ldots d^\nu) \tag{144}$$

$$= \sum_\beta (T_{\beta,\alpha}^\nu(\mathcal{Q}))^* \sum_\mathcal{P} \zeta_\beta(\mathcal{P}) \exp\left\{ i \sum_j k_{\mathcal{P}j} x_{\mathcal{Q}j} \right\} \tag{145}$$

are eigenfunctions of the Hamiltonian by construction.

Consider a permutation operator $\hat{\mathcal{R}}$. Let us define its action on a function so that we permute the coordinates according to $\mathcal{R}^{-1}$:

$$\hat{\mathcal{R}}\varphi(x_1, x_2, \ldots, x_N) = \varphi(x_{\mathcal{R}^{-1}1}, x_{\mathcal{R}^{-1}2}, \ldots x_{\mathcal{R}^{-1}N}) . \tag{146}$$

(This is in analogy with the action of a rotation $\hat{r}$: $\hat{r} f(x) = f(r^{-1}x)$.) Than, since after the action of $\hat{\mathcal{R}}$ the permutation $\mathcal{QR}$ will arrange the koordinates into increasing order,

$$\hat{\mathcal{R}}\varphi_\alpha(x_1, x_2, \ldots, x_N) = \sum_\mathcal{P} a_{\alpha,\mathcal{P}}^{\mathcal{QR}} \exp\left\{ i \sum_j k_{\mathcal{P}j} x_{\mathcal{Q}j} \right\} \tag{147}$$

$$= \sum_\beta (T_{\beta,\alpha}^\nu(\mathcal{QR}))^* \sum_\mathcal{P} \zeta_\beta(\mathcal{P}) \exp\left\{ i \sum_j k_{\mathcal{P}j} x_{\mathcal{Q}j} \right\} \tag{148}$$

This yields

$$\hat{\mathcal{R}}\varphi_\alpha(x_1, x_2, \ldots, x_N) = \sum_\gamma \varphi_\gamma(x_1, x_2, \ldots, x_N)(T_{\gamma,\alpha}^\nu(\mathcal{R}))^* . \tag{149}$$

As we dicussed in Sec. 3.1 in order to have an antisymmetric wavefunction in which the spin part transforms according to $\mathcal{T}^{(l,m)}$, the coordinate dependent part must transform according to $(-1)^{\mathcal{P}}(\mathcal{T}^{(l,m)})^{*}$. This means, that we have to solve Eq. (128) ((136)) with the Y, $\mathcal{T}(\mathcal{C}^{(1 \to N)})$ (and X) operators replaced by

$$Y^{(i,j)}(k_{\alpha}, k_{\beta}) \Longrightarrow \frac{-(k_{\alpha} - k_{\beta})\,\mathcal{T}^{(l,m)}(\mathcal{P}^{(i,j)}) - ic}{(k_{\alpha} - k_{\beta}) + ic} \,, \tag{150}$$

$$\mathcal{T}(\mathcal{C}^{(1 \to N)}) \Longrightarrow (-1)^{(N-1)}\mathcal{T}^{(l,m)}(\mathcal{C}^{(1 \to N)}) \,, \tag{151}$$

and

$$X_{i,j} \Longrightarrow \frac{(k_i - k_j) + ic\,\mathcal{T}^{(l,m)}(\mathcal{P}^{(i,j)})}{(k_i - k_j) + ic} \,, \tag{152}$$

than the functions constructed above (with $\mathcal{T}^{\nu}(\mathcal{Q})$ replaced by $(-1)^{\mathcal{Q}}\mathcal{T}^{(l,m)}(\mathcal{Q})$) will have the right symmetry.

In practice we do not construct the $\mathcal{T}^{(l,m)}$ representation, instead we replace the matrices of the permutations by the permutation operators themselvs[5]:

$$Y^{(i,j)}(k_{\alpha}, k_{\beta}) \Longrightarrow - \,\hat{Y}^{(i,j)}(k_{\alpha}, k_{\beta})$$

$$\hat{Y}^{(i,j)}(k_{\alpha}, k_{\beta}) = \frac{(k_{\alpha} - k_{\beta})\,\hat{\mathcal{P}}^{(i,j)} + ic}{(k_{\alpha} - k_{\beta}) + ic} \,, \tag{153}$$

and

$$X_{i,j} \Longrightarrow \hat{X}_{i,j} = \frac{(k_i - k_j) + ic\hat{\mathcal{P}}^{(i,j)}}{(k_i - k_j) + ic} \,, \tag{154}$$

and solve the resulting equations in the space of spin states. Thus we have to solve the eigenvalue problems

$$e^{ik_j L}S_{\mathcal{P}_0} = \hat{Y}^{(j,j+1)}(k_{j+1}, k_j)\,\hat{Y}^{(j+1,j+2)}(k_{j+2}, k_j) \ldots \hat{Y}^{(N-1,N)}(k_N, k_j) \times$$
$$\times\, \hat{\mathcal{C}}^{(1 \to N)}\hat{Y}^{(1,2)}(k_1, k_j)\,\hat{Y}^{(2,3)}(k_2, k_j) \ldots \hat{Y}^{(j-1,j)}(k_{j-1}, k_j)\, S_{\mathcal{P}_0} \,, \tag{155}$$

or equivalently

$$e^{ik_j L}S_{\mathcal{P}_0} = \hat{X}_{j+1,j}\,\hat{X}_{j+2,j} \ldots \hat{X}_{N,j}\hat{X}_{1,j}\,\hat{X}_{2,j} \ldots \hat{X}_{j-1,j}\,S_{\mathcal{P}_0} \,. \tag{156}$$

It is easy to guess, that the solution $S_{\mathcal{P}_0}$ and the states $S_{\mathcal{P}}$ constructable out of $S_{\mathcal{P}_0}$ will describe the spin state of the system.

We note here, that the $\hat{X}$ operators are the two particle scattering matrices for the spin $1/2$ Fermions with δ-interaction, and that (156) is of a form which can be generalized to treat other cases.

---

[5] Note that the $\hat{Y}$ operators apparently differ in a $-$ sign from the corresponding $Y^{(l,m)}$ matrices. All in the equations of the type (128) these signs cancel with the $(-1)^{(N-1)}$ sign of the ciclic permutation. Of course, in equations of the type (116) the operators $-\hat{Y}$ are to be used.

3.4   BA for the Spin State $S_{\mathcal{P}_0}$

In this subsection we follow the startegy allready known. First we give the solution of (155) for the case, when only one spin is turned down. Actually we show a way how to verify Yangs solution (1967). After that we construct the solution for two turned over spins, and finaly generalize the result. (We find technically easier to treat Eq. (155), but of course the solutions are the same as those for Eq. (156).)

The Case of One Spin Turned Down Consider a chain of N spins, and suppose, all but one of them point up. Suppose also, that the N wavenumbers are in the order $k_1, k_2, \ldots, k_N$ and define the amplitudes

$$S(k_1, k_2, \ldots, k_N | \Lambda; m) = \left\{ \prod_{l=1}^{m-1} \frac{i(k_l - \Lambda) - c/2}{i(k_l - \Lambda) + c/2} \right\} \frac{1}{i(k_m - \Lambda) + c/2} . \tag{157}$$

(For the sake of definitness: if $l = 1$, the 'empty' product is one). According to Yang (1967), if the anplitude of finding the down spin at the site m is $S(k_1, k_2, \ldots, k_N | \Lambda; m)$, the state is an eigenstate of the r.h.s. of (155) provided the Λ has a right value. This can be verified in the following way (see also Fungs work (1981)). First construct the state-vector

$$\left| S^{(1)}(k_1, k_2, \ldots, k_N) \right\rangle = \sum_{m=1}^{N} S(k_1, k_2, \ldots, k_N | \Lambda; m) \sigma_m^- |F\rangle , \tag{158}$$

where $|F\rangle$ is the ferromagnetic state. Since

$$\frac{(k_{j-1} - k_j)}{(k_{j-1} - k_j) + ic} S(\ldots, k_{j-1}, k_j, \ldots | \Lambda; j) +$$

$$+ \frac{ic}{(k_{j-1} - k_j) + ic} S(\ldots, k_{j-1}, k_j, \ldots | \Lambda; j - 1) = S(\ldots, k_j, k_{j-1}, \ldots | \Lambda; j - 1) ,$$

$$\tag{159}$$

and

$$\frac{(k_{j-1} - k_j)}{(k_{j-1} - k_j) + ic} S(\ldots, k_{j-1}, k_j, \ldots | \Lambda; j - 1) +$$

$$+ \frac{ic}{(k_{j-1} - k_j) + ic} S(\ldots, k_{j-1}, k_j, \ldots | \Lambda; j) = S(\ldots, k_j, k_{j-1}, \ldots | \Lambda; j),$$

$$\tag{160}$$

(as can be checked dirrectly,) we have

$$\hat{Y}^{(j-1,j)}(k_{j-1}, k_j) \left| S^{(1)}(\ldots, k_{j-1}, k_j, \ldots) \right\rangle = \left| S^{(1)}(\ldots, k_j, k_{j-1}, \ldots) \right\rangle . \tag{161}$$

In a similar way

$$\hat{Y}^{(j-2,j-1)}(k_{j-2}, k_j) \left| S^{(1)}(\ldots, k_{j-2}, k_j, k_{j-1}, \ldots) \right\rangle =$$
$$\left| S^{(1)}(\ldots, k_j, k_{j-2}, k_{j-1}, \ldots) \right\rangle , \qquad (162)$$

and analogous relations hold for the actions of the next $\hat{Y}$ operators too. This way the $\hat{Y}$ operators standing to the right from the operator $\hat{C}^{(1 \rightarrow N)}$ move the k_j from the position j to the first position in the argument of $\left| S^{(1)}(\{k\}) \right\rangle$. As

$$S(k_j, k_1, \ldots, k_N | \Lambda; 1) = \left\{ \prod_{l \neq j} \frac{i(k_l - \Lambda) - c/2}{i(k_l - \Lambda) + c/2} \right\}^{-1} S(k_1, \ldots, k_N, k_j | \Lambda; N) ,$$
$$(163)$$

and

$$S(k_j, k_1, \ldots, k_N | \Lambda; m \neq 1) = \frac{i(k_j - \Lambda) - c/2}{i(k_j - \Lambda) + c/2} S(k_1, \ldots, k_N, k_j | \Lambda; m-1) , \quad (164)$$

also the relation

$$\hat{C}^{(1 \rightarrow N)} \left| S^{(1)}(k_j, k_1, \ldots, k_N) \right\rangle = \frac{i(k_j - \Lambda) - c/2}{i(k_j - \Lambda) + c/2} \left| S^{(1)}(k_1, \ldots, k_N, k_j) \right\rangle \quad (165)$$

holds, if

$$\prod_{l=1}^{N} \frac{i(k_l - \Lambda) - c/2}{i(k_l - \Lambda) + c/2} = 1 . \qquad (166)$$

That is, the $\hat{C}^{(1 \rightarrow N)}$ operator moves the k_j from the first position to the last one in the argument of $\left| S^{(1)}(\{k\}) \right\rangle$, while $\left| S^{(1)}(\{k\}) \right\rangle$ picks up a factor

$$\frac{i(k_j - \Lambda) - c/2}{i(k_j - \Lambda) + c/2} . \qquad (167)$$

Finally the $\hat{Y}$ operators on the left side of $\hat{C}^{(1 \rightarrow N)}$ move the k_j back to its original jth position in $\left| S^{(1)}(\{k\}) \right\rangle$. This way

$$S_{\mathcal{P}_0} = \left| S^{(1)}(k_1, k_2, \ldots, k_N) \right\rangle , \qquad (168)$$

is an eigenstate of the r.h.s. of (155) with an eigenvalue (167), if (166) holds, so (155) reduces to

$$e^{ik_j L} = \frac{i(k_j - \Lambda) - c/2}{i(k_j - \Lambda) + c/2} . \qquad (169)$$

The Case of Two Spins Turned Down Just as in the case of the Heisenberg chain, we build up the 'two particle' eigenstate as an appropriate combination of the 'one particle' ones. Suppose the two down spins are located at the positions $m_1 < m_2$, and introduce the amplitude

$$S(k_1, k_2, \ldots, k_N \,|\, \Lambda_1, \Lambda_2; m_1, m_2) =$$
$$a(\Lambda_1, \Lambda_2) \; S(k_1, k_2, \ldots, k_N | \Lambda_1; m_1) \, S(k_1, k_2, \ldots, k_N | \Lambda_2; m_2) +$$
$$a(\Lambda_2, \Lambda_1) \; S(k_1, k_2, \ldots, k_N | \Lambda_2; m_1) \, S(k_1, k_2, \ldots, k_N | \Lambda_1; m_2) \; . \quad (170)$$

The supposed form of the eigenstate is

$$\left| S^{(2)}(k_1, k_2, \ldots, k_N) \right\rangle = \sum_{m_1 < m_2} S(k_1, k_2, \ldots, k_N | \Lambda_1, \Lambda_2; m_1, m_2) \, \sigma^-_{m_1} \sigma^-_{m_2} \, |F\rangle \; .$$
$$(171)$$

It is not hard to see, that just as in the 'one particle' case

$$\hat{Y}^{(j-1,j)}(k_{j-1}, k_j) \left| S^{(2)}(\ldots, k_{j-1}, k_j, \ldots) \right\rangle = \left| S^{(2)}(\ldots, k_j, k_{j-1}, \ldots) \right\rangle \; , \quad (172)$$

provided

$$a(\Lambda_1, \Lambda_2) \, S(\ldots, k_{j-1}, k_j, \ldots | \Lambda_1; j-1) \, S(\ldots, k_{j-1}, k_j, \ldots | \Lambda_2; j) +$$
$$a(\Lambda_2, \Lambda_1) \, S(\ldots, k_{j-1}, k_j, \ldots | \Lambda_2; j-1) \, S(\ldots, k_{j-1}, k_j, \ldots | \Lambda_1; j) =$$
$$a(\Lambda_1, \Lambda_2) \, S(\ldots, k_j, k_{j-1}, \ldots | \Lambda_1; j-1) \, S(\ldots, k_j, k_{j-1}, \ldots | \Lambda_2; j) +$$
$$a(\Lambda_2, \Lambda_1) \, S(\ldots, k_j, k_{j-1}, \ldots | \Lambda_2; j-1) \, S(\ldots, k_j, k_{j-1}, \ldots | \Lambda_1; j) \, . $$
$$(173)$$

This requirement is equivalent to the relation

$$\frac{a(\Lambda_1, \Lambda_2)}{a(\Lambda_2, \Lambda_1)} = \frac{i(\Lambda_2 - \Lambda_1) - c}{i(\Lambda_2 - \Lambda_1) + c} \; . \quad (174)$$

If it holds, just like in the 'one particle' case the $\hat{Y}$ opperators standing to the right from the $\hat{C}^{(1 \to N)}$ move the k_j from the position j to the first position in the argument of $\left| S^{(1)}(\{k\}) \right\rangle$. Now the effect of $\hat{C}^{(1 \to N)}$ on $\left| S^{(2)}(k_j, \ldots) \right\rangle$ is

$$\hat{C}^{(1 \to N)} \left| S^{(2)}(k_j, k_1, \ldots, k_N) \right\rangle =$$
$$\frac{i(k_j - \Lambda_1) - c/2}{i(k_j - \Lambda_1) + c/2} \, \frac{i(k_j - \Lambda_2) - c/2}{i(k_j - \Lambda_2) + c/2} \left| S^{(2)}(k_1, \ldots, k_N, k_j) \right\rangle \quad (175)$$

provided

$$a(\Lambda_1, \Lambda_2) \, S(k_j, k_1, \ldots, k_N | \Lambda_1; 1) \, S(k_j, k_1, \ldots, k_N | \Lambda_2; m) +$$
$$a(\Lambda_2, \Lambda_1) \, S(k_j, k_1, \ldots, k_N | \Lambda_2; 1) \, S(k_j, k_1, \ldots, k_N | \Lambda_1; m) =$$
$$\frac{i(k_j - \Lambda_1) - c/2}{i(k_j - \Lambda_1) + c/2} \, \frac{i(k_j - \Lambda_2) - c/2}{i(k_j - \Lambda_2) + c/2} \times \quad (176)$$

$$\Big(a(\Lambda_1, \Lambda_2) \, S(k_1, \ldots, k_N, k_j | \Lambda_1; m-1) \, S(k_1, \ldots, k_N, k_j | \Lambda_2; N) +$$
$$a(\Lambda_2, \Lambda_1) \, S(k_1, \ldots, k_N, k_j | \Lambda_2; m-1) \, S(k_1, \ldots, k_N, k_j | \Lambda_1; N) \Big) \, . $$

This condition reduces to the equations

$$\prod_{\beta=1}^{2} \frac{i(\Lambda_\alpha - \Lambda_\beta) + c}{i(\Lambda_\alpha - \Lambda_\beta) - c} = -\prod_{l=1}^{N} \frac{i(\Lambda_\alpha - k_l) + c/2}{i(\Lambda_\alpha - k_l) - c/2}, \quad (\alpha = 1, 2) . \qquad (177)$$

As also in this case the $\hat{Y}$ operators on the left side of $\hat{C}^{(1 \to N)}$ move the k_j back to its original jth position in $\left| S^{(2)}(\{k\}) \right\rangle$, we conclude that (155) is satisfied by

$$S_{\mathcal{P}_0} = \left| S^{(2)}(k_1, k_2, \ldots, k_N) \right\rangle \qquad (178)$$

under the conditions (177) and

$$e^{ik_j L} = \prod_{\beta=1}^{2} \frac{i(k_j - \Lambda_\beta) - c/2}{i(k_j - \Lambda_\beta) + c/2} , \quad (j = 1, 2, \ldots, N) . \qquad (179)$$

The General Case Suppose the $M(< N/2)$ turned down spins[6] are located at the positions $m_1 < m_2 < \ldots < m_M$. Now we introduce the amplitude

$$S(k_1, \ k_2, \ldots, k_N | \Lambda_1, \Lambda_2, \ldots, \Lambda_M; m_1, m_2, \ldots, m_M) =$$
$$\sum_{\mathcal{R}} a(\Lambda_{\mathcal{R}1}, \Lambda_{\mathcal{R}2}, \ldots, \Lambda_{\mathcal{R}M}) \prod_{l=1}^{M} S(k_1, k_2, \ldots, k_N | \Lambda_{\mathcal{R}l}; m_l) . \qquad (180)$$

Here $\mathcal{R}$ is a permutation of M elements and the summation is extended over all possible permutations. The 'Ansatz' for the M particle eigenstates is

$$\left| S^{(M)}(k_1, k_2, \ldots, k_N) \right\rangle = \qquad (181)$$
$$\sum_{m_l < m_{l+1}} S(k_1, k_2, \ldots, k_N | \Lambda_1, \Lambda_2, \ldots, \Lambda_M; m_1, m_2, \ldots, m_M) \prod_{q=1}^{M} \sigma_{m_q}^{-} |F\rangle ,$$

where the summation is extended over all possible configurations.

To find under what conditions does

$$S_{\mathcal{P}_0} = \left| S^{(M)}(k_1, k_2, \ldots, k_N) \right\rangle , \qquad (182)$$

satisfy (155) needs no new trick. Actually we find, that the first $\hat{Y}$ operators move the k_j left in the argument of $\left| S^{(M)}(\{k\}) \right\rangle$, if

$$\frac{a(\ldots, \Lambda_{\mathcal{R}1}, \Lambda_{\mathcal{R}2}, \ldots)}{a(\ldots, \Lambda_{\mathcal{R}2}, \Lambda_{\mathcal{R}1}, \ldots)} = \frac{i(\Lambda_{\mathcal{R}2} - \Lambda_{\mathcal{R}1}) - c}{i(\Lambda_{\mathcal{R}2} - \Lambda_{\mathcal{R}1}) + c} . \qquad (183)$$

---

[6] The $M < N/2$ condition does not restrict generality as we can choose which is the up (or down) dirrection.

The operator $\hat{C}^{1 \to N}$ moves the k_j from the first to the last position while $\left|S^{(M)}(\{k\})\right\rangle$ picking up a constant coefficient

$$\prod_{\beta=1}^{M} \frac{i(k_j - \Lambda_\beta) - c/2}{i(k_j - \Lambda_\beta) + c/2}, \tag{184}$$

if

$$\prod_{\beta=1}^{M} \frac{i(\Lambda_\alpha - \Lambda_\beta) + c}{i(\Lambda_\alpha - \Lambda_\beta) - c} = -\prod_{l=1}^{N} \frac{i(\Lambda_\alpha - k_l) + c/2}{i(\Lambda_\alpha - k_l) - c/2}, \quad (\alpha = 1, 2, \ldots, M). \tag{185}$$

The remaining $\hat{Y}$ operators move the k_j back to its original position. This way for the case of M spins turned down (155) is equivalent to (185) and

$$e^{ik_j L} = \prod_{\beta=1}^{M} \frac{i(k_j - \Lambda_\beta) - c/2}{i(k_j - \Lambda_\beta) + c/2}, \quad (j = 1, 2, \ldots, N). \tag{186}$$

Taking the logarithm of (186) and (185) we conclude that $\left|S^{(M)}(\{k\})\right\rangle$ is a solution of (155), if

$$Lk_j = 2\pi I_j - \sum_{\beta}^{M} 2\tan^{-1}\left(\frac{k_j - \Lambda_\beta}{c/2}\right),$$

$$\left(I_j = \frac{M}{2} \,(\text{mod}\,1)\right), \tag{187}$$

and

$$\sum_{l}^{N} 2\tan^{-1}\left(\frac{\Lambda_\alpha - k_l}{c/2}\right) = 2\pi J_\alpha + \sum_{\beta}^{M} 2\tan^{-1}\left(\frac{\Lambda_\alpha - \Lambda_\beta}{c}\right),$$

$$\left(J_\alpha = \frac{N + M + 1}{2} \,(\text{mod}\,1)\right). \tag{188}$$

The Spin of the $\left|S^{(M)}(\{k\})\right\rangle$ States In the previous sections we solved the eigenvalue problem (155) in a reducible representation of the permutation group, namely in the representation generated by the spin functions describing the $S^z = N/2 - M$ states of N spins, each of length 1/2. On the other hand we know, that the solution should transform according to a single irreducible representation. What we are interested in is the length of the total spin S^2 connected with this irreducible representation. Actually it is generally true, that *any* solution of Eqs. (187) and (188) gives a state $\left|S^{(M)}(\{k\})\right\rangle$ with $S^2 = S^z(S^z + 1)$ provided all the Λs are finite. This can be proved – just as in the case of isotropic Heisenberg chain (Sec. 2.4) – by showing that

$$\sigma^+ \left|S^{(M)}(\{k\})\right\rangle = 0. \tag{189}$$

Here we do not prove this statement, but give a verification of it for the case $c > 0$.

Consider a solution of Eqs. (187) and (188)! For $c > 0$ in any solution all the k_j are expected to be real. For this wavenumbers $k_j/c \to 0$ as $|c| \to \infty$, but $2\Lambda_\alpha/c \to x_\alpha = finite$ as it can be seen from the $|c| \to \infty$ limit of Eqs. (188):

$$2\tan^{-1}(x_\alpha) = 2\pi J'_\alpha + \sum_\beta^M 2\tan^{-1}\left(\frac{x_\alpha - x_\beta}{2}\right) ,$$

$$\left(J'_\alpha = \frac{N + M + 1}{2} \,(\mathrm{mod}\, 1)\right) . \tag{190}$$

It is important, that this equation is equivalent to the Eq. (48) for $\rho = 1$ (see (47)) with

$$\cot \frac{k_\alpha}{2} = -\lim_{|c| \to \infty} \frac{2\Lambda_\alpha}{c} = -x_\alpha . \tag{191}$$

To see the $|c| \to \infty$ limit of the $\left|S^{(M)}(\{k\})\right\rangle$ we write

$$S(k_1,\ k_2,\ldots,k_N|\Lambda_1,\Lambda_2,\ldots,\Lambda_M;m_1,m_2,\ldots,m_M) = \tag{192}$$

$$C \sum_{\mathcal{R}} \left(\prod_{\alpha<\beta} \left(\frac{i(\Lambda_{\mathcal{R}\alpha} - \Lambda_{\mathcal{R}\beta}) + c}{i(\Lambda_{\mathcal{R}\alpha} - \Lambda_{\mathcal{R}\beta}) - c}\right)^{1/2}\right) \prod_{\alpha=1}^M S(k_1, k_2,\ldots,k_N|\Lambda_{\mathcal{R}\alpha};m_\alpha),$$

where we used, that the solution of (183) is of the form

$$a(\Lambda_{\mathcal{R}1}, \Lambda_{\mathcal{R}2},\ldots,\Lambda_{\mathcal{R}M}) = C \prod_{\alpha<\beta} \left(\frac{i(\Lambda_{\mathcal{R}\alpha} - \Lambda_{\mathcal{R}\beta}) + c}{i(\Lambda_{\mathcal{R}\alpha} - \Lambda_{\mathcal{R}\beta}) - c}\right)^{1/2} . \tag{193}$$

It is clear, that if we choose the constatnt C

$$C \propto (-1)^M \prod_\alpha (i\Lambda_\alpha + c/2) . \tag{194}$$

than

$$\lim_{|c| \to \infty} S(k_1,\ k_2,\ldots,k_N|\Lambda_1,\Lambda_2,\ldots,\Lambda_M;m_1,m_2,\ldots,m_M) =$$

$$\sum_{\mathcal{R}} \left(\prod_{\alpha<\beta} \left(\frac{i(x_{\mathcal{R}\alpha} - x_{\mathcal{R}\beta}) + 2}{i(x_{\mathcal{R}\alpha} - x_{\mathcal{R}\beta}) - 2}\right)^{1/2}\right) \prod_{\alpha=1}^M \left(\frac{ix_\alpha + 1}{ix_\alpha - 1}\right)^{m_\alpha} \tag{195}$$

This is, however, a Bethe Ansatz eigenfunction of M turned down spins in an isotropic Heisenberg chain of length N (see (45)), what describes indeed an $S^2 = S^z(S^z + 1)$ state (see Sec. 2.4). This means, that

$$\hat{S}^2 \lim_{|c| \to \infty} \left|S^{(M)}(\{k\})\right\rangle = S^z(S^z + 1) \lim_{|c| \to \infty} \left|S^{(M)}(\{k\})\right\rangle . \tag{196}$$

The symmetry of an eigenstate can change only at a singularity of the solution. Since, however, the solution is continuous in c, the symmetry of the solution is expected to be independent of the actual value of c, i.e.

$$\hat{S}^2 \left| S^{(M)}(\{k\}) \right\rangle = S^z (S^z + 1) \left| S^{(M)}(\{k\}) \right\rangle . \tag{197}$$

3.5 Construction of the Complete Wavefunction

In Sect. 3.4 for an easy 'book-keeping' we used the state-vectors $\left| S^{(M)}(\{k\}) \right\rangle$ instead of the amplitudes of the different spin configurations. To apply, however, the results of Sect. 3.3, specially the construction (145), we need the $\zeta(\mathcal{P})$ vectors. Before constructing them, for the sake of easy reference we define some notations.

We denote a (spin-)configuration by $\{m\}$. (Here m refers to the $m_1 < m_2 < \ldots < m_M$ positions of the down spins in the series $\sigma_1, \sigma_2, \ldots, \sigma_N$ (i.e. in $\{\sigma\}$.) The corresponding state is

$$|\{m\}\rangle = \prod_{q=1}^{M} \sigma_{m_q}^- |F\rangle . \tag{198}$$

We denote the configuration obtained by permuting the series $\sigma_1, \sigma_2, \ldots, \sigma_N$ according to $\mathcal{P}$ by $\{\mathcal{P}m\}$. Clearly

$$\hat{\mathcal{P}}|\{m\}\rangle = |\{\mathcal{P}m\}\rangle . \tag{199}$$

Finally we choose the basis in the space of spin functions (defined as functions of the spin-configurations)

$$S_{\{m\}}^0(\{m'\}) = \delta_{\{m\},\{m'\}} . \tag{200}$$

It is clear, that the set $S_{\{m\}}^0$ forms a basis for an (in general reducible) representation of the permutation group: if

$$\hat{\mathcal{P}} S_{\{m'\}}^0(\{m\}) = S_{\{m'\}}^0(\{\mathcal{P}^{-1}m\}) , \tag{201}$$

as it should be according to (199), than

$$\hat{\mathcal{P}} S_{\{m'\}}^0(\{m\}) = \sum_{\{m''\}} S_{\{m''\}}^0(\{m\}) \mathcal{T}_{\{m''\},\{m'\}}^S(\mathcal{P}) \tag{202}$$

with

$$\mathcal{T}_{\{m''\},\{m'\}}^S(\mathcal{P}) = \delta_{\{m''\},\{\mathcal{P}m'\}} . \tag{203}$$

If $\mathcal{T}^S$ is reducible, it can be transformed into a block-diagonal form T^S by a unitary transformation V

$$V^{-1} \mathcal{T}^S V = T^S \tag{204}$$

so, that the blocks in the diagonal are some of the irreducible representations.

Due to (199)

$$\hat{\mathcal{P}} = \sum_{\{m''\},\{m'\}} |\{m''\}\rangle \mathcal{T}^S_{\{m''\},\{m'\}}(\mathcal{P})\langle\{m'\}| \, . \tag{205}$$

where $\mathcal{T}^S$ is the same as above. If we put this into (155) we see, that the vector with components (labelled by the configurations)

$$S_{\{m\}}(\mathcal{P}_0) = S(k_1, k_2, \ldots, k_N | \Lambda_1, \Lambda_2, \ldots, \Lambda_M; m_1, m_2, \ldots, m_M) \tag{206}$$

satisfies (155) with $\hat{\mathcal{P}}$ replaced by $\mathcal{T}^S(\mathcal{P})$. If we transform this representation into a blockdiagonal form, we have to transform the $S_{\{m\}}(\mathcal{P}_0)$ too. The result is

$$\zeta_\alpha(\mathcal{P}_0) = \sum_{\{m\}} V^*_{\{m\}\alpha} S_{\{m\}}(\mathcal{P}_0) \, . \tag{207}$$

If the spin part transforms according to one single irreducible representation, only those components ζ_α are nonzero, which correspond to the block of the given representation in $\mathcal{T}^S$. The same holds for

$$\zeta_\alpha(\mathcal{P}) = (-1)^{\mathcal{P}} \sum_{\{m\}} V^*_{\{m\}\alpha} S_{\{m\}}(\mathcal{P}) \, , \tag{208}$$

where

$$S_{\{m\}}(\mathcal{P}) = S(k_{\mathcal{P}1}, k_{\mathcal{P}2}, \ldots, k_{\mathcal{P}N} | \Lambda_1, \Lambda_2, \ldots, \Lambda_M; m_1, m_2, \ldots, m_M) \, . \tag{209}$$

It is clear, that due to relations of the type (159) and (160), the application of the appropriate combinations of the Y matrices on $\zeta(\mathcal{P}_0)$, leads to $\zeta(\mathcal{P})$ of (208)[7], thus these are the vectors to be used in (145). Keeping in mind, that only the components corresponding to one single block are nonzero in ζ, we may use $\mathcal{T}^S$ instead of that single block in (145). Thus we have

$$\varphi_\alpha(x_1, x_2, \ldots, x_N) = \sum_\beta (-1)^{\mathcal{Q}} (T^S_{\beta,\alpha}(\mathcal{Q}))^* \sum_{\mathcal{P}} \zeta_\beta(\mathcal{P}) \exp\left\{ i \sum_j k_{\mathcal{P}j} x_{\mathcal{Q}j} \right\} \, , \tag{210}$$

(where only those φ_α are nonzero, which correspond to that block, for which ζ_α are nonzero.) In order to obtain the complete wavefunction we have to calculate the sum

$$f(\ldots, x_j, \sigma_j, \ldots) = \sum_\alpha \varphi_\alpha(\ldots, x_j, \ldots) S_\alpha(\ldots, \sigma_j, \ldots) \, , \tag{211}$$

with

$$S_\alpha(\ldots, \sigma_j, \ldots) = \sum_{\{m'\}} S^0_{\{m'\}}(\{m\}) V_{\{m'\}\alpha} \tag{212}$$

---

[7] The coefficient $(-1)^{\mathcal{P}}$ in (208) has its origin in the different signs of the Y matrices and the corresponding $\hat{Y}$ operators.

where $\{m\}$ refers to the configuration $\ldots \sigma_j \ldots$. Substituting (210) and (212) into (211), after some straightforward manipulations we arrive at

$$f(\ldots, x_j, \sigma_j, \ldots) = \sum_{\mathcal{P}} (-1)^{\mathcal{Q}} (-1)^{\mathcal{P}} S_{\{\mathcal{Q}m\}}(\mathcal{P}) \exp\left\{ i \sum_j k_{\mathcal{P}j} x_{\mathcal{Q}j} \right\} . \quad (213)$$

3.6 Summary and Comments

Summary of the Results As the diagonalisation of the Hamiltonian (70) involved several nontrivial steps it is worth to summarise them.

- First it has been observed, that the eigenstates of the Hamiltonian should transform according to the different irreducible representations of the permutation group. It has been seen also that in order to have a wavefunction describing N spin $1/2$ Fermions in a certain spin state, the coordinate dependent part of the wavefunction should transform according the representation $(-1)^{\mathcal{P}} \left(\mathcal{T}^{(l,m)}(\mathcal{P}) \right)^*$ with $\mathcal{T}^{(l,m)}(\mathcal{P})$ being the representation according to which the spin part transforms.
- The Ansatz (112) for the wavfunction reflects the fact that the particles move as free ones, if they are apart. The coefficients $a_{\mathcal{P}}^{\mathcal{Q}}$ should satisfy the conditions (115) ((116)) in order to have the wavefunction continuous, with the right discontinuities (required by the Hamiltonian) in the derivatives. These conditions are compatible to each other, if the so called Yang-Baxter equations (123) for the Ys defined by (118) hold.
- Imposing the periodic boundary conditions on the system leads to the equations (128) or to the equivalent set of equations (136). These equations are sets of eigenvalue equations which must hold simultanously. (This is possible as the matrices involved commute due to the Yang-Baxter equations (138) written up in terms the X matrices.) It should be noted, that these equations must hold for any solution regardless of its symmetry.
- For the eigenfunctions corresponding to a system of spin $1/2$ Fermions one has to solve the equations (128) ((136)) with the matrices of the regular representation of the permutation group replaced by the matrices of the representation $(-1)^{\mathcal{P}} \mathcal{T}^{(l,m)}(\mathcal{P})$. In practice one replaces the matrices of the representations $\mathcal{T}^{(l,m)}$ by the permutation operators themselvs, but solves the equations (155) ((156)) in the space of spin functions.
- By a second Bethe Ansatz (155) ((156)) can be reduced to two sets of algebraic equations, and the complete wavefunction can be reconstructed.

As a result we arrived at the following. To describe a system of N electrons being in a spin state $S^2 = S^z(S^z + 1)$, $S^z = N/2 - M$, the equations

$$Lk_j = 2\pi I_j - \sum_{\beta}^{M} 2\tan^{-1}\left(\frac{k_j - \Lambda_\beta}{c/2} \right) ,$$

$$\left(I_j = \frac{M}{2} \,(\mathrm{mod}\,1) \right) , \quad (214)$$

and

$$\sum_l^N 2\tan^{-1}\left(\frac{\Lambda_\alpha - k_l}{c/2}\right) = 2\pi J_\alpha + \sum_\beta^M 2\tan^{-1}\left(\frac{\Lambda_\alpha - \Lambda_\beta}{c}\right) ,$$

$$\left(J_\alpha = \frac{N + M + 1}{2} \,(\mathrm{mod}\,1)\right) . \qquad (215)$$

have to be solved. Once these equations are solved, the amplitude of finding the electrons at the positions $x_1, x_2, \ldots, x_N$ with spins $\sigma_1, \sigma_2, \ldots, \sigma_N$ is

$$f(\ldots, x_j, \sigma_j, \ldots) = C \sum_{\mathcal{P}} (-1)^{\mathcal{Q}}(-1)^{\mathcal{P}} S_{\{\mathcal{Q}m\}}(\mathcal{P}) \exp\left\{i\sum_j k_{\mathcal{P}j} x_{\mathcal{Q}j}\right\} . \qquad (216)$$

Here the permutation $\mathcal{Q}$ arranges the coordinates in increasing order

$$x_{\mathcal{Q}1} \le x_{\mathcal{Q}2} \le x_{\mathcal{Q}3} \le x_{\mathcal{Q}} \le \ldots \le x_{\mathcal{Q}N}; \qquad (217)$$

the spin function is

$$S_{\{\mathcal{Q}m\}}(\mathcal{P}) = \sum_{\mathcal{R}} \left(\prod_{\alpha<\beta} \left(\frac{i(\Lambda_{\mathcal{R}\alpha} - \Lambda_{\mathcal{R}\beta}) + c}{i(\Lambda_{\mathcal{R}\alpha} - \Lambda_{\mathcal{R}\beta}) - c}\right)^{1/2}\right) \times$$

$$\prod_{\gamma=1}^M S(k_{\mathcal{P}1}, k_{\mathcal{P}2}, \ldots, k_{\mathcal{P}N} | \Lambda_{\mathcal{R}\gamma}; \mathcal{Q}m_\gamma) , \qquad (218)$$

with

$$S(k_1, k_2, \ldots, k_N | \Lambda; m) = \left\{\prod_{l=1}^{m-1} \frac{i(k_l - \Lambda) - c/2}{i(k_l - \Lambda) + c/2}\right\} \frac{1}{i(k_m - \Lambda) + c/2} , \qquad (219)$$

and $\mathcal{Q}m_\gamma$ denoting the positions of the down spins in the series $\sigma_{\mathcal{Q}1}, \sigma_{\mathcal{Q}2}, \ldots, \sigma_{\mathcal{Q}N}$ in increasing order; the summations are extended over all possible $\mathcal{P}$ and $\mathcal{R}$ permutations of the wavenumbers k and rapidities Λ respectively; and the C is a normalisation constant. It is clear, that this wavefunction is nonzero, if all the ks and all the Λs are not equal. The energy and the momentum is

$$E = \sum_{j=1}^N k_j^2 , \quad P = \sum_{j=1}^N k_j . \qquad (220)$$

It is worth to note, that this solution is Galillei-invariant: shifting all momenta and rapidities by a given value $2\pi n/L$ ($n = $integer) will result in an other eigenstate with the total momentum shifted by $N2\pi n/L$.

Possible Generalisations In the previous sections we treated the δ-gas of identical particles. Our final aim was to find those equations, which correspond to spin 1/2 Fermions, but we tried to be as general as possiblle, and we specified the symmetry of the particles only if it was needed. Actually Eqs. (128) and (136) must hold for solutions corresponding to particles of any internal degrees of freedom. To find the solutions describing particles of a given symmetry, we have to replace in these equations the regular representation of the permutation group by an appropriatelly choosen irreducible one. This irreducible representation can be found by a reasoning analogous to that presented in Sects. 3.1 and 3.3. This way the procedure given in the previous sections is appropriate to discribe particles of different internal symmetries.

A further generalisation is based on the observation, that the $\hat{X}$ operators entering Eq. (156) are the scattering matrices of the particles. To see this, let us consider first the two particle case and write up the Ansatz for the complete wavefunction in the form

$$
\begin{aligned}
f(x_1,\sigma_1,\ x_2,\sigma_2) &= \mathcal{A}e^{ik_1x_1 + ik_2x_2}\left(A^{(12)}_{\sigma_1,\sigma_2}\theta(x_2-x_1) + A^{(21)}_{\sigma_1,\sigma_2}\theta(x_1-x_2)\right) \\
&= e^{ik_1x_1 + ik_2x_2}\left(A^{(12)}_{\sigma_1,\sigma_2}\theta(x_2-x_1) + A^{(21)}_{\sigma_1,\sigma_2}\theta(x_1-x_2)\right) \\
&- e^{ik_1x_2 + ik_2x_1}\left(A^{(12)}_{\sigma_2,\sigma_1}\theta(x_1-x_2) + A^{(21)}_{\sigma_2,\sigma_1}\theta(x_2-x_1)\right)\ ,
\end{aligned}
\tag{221}
$$

where $\mathcal{A}$ antisymmetrizes the function behind and θ is the step-function. (Note, that these coefficients are not identical to the coefficients a^Q_P.) It is not hard to see that to satisfy the continuity and discontinuity prescriptions (85), the coefficients should satisfy the equations

$$
A^{(21)}_{\sigma_1\sigma_2} = S^{\sigma'_1\sigma'_2}_{\sigma_1\sigma_2}A^{(12)}_{\sigma'_1\sigma'_2}\ ,
\tag{222}
$$

with the S-matrix

$$
S^{\sigma'_1\sigma'_2}_{\sigma_1\sigma_2} = \frac{(k_1-k_2)I^{\sigma'_1\sigma'_2}_{\sigma_1\sigma_2} + icP^{\sigma'_1\sigma'_2}_{\sigma_1\sigma_2}}{(k_1-k_2)+ic}\ ,\quad I^{\sigma'_1\sigma'_2}_{\sigma_1\sigma_2} = \delta^{\sigma'_1}_{\sigma_1}\delta^{\sigma'_2}_{\sigma_2},\quad P^{\sigma'_1\sigma'_2}_{\sigma_1\sigma_2} = \delta^{\sigma'_1}_{\sigma_2}\delta^{\sigma'_2}_{\sigma_1}\ .
\tag{223}
$$

It is apparent, that the S-matrix is actually *identical* to the operator $\hat{X}_{1,2}$ (as we have noted allready):

$$
S^{\sigma'_1\sigma'_2}_{\sigma_1\sigma_2} = \langle\sigma_1\sigma_2|\hat{X}_{1,2}|\sigma'_1\sigma'_2\rangle.
\tag{224}
$$

We look for the complete eigenfunction describing N particles in the form

$$
f(\ldots,x_j,\sigma_j,\ldots) = \mathcal{A}e^{i\sum k_jx_j}\sum_Q A^Q_{\sigma_1,\sigma_2,\ldots,\sigma_N}\theta(Q),
\tag{225}
$$

(where $\theta(Q)$ is one, if Q arranges the coordinates into increasing order, and zero otherwise). It should not be hard to show, that the continuity and discontinuity prescriptions (111) will connect the different coefficients by equations analogous to (223) (and also to (115) and (116)). The periodic boundary conditions will lead to the equations (156) ([Andrei 1992]). Based on the observation, that in all this procedure the actual form of the S-matrix (what in the case of the δ-gas of spin 1/2 Fermions happened to be the $\hat{X}_{i,j}$) does not play a role, we may say the following: Suppose, that in a problem we have the S-matrices characterising the scattring of the particles (of some symmetry, not necessarily Fermions). If the many-particle scattrings factorize into products of two-particle scatterings, and the two particle S-matrices satisfy the Yang-Baxter equations

$$S_{j,k}S_{i,k}S_{i,j} = S_{i,j}S_{i,k}S_{j,k}\ , \quad (S_{i,j}S_{j,i} = 1)\ , \tag{226}$$

than the periodic boundary conditions lead to equations of the type (156).

$$e^{ik_j L}S_{j-1,j}\,S_{j,j-2}\ldots S_{j,1}S_{j,N}\,S_{N,N-1}\ldots S_{j,j+1}\,A = A\ . \tag{227}$$

This equation has the meaning: taking a particle around the ring the wavefunction picks up a phase due to the momentum, and it changes also due to the scattering of the particle on the others, newerthless the total change must be zero.

4   Treatment of the BA Equations of the Heisenberg Chain

In the previous sections we have seen the method – the coordinate space Bethe Ansatz – by which the diagonalization of certain 1D Hamiltonians can be reduced to the solution of certain sets of nonlinear equations, the so called Bethe Ansatz equations. In this section we present a method to solve these BA equations, and analyze the obtained solution. The application of this method may need model-specific considerations (just as the BA itself) but we belive the main ideas can be illuminated by following the sulution of the BA equations for the *isotropic* (XXX) *Heisenberg chain*. First we discuss the solutions corresponding to a small (non-macroscopic) number of turned-down spins (particles), than we describe the ground state what is a state with no magnetization (the number of 'particles' is half the number of the sites), and finally we discuss the low energy excited states. (References for the treatment of the anisotropic case are: [Johnson et al 1973, Yang and Yang 1966, Woynarovich 1982, Babelon et al 1983, Virosztek and Woynarovich 1984]. An alternative treatment to find the spectrum of the XYZ chain is presented by Klümper and Zittartz (1988).)

For the $\rho = 1$ isotropic case the BA equations, the energy and the momentum corresponding to r turned-down spins (particles) read:

$$Nk_\alpha = 2\pi\lambda_\alpha + \sum_{\beta(\neq\alpha)}\psi_{\alpha\beta}\,, \quad (\alpha = 1, 2, \ldots, r)\,, \quad (\lambda_\alpha = \text{integers})\,, \tag{228}$$

$$\cot\frac{1}{2}\psi_{\alpha\beta} = \frac{1}{2}\left(\cot\frac{1}{2}k_\alpha - \cot\frac{1}{2}k_\beta\right)\,, \tag{229}$$

$$E^{(r)}(k_1, k_2, \ldots, k_r) = \sum_{\alpha}^{r}(\cos k_\alpha - 1), \quad P^{(r)}(k_1, k_2, \ldots, k_r) = \sum_{\alpha}^{r} k_\alpha. \quad (230)$$

It is convenient to parametrize the wavenumbers k by the rapidities x defined through

$$x = \cot \frac{k}{2}. \quad (231)$$

Fixing $0 \leq \operatorname{Re} k < 2\pi$ and $-\pi \leq \operatorname{Re} \psi \leq \pi$ we have

$$k_\alpha = \pi - 2\tan^{-1} x_\alpha, \quad \psi_{\alpha\beta} = \pi\operatorname{sgn}\left(\operatorname{Re}(x_\alpha - x_\beta)\right) - 2\tan^{-1}\frac{x_\alpha - x_\beta}{2}, \quad (232)$$

(where we understand $-\pi \leq \operatorname{Re}\left(\tan^{-1} x\right) \leq \pi$). Due to (232) the BA equatins read

$$N2\tan^{-1} x_\alpha = 2\pi J_\alpha + \sum_{\beta}^{r} 2\tan^{-1}\frac{x_\alpha - x_\beta}{2},$$

$$\left(J_\alpha = \frac{N - r + 1}{2} \,(\operatorname{mod} 1)\right), \quad (233)$$

and

$$E^{(r)} = -\sum_{\alpha}^{r}\frac{2}{1 + x_\alpha^2}, \quad P^{(r)} = r\pi - \sum_{\alpha}^{r} 2\tan^{-1} x_\alpha. \quad (234)$$

4.1   States with a Non-macroscopic r

$r = 1$. The simplest not completely ferromagnetic states are those with one spin turned down, i.e. with one particle (spin-wave). The rapidity of this particle is real, corresponding to a simple plain-wave.

$r = 2$. The simplest two-particle states are those with two real rapidities. These are scattering-states of the two particles. Eqs. (48) have, however, complex solutions too. In these solutions the two rapidities must be (in order to have the energy and momentum real) the complex conjugate of each other: $x_1 = x$, $x_2 = x^*$. Their equations read:

$$N2\tan^{-1} x = 2\pi J_1 + 2\tan^{-1}\frac{x - x^*}{2}, \quad (235)$$

$$N2\tan^{-1} x^* = 2\pi J_2 + 2\tan^{-1}\frac{x^* - x}{2}, \quad (J_{1,2} = \text{half} - \text{integers}). \quad (236)$$

(Here and in the following we suppose, that N is even.) Taking the sum and the difference of these equations we have

$$N\left(2\tan^{-1}\frac{\operatorname{Re} x}{1 + \operatorname{Im} x} + 2\tan^{-1}\frac{\operatorname{Re} x}{1 - \operatorname{Im} x}\right) = 2\pi\left(J_1 + J_2\right), \quad (237)$$

and

$$N \frac{1}{i} \ln \frac{(1 - \mathrm{Im}\,x)^2 + (\mathrm{Re}\,x)^2}{(1 + \mathrm{Im}\,x)^2 + (\mathrm{Re}\,x)^2} = 2\pi \left(J_1 - J_2\right) + \frac{2}{i} \ln \frac{1 - \mathrm{Im}\,x}{1 + \mathrm{Im}\,x}. \tag{238}$$

Exponentiating (238) yields

$$\frac{1 - \mathrm{Im}\,x}{1 + \mathrm{Im}\,x} = e^{i\pi(J_1 - J_2)} \left[\frac{(1 - \mathrm{Im}\,x)^2 + (\mathrm{Re}\,x)^2}{(1 + \mathrm{Im}\,x)^2 + (\mathrm{Re}\,x)^2} \right]^{N/2}. \tag{239}$$

From here its obvious, that

$$\mathrm{Im}\,x = 1 + \delta \tag{240}$$

with

$$\delta = 2 e^{i\pi(J_1 - J_2)} \left[\frac{(\mathrm{Re}\,x)^2}{4 + (\mathrm{Re}\,x)^2} \right]^{N/2} \tag{241}$$

in leading order. The $\mathrm{Re}\,x$ is determined by the equation

$$N 2\tan^{-1} \frac{\mathrm{Re}\,x}{2} = 2\pi \left(J_1 + J_2 + \frac{N}{2} \mathrm{sgn}(\delta \cdot \mathrm{Re}\,x) \right). \tag{242}$$

Such a complex is called *2-string*. It corresponds to a bound pair of the two particles:

$$a(n_1, n_2) \propto \begin{cases} e^{i \mathrm{Re}\,k(n_1 + n_2)} \cosh \left(\mathrm{Im}\,k \left(N/2 - (n_2 - n_1) \right) \right) \ (\delta > 0), \\ e^{i \mathrm{Re}\,k(n_1 + n_2)} \sinh \left(\mathrm{Im}\,k \left(N/2 - (n_2 - n_1) \right) \right) \ (\delta < 0). \end{cases} \tag{243}$$

The energy and momentum of such a bound pair is

$$E = \frac{1}{2}(\cos p - 1), \quad p = \pi - 2\tan^{-1} \frac{\mathrm{Re}\,x}{2}. \tag{244}$$

(It is worth to note, that the energy of a 2-string is allways higher, than the energy of two free particles with the same total momentum. Another piont is, that these bound pairs exist although the interaction is repulsive: this is possible as the effective mass of the particles can be negative.)

$r = 3$. The states with three particles can be of three type. One is the scattering state of the particles. The other consist of a free particle and a bound pair. Of the third type are the bound states of the three particles. These are described by the so called *3-strings*. The three rapidities in a 3-string are given as

$$x_1 = x + 2i + \delta, \quad x_2 = x, \quad x_3 = x - 2i + \delta^*. \tag{245}$$

Here x is real and δ is a complex number exponentially small in N. The energy and momentum of a 3-string is

$$E = \frac{1}{3}(\cos p - 1), \quad p = \pi - 2\tan^{-1} \frac{x}{3}. \tag{246}$$

m-strings. Complex rapidity configurations of the form

$$x_l = x + i(m + 1 - 2l) + \delta_l, \quad (l = 1, 2, \ldots, m) \tag{247}$$

with x being real and the δ_l being exponentially small in N, are called m-strings. Such complexes correspond to bound states of m particles with an energy and momentum

$$E = \frac{1}{m}(\cos p - 1), \quad p = \pi - 2\tan^{-1}\frac{x}{m}. \tag{248}$$

(Its a matter of elementary calculations to see, that two strings of lengths m_1 and m_2, and momenta p_1 and p_2 are of lower energy than one $(m_1 + m_2)$-string with a momentum $p_1 + p_2$.) In a solution corresponding to $r \ll N$ all kind of strings can occour whichs lengths are compatible with r: all combinations of different strings can occur for which

$$r = \sum m\, n(m), \tag{249}$$

with $n(m)$ being the number of m-strings. We have to note, that the complex rapidities (the complex elements of an x set obtained as a solution of (233)) form certainly strings of different length and centers as long as the number of particles are not macroscopic. If a string consists of a macroscopic number of rapidities ($m \propto N$), its form will deviate from (247) ([Sutherland 1995]). If a state is in the vicinity of the ground-state (in the sense, that there are a nonmacroscopic number of *excitations*) the longest strings to occur are the 2-strings (see Sec. (4.2)). Strings of different length with finite density are present at finite temperatures ([Takahashi 1971, Gaudin 1971, Takahashi and Suzuki 1972]).

4.2   States Near to the Ground-state

The states discussed in the previous subsection are of relatively simple structure, but of high energy. Now we want to find the ground state and the low energy excited states. We have seen, that states with real wavenumbers (1-strings) are of lower energy, than those with different longer strings. This makes us think (but does not prove), that in the ground-state all the rapidities are real. Actually Yang and Yang (1966) proved, that the ground state is the state with $r = N/2$ real rapidities. First we describe the solution corresponding to this state, than we show a systematic way, how the states 'slightly' deviating from the ground-state can be found.

The Ground-State We write the BA equations (233) in the form

$$\frac{J_\alpha}{N} = \frac{1}{2\pi}\left\{ 2\tan^{-1}x_\alpha - \frac{1}{N}\sum_{\beta}^{r} 2\tan^{-1}\frac{x_\alpha - x_\beta}{2} \right\}. \tag{250}$$

The set J_α corresponding to the ground-state is a set of consecutive integers (or half-integers) centered around the origin:

$$J_{\alpha+1} = J_\alpha + 1, \quad J_1 = -J_r, \quad \left(J_r = \frac{r-1}{2} \right).$$ (251)

Introducing the 'counting function'

$$z(x) = \frac{1}{2\pi} \left\{ 2\tan^{-1} x - \frac{1}{N} \sum_\beta^r 2\tan^{-1} \frac{x - x_\beta}{2} \right\},$$ (252)

(250) reads

$$z(x_\alpha) = \frac{J_\alpha}{N}.$$ (253)

We suppose, this relation can be inverted making possible to consider x_α as a function of J_α/N:

$$x_\alpha = x \left(\frac{J_\alpha}{N} \right).$$ (254)

Next we use, that

$$\frac{1}{N} \sum_{n=n_1}^{n_2} f \left(\frac{n}{N} \right) = \int_{(n_1 - \frac{1}{2})/N}^{(n_2 + \frac{1}{2})/N} f(z)\, dz$$ (255)

in leading order in N, provided f is a smuth function. This applied to our case reads:

$$\frac{1}{N} \sum_\alpha g(x_\alpha) = \int_{-B}^{B} g(x)\rho(x)\, dx, \quad \text{with} \quad \rho(x) = \frac{dz}{dx},$$ (256)

and

$$z(\pm B) = \pm \frac{r}{2N}, \quad \text{i.e.:} \quad \int_{-B}^{B} \rho = \frac{r}{N}.$$ (257)

Taking the derivative of (252) and applying (256) leads to the integral equation

$$\rho(x) = \frac{1}{2\pi} \left\{ \frac{2}{1+x^2} - \int_{-B}^{B} \frac{4}{4 + (x - x')^2} \rho(x')\, dx' \right\}.$$ (258)

Integrating this equation from $-\infty$ to ∞ yields

$$\int_{B}^{\infty} \rho = \frac{1}{2} - \frac{r}{N} = \frac{S^z}{N}.$$ (259)

As in the ground-state $r = N/2$ ($S^z = 0$), B should be taken to ∞ and the ground-state distribution of the rapidities $\rho_0(x)$ is determined by the equation

$$\rho_0(x) = \frac{1}{2\pi}\left\{\frac{2}{1+x^2} - \int_{-\infty}^{\infty} \frac{4}{4+(x-x')^2}\rho_0(x')\,dx'\right\}. \tag{260}$$

This integral equation is solved by Fourier transformation:

$$\rho_0(x) = \frac{1}{4}\frac{1}{\cosh\frac{\pi x}{2}}, \quad \text{i.e.:} \quad z_0(x) = \frac{1}{\pi}\tan^{-1}\left(\tanh\frac{\pi x}{4}\right). \tag{261}$$

The ground-state energy is

$$E_0 = -\sum_{\alpha}\frac{2}{1+x_\alpha} = -N\int_{-\infty}^{\infty}\frac{2}{1+x^2}\rho(x) = -N2\pi\ln 2. \tag{262}$$

Excited States The ground-state is the only state, in which $r = N/2$ and all of the rapidities are real. We look for the solutions corresponding to excited states by allowing for *holes* in the real rapidity disztribution and introducing *complex rapidities* too. As in the excited states the magnetization can be different from zero, the number of rapidities r can be smaller than $N/2$. Our strategy is as follows: we write up and solve an integral equation for the distribution of the real rapidities. In the obtained real-root-density the positions of the holes and complex rapidities appear as parameters. By means of this density we eliminate the real rapidities from the equations of the holes and complex rapidities, and this way we arrive at a set of equations, which relates the parameters of the excitations only. For an easy reference in the following we denote by x with Greek indeces the real rapidities only, and the complex rapidities will be denoted by y with Latin indeces. Because of the analytic properties of the $\tan^{-1}$ function, we will have to distinguish between the ys with $|\text{Im}\,y| < 2$ and $|\text{Im}\,y| > 2$. The first we call 'close', the second we call 'wide' roots.

The distribution of real rapidities. Similarly to the treatment of the ground-state, we define the counting function $z(x)$. We have to be, however, carefull. As the $\tan^{-1}$ function for complex arguments has branching points and cuts, in the equations of the real rapidities, i.e. in

$$\frac{J_\alpha}{N} = \frac{1}{2\pi}\left\{2\tan^{-1}x_\alpha - \frac{1}{N}\sum_\beta 2\tan^{-1}\frac{x_\alpha - x_\beta}{2} - \frac{1}{N}\sum_l 2\tan^{-1}\frac{x_\alpha - y_l}{2}\right\} \tag{263}$$

some of the J_α can be equal. (This can be seen by observing that the function

$$\frac{1}{2\pi}\left\{2\tan^{-1}x - \frac{1}{N}\sum_\beta 2\tan^{-1}\frac{x - x_\beta}{2} - \frac{1}{N}\sum_l 2\tan^{-1}\frac{x - y_l}{2}\right\} \tag{264}$$

has $-2\pi/N$ jumps at $\mathrm{Re}\, y$ if $\mathrm{Im}\, y > 2$. (Actually the value of these jumps are $-4\pi/N$ as the ys form complex conjugate pairs.)) To resolve this problem we define the counting function as

$$z(x) = \frac{1}{2\pi}\left\{ 2\tan^{-1}x - \frac{1}{N}\sum_{\beta} 2\tan^{-1}\frac{x-x_{\beta}}{2} - \frac{1}{N}\sum_{l} 2\tan_{c}^{-1}\frac{x-y_{l}}{2}\right\}, \quad (265)$$

where

$$2\tan_{c}^{-1}y = \begin{cases} 2\tan^{-1}y, & \text{if } |\mathrm{Im}\, y| < 1, \\ 2\tan^{-1}y - \pi\mathrm{sgn}(\mathrm{Re}\, y), & \text{if } |\mathrm{Im}\, y| > 1. \end{cases} \quad (266)$$

By this definitions (263) reads:

$$\frac{J'_{\alpha}}{N} = z(x_{\alpha}), \quad \text{with} \quad J'_{\alpha} = J_{\alpha} + \sum_{|\mathrm{Im}y|>2}\frac{1}{2}\mathrm{sgn}(x_{\alpha} - \mathrm{Re}\, y). \quad (267)$$

(Note, that $J'_{\alpha} - J_{\alpha} =$ integer as the ys form complex conjugate pairs.) If the number of excitations (the number of ys) is non-macroscopic, the $z(x)$ is a monotonic function, and to have all x_{α} different, all of the J'_{α} should be different. As $N\{z(\infty) - z(-\infty)\} > n(x)$ (here $n(x)$ is the number of the real rapidities) after choosing the J'_{α} set, there will remain numbers, J'_{h}, which are integers or half-integers as required by r, $Nz(\infty) > J'_{h} > Nz(-\infty)$, but are not elements of the J'_{α} set. The holes are defined by this numbers:

$$\frac{J'_{h}}{N} = z(x_{h}). \quad (268)$$

As the union of the J'_{α} and J'_{h} sets is an equidistant series of integers or half-integers, now the summation formula is:

$$\frac{1}{N}\sum_{\alpha} g(x_{\alpha}) = \int_{-\infty}^{\infty} g(x)\rho(x)\,dx - \frac{1}{N}\sum_{h} g(x_{h}), \quad (269)$$

where again

$$\rho(x) = \frac{dz}{dx}. \quad (270)$$

(Note, that now we took for the integration limits $\pm\infty$, later we shall determine the number of holes $n(h)$ accordingly.) Again we obtain the integral equation for the ρ by taking the derivative of (265) and applying (269):

$$\rho(x) = \frac{1}{2\pi}\left\{ \frac{2}{1+x^2} - \int_{-\infty}^{\infty}\frac{4}{4+(x-x')^2}\rho_0(x')\,dx' \right.$$

$$\left. + \frac{1}{N}\sum_{h}\frac{4}{4+(x-x_h)^2} - \frac{1}{N}\sum_{l}\frac{4}{4+(x-y_l)^2}\right\}. \quad (271)$$

Just as in the case of the ground-state, also this equation can be solved by Fourier transformation:

$$\rho(x) = \rho_0(x) + \frac{1}{N} \sum_h \rho_1(x - x_h) + \frac{1}{N} \sum_l \rho_2(x - y_l), \qquad (272)$$

with

$$\rho_1(x) = \frac{1}{2\pi} \int\limits_{-\infty}^{\infty} \frac{e^{-|\omega|}}{2 \cosh \omega} e^{i\omega x}, \qquad (273)$$

and

$$\rho_2(x - y) = -\frac{1}{2\pi} \int\limits_{-\infty}^{\infty} \frac{t(\omega, y) e^{+|\omega|}}{2 \cosh \omega} e^{i\omega(x-y)}, \qquad (274)$$

where

$$t(\omega, y) = \begin{cases} 2 \sinh 2\omega \, \theta(-\omega), & \text{if } \operatorname{Im} y > 2, \\ e^{-2|\omega|}, & \text{if } |\operatorname{Im} y| < 2, \\ -2 \sinh 2\omega \, \theta(\omega), & \text{if } \operatorname{Im} y < -2, \end{cases} \qquad (275)$$

(with $\theta(\omega)$ being the step-function). Now we can deteremine the number of holes. On one side we know that $N \int \rho$ must be the sum $n(x) + n(h)$, on the other side we can calculate this integral. This way we get

$$n(h) = N - 2n(x) - n(c) = 2S^z + n(c) + 2n(w), \qquad (276)$$

where $n(c)$ and $n(w)$ are the numbers of the close and wide roots, respectively. *The complex rapidities.* The complex rapidities are determined by the equations

$$N \tan^{-1} y_l = J_h + \sum_\beta 2 \tan^{-1} \frac{y_l - x_\beta}{2} + \frac{1}{N} \sum_m 2 \tan^{-1} \frac{y_l - y_m}{2}. \qquad (277)$$

If we replace the summation over the real x by the appropriate integral, we obtain the relations

$$\left. \begin{array}{l} N 2 \tan^{-1} \left(\tanh \frac{\pi y_l}{4} \right) \ (\bmod 2\pi) \ (\text{if } |\operatorname{Im} y_l| < 2) \\ \pi(N - r) \operatorname{sgn}(\operatorname{Re} y_l) \ (\bmod 2\pi) \quad (\text{if } |\operatorname{Im} y_l| > 2) \end{array} \right\} = 2\pi J_l +$$

$$+ \int\limits_{-\infty}^{\infty} 2 \tan_c^{-1} \frac{y_l - x}{2} \left(\sum_h \rho_1(x - x_h) + \sum_m \rho_2(x - y_m) \right)$$

$$- \sum_h 2 \tan^{-1} \frac{y_l - x_h}{2} + \sum_m 2 \tan^{-1} \frac{y_l - y_m}{2} . \qquad (278)$$

It is apparent, that for the close roots the imaginary part of the l.h.s. of (278) is proportional to N. For this (278) can be satisfied only if the close roots form pairs $y^\pm$ with the property

$$y^+ - y^- = 2i + 2\delta, \quad \delta \propto \exp\left\{ -N \operatorname{Im} \left(2 \tan^{-1} \left(\tanh \frac{\pi y_l}{4} \right) \right) \right\}. \qquad (279)$$

Note, that the members of such a close pair need not be complex conjugated to each other. If y^+ and y^- are the complex conjugate of each other, they form a 2-string. If y^+ and y^- is not a complex conjugate pair, there must be another close pair consisting of ys, which are complex conjugated to the y^+ and y^-. Such a configuration is often called a 'quartet'. The wide roots need not form special configurations. Actually (278) can be transformed into a more tractable form: we can eliminate the terms with imaginary part $\propto N$ by summing the two equations belonging to the two members of a close pair. The resulting set of equations take an extremly symple form, if we introduce a set of new variables ξ to represent the y set. If

$$
\begin{aligned}
y^\pm &= \xi \pm i && \text{if } |\mathrm{Im}\,\xi| < 1\,, \\
y &= \xi + i\mathrm{sgn}(\mathrm{Im}\,\xi) && \text{if } |\mathrm{Im}\,\xi| > 1\,,
\end{aligned} \tag{280}
$$

and we neglect the exponentially small δs, after a lengthy but straightforward manipulation we find, that this set of new variables satisfys the equations

$$
\sum_h 2\tan^{-1}(\xi_n - x_h) = 2\pi J_n + \sum_m 2\tan^{-1}\frac{\xi_n - \xi_m}{2}\,,
$$
$$
\left(J_n = \frac{n(h) + n(\xi) + 1}{2} \,(\mathrm{mod}\,1) \right)\,. \tag{281}
$$

Here $n(\xi)$ is the number of ξs: $n(\xi) = n(c)/2 + n(w)$, and the new quantumnumbers J_n are constrtucted by absorbing all terms, which are integer multiples of π, into the term $2\pi J_n$.

Equations for the holes. It is not hard to see, that in (272)

$$
\sum_l \rho_2(x - y_l) = -\frac{1}{2\pi} \sum_m \frac{2}{1 + (x - \xi_m)^2}\,. \tag{282}
$$

The density (280) can be integrated in a closed form yielding

$$
z(x) = \frac{1}{\pi}\tan^{-1}\left(\tanh\frac{\pi x}{4} \right) + \frac{1}{2\pi N}\sum_{h'}\varphi(x - x_{h'}) - \frac{1}{2\pi N}\sum_m 2\tan_c^{-1}(x - \xi_m)\,, \tag{283}
$$

with

$$
\varphi(x) = \frac{1}{i}\ln\frac{\Gamma\left(\frac{1}{2} - \frac{ix}{2}\right)\Gamma\left(1 + \frac{ix}{2}\right)}{\Gamma\left(\frac{1}{2} + \frac{ix}{2}\right)\Gamma\left(1 - \frac{ix}{2}\right)}\,, \tag{284}
$$

where Γ is the gamma function. Eq. (268) is than equivalent to

$$
N2\tan^{-1}\left(\tanh\frac{\pi x_h}{4} \right) = 2\pi J_h - \sum_{h'}\varphi(x_h - x_{h'}) + \sum_m 2\tan^{-1}(x_h - \xi_m)\,,
$$
$$
\left(J_h = \frac{N/2 + n(h)/2 - n(\xi) + 1}{2} \right)\,, \tag{285}
$$

what, for later purposes, we write in the form

$$Np_h = 2\pi I_h - \sum_{h'} \varphi(x_h - x_{h'}) + \sum_m 2\tan^{-1}(x_h - \xi_m),$$

$$\left(I_h = \frac{n(h)/2 + n(\xi) + 1}{2} \right), \tag{286}$$

with p_h defined below by (291).

The energy and the momentum. The energy of the state is calculated by the formula

$$E = -\sum_\alpha \frac{2}{1 + x_\alpha^2} - \sum_l \frac{2}{1 + y_l^2}. \tag{287}$$

Replacing the sum over the x_α by the appropriate integral, and substituting (280) for the y_l variables, after a straightforward calculation we find, that to the energy only the holes contribute:

$$E = E_0 + \sum_h \frac{\pi}{2} \frac{1}{\cosh(\pi x_h/2)}. \tag{288}$$

After a similar manipulation the formula

$$P = r\pi - \sum_\alpha 2\tan^{-1}x_\alpha - \sum_l 2\tan^{-1}y_l \tag{289}$$

for the momentum leads to the expression

$$P = \sum_h p_h + \frac{N\pi}{2} \pmod{2\pi}, \tag{290}$$

with

$$0 < p_h = 2\tan^{-1}\left(\tanh\frac{x_h\pi}{4}\right) + \frac{\pi}{2} < \pi. \tag{291}$$

Realising that $N\pi/2$ is actually P_0, the momentum of the ground-state, and that the energy associated with a hole

$$\epsilon(x_h) = \frac{\pi}{2} \frac{1}{\cosh(\pi x_h/2)} = \frac{\pi}{2} \sin p_h = \epsilon_h, \tag{292}$$

we can write

$$E - E_0 = \sum_h \epsilon_h, \quad P - P_0 = \sum_h p_h \tag{293}$$

Comments on the simplest excited states. The low energy excited states are described by the set of equations (281) and (286). In deriving these equations we have not fixd the quantum numbers connected with the holes and complex rapidities. Actually we have to chose the J_m and I_h quantum numbers so that (281) and (286) should have a solution. Once a solution is found the energy and momentum of the corresponding state is given by (291), (292) and (293). It is remarcable, that in the energy and the momentum only the holes have a dirrect contribution. It is also worth to note, that although the momentum contribution of one hole can not be larger than π, as the number of holes is allways even (for N even), the full $\{0, 2\pi\}$ region is covered by the possible momenta of the excited states. (Actually our expressions define the momenta up to a constant equal to π only: if we shift *all* momenta by π the total momentum will not be changed (mod 2π).) The number of holes $n(h)$, and the number of ξ variables $n(\xi)$ are connected to the magnetization S^z through the the relation

$$S^z = \frac{1}{2}n(h) - n(\xi), \qquad (294)$$

(see (276)) and we should also recall, that the solutions of the BA equations correspond to $S^2 = S^z(S^z + 1)$ states.

The simplest excited states are those with two holes. They are triplets if there is no ξ. Such solutions exist, if the two quantum numbers are $0 \leq I_1 < I_2 \leq N/2$. The states with one ξ are singlets. For two holes and one single ξ (281) has a solution only if the corresponding J is equal to zero. Than $\xi = (x_1 + x_2)/2$. Eq. (286) has a solution, if $1/2 \leq I_1 < I_2 \leq N/2 - 1/2$. For both the singlet and triplet states the energy is of the form

$$E - E_0 = \frac{\pi}{2}(\sin p_1 + \sin p_2). \qquad (295)$$

In the thermodynamic limit the two momenta become continuous parameters in the $\{0, \pi\}$ interwall, and the two continua of excitations become degenerated.

The states with four holes can be singlets (those with two ξs), triplets (if there is one ξ) and quadruplets (if there is no ξ). Also these continua become degenerated in the thermodynamic limit.

In general, if the number of holes is $n(h) = 2n$, S^z can take any integer value $0 \leq S^z \leq n$ depending on the number os the ξs.

We should note that if there are more ξs, we can not prescribe that these variables are real or complex. This is determined by the values of the x_h variables through the Eqs. (286). One can see, however, that if the number of holes is not large, the solutions for the ξs will take generic values, thus the possible y configurations are the 2-strings (real ξs), quartets (complex conjugate pairs of ξs with $|\text{Im}\xi| < 1$), and wide pairs (complex conjugate pairs of ξs with $|\text{Im}\xi| > 1$).

Nature of Excitations and the 'Higher Level Bethe Ansatz' As we have seen, the excited states are determined by the solutions of the equations

$$Np_h = 2\pi I_h - \sum_{h'} \varphi(x_h - x_{h'}) + \sum_m 2\tan^{-1}(x_h - \xi_m) \ ,$$

$$\left(I_h = \frac{n(h)/2 + n(\xi) + 1}{2}\right) \ , \qquad (296)$$

and

$$\sum_h 2\tan^{-1}(\xi_n - x_h) = 2\pi J_n + \sum_m 2\tan^{-1}\frac{\xi_n - \xi_m}{2},$$

$$\left(J_n = \frac{n(h) + n(\xi) + 1}{2} \,(\bmod\, 1)\right). \tag{297}$$

Once these equations are solved, the energy and the momentum of the state are given by

$$E - E_0 = \sum \epsilon_h, \quad \epsilon_h = \frac{\pi}{2}\sin p_h,$$

$$P - P_0 = \sum p_h, \quad p_h = 2\tan^{-1}\left(\tanh\left(\frac{\pi x_h}{4}\right)\right). \tag{298}$$

The variables ξ do not appear dirrectly in the energy and momentum, but they are needed to distinguish between the states of different internal symmetry: the spin of a state is given by

$$S^z = \frac{1}{2}n(h) - n(\xi), \quad S^2 = S^z(S^z + 1). \tag{299}$$

As only the holes contribute to the energy and momentum, and these quantities are the sums of the contributions of the individual holes, it is consitent to consider the holes as the *dressed particles* of the system. These particles are of spin 1/2 ([Faddeev and Takhtajan 1981]), and in a state the spin of $n(h) - n(\xi)$ particles point up, and the spin of the other $n(\xi)$ particles point down. The secular equations of the dressed particles are the Eqs. (296) and (297). The apparent similarity of these equations to (214) and (215) suggest to consider (296-297) as the Bethe Ansatz equations of the dressed particles. Equations of this type are often called the *higher level Bethe Ansatz* equations.

The above interpretation of (296-297) makes possible to reconstruct the scattering matrix of the dressed particles ([Andrei and Lowenstein 1980b, Korepin 1980, Destri and Lowenstein 1982]). We recall (227), according to which the scattering matrix $S_{\sigma_1\sigma_2}^{\sigma_1'\sigma_2'}$ of two particles of momenta p_1 and p_2 in case of periodic boundary conditions satifys the relations

$$e^{ip(x_1)L}S_{\sigma_1\sigma_2}^{\sigma_1'\sigma_2'}(x_1, x_2)\,A_{\sigma_1'\sigma_2'} = A_{\sigma_1\sigma_2},$$

$$e^{ip(x_2)L}S_{\sigma_1\sigma_2}^{\sigma_1'\sigma_2'}(x_2, x_1)\,A_{\sigma_1'\sigma_2'} = A_{\sigma_1\sigma_2}, \tag{300}$$

where now the σ are spin variables. Diagonalizing these relations and taking the logarithm we have that in a singlet/triplet state the momenta and the phaseshifts $\delta^{s/t}$ satisfy the equations

$$Np_i + \delta_{ij}^{s/t} = 2\pi\lambda_i, \quad (i, j = 1, 2 \text{ or } 2, 1). \tag{301}$$

In a triplet state there is no ξ, and the I_j numbers are integers. Comparing (296) for such a state and (301) yields

$$\delta_{12}^t = \varphi(x_1 - x_2). \tag{302}$$

In the two particle singlet states there is one ξ which has the value $\xi = (x_1 + x_2)/2$, and the I_j quantumnumbers are half integers. In this case the comparing (296) and (301) yields

$$\delta_{12}^s = \pi + \varphi(x_1 - x_2) - 2\tan^{-1}\frac{x_1 - x_2}{2} . \qquad (303)$$

From these phaseshifts the scattering matrix can be reconstructed:

$$S(x_1, x_2) = \frac{1}{2}(I + \mathcal{P})e^{i\delta^t} + \frac{1}{2}(I - \mathcal{P})e^{i\delta^s} , \qquad (304)$$

with I and $\mathcal{P}$ being, just as in (223), the unit and permutation matrices, respectively. Thus S written in a more usual form reads

$$S(x_1, x_2) = e^{i\varphi(x_1 - x_2)} \frac{(x_1 - x_2)I + 2i\mathcal{P}}{(x_1 - x_2) + 2i} . \qquad (305)$$

4.3   Finite Size Corrections

The fact, that the one particle energy-momentum dispersion (292) startes at 0 and terminates at π with finite slope $(\pm\pi/2)$ indicates that the model has a conformally invariant continuum limit. The conformal properties (the central charge and the conform dimensions) of the limiting model can be found by calculating the finite size correction to the ground-state energy and the fine structure of the excitation spectrum arising due to the quantisation of the momenta. The spectrum is of the *tower* structure

$$E - E_0 = -\frac{\pi v}{6N}c + \frac{2\pi v}{N}(x + n^+ + n^-), \quad P - P_0 = \frac{2\pi}{N}(s + n^+ - n^-), \quad (306)$$

where c is the central charge, v is the slope of the one particle dispersion (for our case $v = \pi/2$), x and s are related to the conform dimensis $\Delta^\pm$ by $\Delta^\pm = x \pm s$, and $n^\pm$ are nonnegative integers ([Cardy 1987]). In the following we show a method, by which the energy and momentum of the states corresponding to the apexes of the towers (the states for which n^+ and n^- are zero) can be calculated.

Our starting point is given by (233) and (234), and we look for the simplest states with r real rapidities. We choose for the J_α set the equidistant series of integers (or half-integers, as required by r) between $J^+ \leq (N - r - 1)/2$ and $J^- \geq -(N - r - 1)/2$ (with $J^+ - J^- = r - 1$), i.e.:

$$J^- = J_1, \quad J_\alpha + 1 = J_{\alpha+1}, \quad J_r = J^+ . \qquad (307)$$

The counting function and the density are introduced just as in the case of the ground-state:

$$z(x) = \frac{1}{2\pi}\left\{2\tan^{-1}x - \frac{1}{N}\sum_{\alpha=1}^{r}2\hat{T}_n^a{}^{-1}\frac{x - x_\alpha}{2}\right\}, \quad \rho(x) = \frac{dz}{dx} . \qquad (308)$$

Now we introduce

$$S(x) = \frac{1}{N} \sum_\alpha \delta(x - x_\alpha) - \rho(x), \tag{309}$$

and write

$$\rho(x) = \frac{1}{2\pi} \left\{ \frac{2}{1 + x^2} - \int_{-\infty}^{+\infty} \frac{4}{4 + (x - x')^2} (\rho(x') + S(x')) \right\}. \tag{310}$$

This equation, after some straightforward algebra leads to

$$\rho(x) = \rho_0(x) - \int_{-\infty}^{+\infty} K(x - x')S(x'), \tag{311}$$

with ρ_0 given by (261) and

$$K(x) = \frac{1}{2\pi} \int_{-\infty}^{+\infty} \frac{e^{i\omega x}}{1 + e^{2|\omega|}}. \tag{312}$$

The energy

$$E = -\sum_\alpha \frac{2}{1 + x_\alpha^2} \tag{313}$$

after some manipulation turns out to be

$$E = E_0 - N \int_{-\infty}^{+\infty} \epsilon(x)S(x), \tag{314}$$

with E_0 given by (262) and $\epsilon(x)$ given by (292). To calculate integrals of $S(x)$ the Euler-Maclaurin type formula

$$\frac{1}{2N} f\left(\frac{J_1}{N}\right) + \frac{1}{N} \sum_{\alpha=2}^{r-1} f\left(\frac{J_\alpha}{N}\right) + \frac{1}{2N} f\left(\frac{J_r}{N}\right) =$$

$$\int_{J_1/N}^{J_r/N} f(x) + \frac{1}{12N^2} \left(f'\left(\frac{J_n}{N}\right) - f'\left(\frac{J_1}{N}\right)\right) \tag{315}$$

(with f' being the derivative of f) can be used. By this formula

$$\int_{-\infty}^{\infty} g(x) S(x) = -\int_{B^+}^{\infty} g(x)\rho(x) + \frac{g(B^+)}{2N} + \frac{g'(B^+)}{12N^2\rho(B^+)}$$

$$- \int_{-\infty}^{B^-} g(x)\rho(x) + \frac{g(B^-)}{2N} - \frac{g'(B^-)}{12N^2\rho(B^-)}. \tag{316}$$

Here B^+ and B^- are the largest and smallest rapidities:

$$B^+ = x_r, \quad \text{i.e.:} \quad \frac{J^+}{N} = z(B^+), \quad \left(\frac{N-r}{N} - \frac{J^+}{N} = \int\limits_{B^+}^{+\infty} \rho(x') \right), \qquad (317)$$

and analogous expressions hold for B^-. The application of (316) to the energy (314) leads to

$$E - E_0 = N\epsilon^+ + N\epsilon^-, \qquad (318)$$

with

$$\epsilon^+ = \int\limits_{B^+}^{+\infty} \epsilon(x)\rho(x) - \frac{\epsilon(B^+)}{2N} - \frac{\epsilon'(B^+)}{12N^2\rho(B^+)}, \qquad (319)$$

and a symmetric expression for ϵ^-. The equation for the density (311) reads

$$\rho(x) = \rho_0(x) + \int\limits_{B^+}^{+\infty} K(x - x')\rho(x') - \frac{1}{2N}K(x - B^+) + \frac{K'(x - B^+)}{12N^2\rho(B^+)} +$$

$$+ \int\limits_{-\infty}^{B^-} K(x - x')\rho(x') - \frac{1}{2N}K(x - B^-) - \frac{K'(x - B^-)}{12N^2\rho(B^-)}. \quad (320)$$

It is obvious, that to calculate ϵ^+ it is enough to know $\rho(x)$ with a sufficient accuracy in the region $B^+ \leq x < \infty$. Consequently, when calculating ϵ^+, in (320) we may neglect the terms containing B^-, as they contribute to $\rho(x)$ in the region $B^+ \leq x < \infty$ terms of the order $o(1/N)$ only. The resulting equation

$$\rho(x) = \rho_0(x) + \int\limits_{B^+}^{+\infty} K(x - x')\rho(x') - \frac{1}{2N}K(x - B^+) + \frac{K'(x - B^+)}{12N^2\rho(B^+)} \qquad (321)$$

is of Wiener-Hoppf type, and can be solved by standard methods (as described by Woynarovich and Eckle (1987), and Hamer at al (1987)).

First we introduce the functions

$$\sigma^+(x) = \begin{cases} \rho(x + B^+) & \text{if } x > 0, \\ 0 & \text{if } x < 0, \end{cases} \quad \sigma^-(x) = \begin{cases} 0 & \text{if } x > 0, \\ \rho(x + B^+) & \text{if } x < 0. \end{cases} \qquad (322)$$

Note, that we can express all the quantities we need by the Fourier transforms of the $\sigma^\pm$, what we denote by $\tilde{\sigma}^\pm$. As for large arguments $\epsilon(x) \simeq \pi \exp\{-\pi x/2\}$,

$$\epsilon^+ = 2\pi^2 e^{-\pi B^+/2} \left\{ \tilde{\sigma}^+ \left(-\frac{i\pi}{2} \right) - \frac{1}{2\pi} \left(\frac{1}{2N} - \frac{\pi}{24N^2\rho(B^+)} \right) \right\}. \qquad (323)$$

The equation (317) determining B^+ is simply

$$2\pi\tilde{\sigma}^+(0) = \frac{N-r}{2N} - \frac{J^+}{N}. \qquad (324)$$

Finally

$$\rho(B^+) = \int\limits_{-\infty}^{+\infty} \left(\tilde{\sigma}^+(\omega) + \tilde{\sigma}^-(\omega)\right). \tag{325}$$

Substituting (322) into (321), it reads

$$\sigma^+(x) + \sigma^-(x) = \rho_0(x + B^+) + \int\limits_{-\infty}^{+\infty} K(x - x')\sigma^+(x') - \frac{K(x)}{2N} + \frac{K'(x)}{12N^2\rho(B^+)}. \tag{326}$$

After Fourier transformation we have

$$(1 - \tilde{K}(\omega))\tilde{\sigma}^+(\omega) + \tilde{\sigma}^-(\omega) = \tilde{\rho}_0(\omega)e^{i\omega B^+} - \left(\frac{1}{2N} - \frac{i\omega}{12N^2\rho(B^+)}\right)\tilde{K}(\omega). \tag{327}$$

In finding the solution an important point is that the $\tilde{\sigma}^+$ is analytic if $\mathrm{Im}\,\omega \leq 0$, while the $\tilde{\sigma}^-$ is analytic if $\mathrm{Im}\,\omega \geq 0$. The $(1 - \tilde{K}(\omega))$ can bee factorized into a product:

$$1 - \tilde{K}(\omega) = \frac{1}{1 + e^{-2|\omega|}} = \frac{1}{G(\omega)G(-\omega)}, \tag{328}$$

with

$$G(\omega) = \sqrt{2\pi} \frac{\exp\left\{-\frac{i\omega}{\pi}\left(1 - \ln\frac{i\omega}{\pi}\right)\right\}}{\Gamma\left(\frac{1}{2} + \frac{i\omega}{\pi}\right)}. \tag{329}$$

The $G(\omega)$ is analytic and free of zeroes in the lower half of the ω plain. After substituting $\tilde{\rho}_0$ and $\tilde{K}$, and writing (327) in the form

$$\frac{\tilde{\sigma}^+(\omega)}{G(\omega)} + G(-\omega)\tilde{\sigma}^-(\omega) = \tag{330}$$

$$= \frac{1}{2\pi}\frac{G(-\omega)e^{i\omega B^+}}{2\cosh\omega} - \frac{1}{2\pi}\left(\frac{1}{2N} - \frac{i\omega}{12N^2\rho(B^+)}\right)\left(G(-\omega) - \frac{1}{G(\omega)}\right),$$

it is clear, that the task is to split up the r.h.s. of this equation into parts analytic in the upper and lower half of the ω plain. It can be done, using, that the functions $f^\pm(\omega)$ defined by the integrals

$$f^\pm(\omega) = \mp\frac{1}{2\pi i}\int\limits_{-\infty}^{+\infty}\frac{f(\omega)}{\omega' - \omega \pm i0}d\omega' \tag{331}$$

have the required analytic properties, and their sum on the real axis is just $f(\omega)$. This way we get

$$\tilde{\sigma}^+(\omega) = -\frac{G(\omega)}{2\pi i} \int\limits_{-\infty}^{+\infty} \frac{G(-\omega')e^{i\omega' B^+}}{4\pi \cosh \omega'} \frac{1}{\omega' - \omega + i0} \, d\omega' +$$

$$+ \frac{G(\omega)}{2\pi i} \int\limits_{-\infty}^{+\infty} \frac{1}{2\pi} \left(\frac{1}{2N} - \frac{i\omega'}{12N^2\rho(B^+)} \right) \times$$

$$\left(G(-\omega') - \frac{1}{G(\omega')} \right) \frac{1}{\omega' - \omega + i0} \, d\omega' , \tag{332}$$

and

$$\tilde{\sigma}^-(\omega) = \frac{1}{2\pi i G(-\omega)} \int\limits_{-\infty}^{+\infty} \frac{G(-\omega')e^{i\omega' B^+}}{4\pi \cosh \omega'} \frac{1}{\omega' - \omega - i0} \, d\omega'$$

$$- \frac{1}{2\pi i G(-\omega)} \int\limits_{-\infty}^{+\infty} \frac{1}{2\pi} \left(\frac{1}{2N} - \frac{i\omega'}{12N^2\rho(B^+)} \right) \times$$

$$\left(G(-\omega') - \frac{1}{G(\omega')} \right) \frac{1}{\omega' - \omega - i0} \, d\omega' . \tag{333}$$

In evaluating $\tilde{\sigma}^\pm$ one has to calculate integrals of the type

$$\int\limits_{-\infty}^{+\infty} \frac{a + b\omega}{c + d\omega} \left(G(-\omega) - \frac{1}{G(\omega)} \right) . \tag{334}$$

This can be done using the assymptotic form of $G(\omega)$:

$$G(\omega) \simeq \exp\left\{ \frac{1}{24} \frac{\pi}{i\omega} \right\} + O\left(\frac{1}{\omega^3} \right) , \quad \text{if } \operatorname{Im}\omega \le 0 , \text{ and } |\omega|\gamma 1 . \tag{335}$$

As on the real axis $G(-\omega) - 1/G(\omega)$ decays exponentially for large $|\omega|$ (being equal to $G(\omega)\tilde{K}(\omega)$) we may cut the integral to run from $-R$ to R, where R is a large number. Than the integrand can be cut into two parts and in the two parts the integration contour can be deformed:

$$\int\limits_{-R}^{+R} \frac{a + b\omega}{c + d\omega} \left(G(-\omega) - \frac{1}{G(\omega)} \right) = -\int\limits_{C_1} \frac{a + b\omega}{c + d\omega} G(-\omega) - \int\limits_{C_2} \frac{a + b\omega}{c + d\omega} \frac{1}{G(\omega)} + p.c.$$

$$\tag{336}$$

where C_1 and C_2 are semicircles of radii R on the uper and lower half plain, respectively, and p.c. stands for the possible pole contribution. Now substituting

(335) and expanding in Taylor series of $1/\omega$ we obtain

$$\int_{-\infty}^{+\infty} \frac{a+b\omega}{c+d\omega}\left(G(-\omega) - \frac{1}{G(\omega)}\right) = -\oint_C \frac{a+b\omega}{c+d\omega}\left(1 - \frac{\pi}{24i\omega} + \frac{1}{2}\left(\frac{\pi}{24i\omega}\right)^2\right) + p.c.,$$

(337)

where the contour C is a circle of radius R. This integral can be calculated in an elementary way. Note, that although we used the assymptotic form of G, the result is exact.

Finally we arrive at the expressions

$$\tilde{\sigma}^+\left(-\frac{i\pi}{2}\right) = \frac{1}{(2\pi)^2}\left(G\left(-\frac{i\pi}{2}\right)\right)^2 e^{-\pi B^+/2} - \frac{G\left(-\frac{i\pi}{2}\right)}{(24)^2 N^2 \rho(B^+)} +$$
$$+ \frac{1}{2\pi}\left(\frac{1}{2N} - \frac{\pi}{24N^2\rho(B^+)}\right)\left(1 - G\left(-\frac{i\pi}{2}\right)\right),$$

(338)

$$2\pi\tilde{\sigma}^+(0) = \frac{1}{2N} + 2G(0)\left\{\frac{1}{2\pi}G\left(-\frac{i\pi}{2}\right)e^{-\pi B^+/2} - \left(\frac{1}{4N} + \frac{\pi}{(24)^2 N^2 \rho(B^+)}\right)\right\},$$

(339)

and

$$\rho(B^+) = \frac{1}{2}G\left(-\frac{i\pi}{2}\right)e^{-\pi B^+/2} - \frac{\pi}{24}\left(\frac{1}{2N} + \frac{\pi}{(24)^2 N^2 \rho(B^+)}\right).$$

(340)

(340) is equivalent to

$$\frac{1}{2\pi}G\left(-\frac{i\pi}{2}\right)e^{-\pi B^+/2}\frac{\pi}{24N^2\rho(B^+)} - \left(\frac{1}{4N} + \frac{\pi}{24N^2\rho(B^+)}\right)^2 = -\frac{1}{48N^2}.$$

(341)

This combined with (323) and (338) yields

$$\epsilon^+ = 2\pi^2\left\{\left(\frac{1}{2\pi}G\left(-\frac{i\pi}{2}\right)e^{-\pi B^+/2} - \left(\frac{1}{4N} + \frac{\pi}{24N^2\rho(B^+)}\right)\right)^2 - \frac{1}{48N^2}\right\}$$

(342)

Comparing (342), (339) and (324) we arrive at

$$\epsilon^+ = 2\pi^2\left\{\frac{(h^+)^2}{8N^2} - \frac{1}{48N^2}\right\}, \quad \left(h^+ = \frac{N-r-1}{2} - J^+\right).$$

(343)

(Here we used that $G(0) = \sqrt{2}$.)

An analogous calculation for ϵ^- yields

$$\epsilon^- = 2\pi^2\left\{\frac{(h^-)^2}{8N^2} - \frac{1}{48N^2}\right\}, \quad \left(h^- = \frac{N-r-1}{2} + J^-\right),$$

(344)

thus, using that $v = \pi/2$, the finite size correction to the energy is

$$E - E_0 = -\frac{\pi v}{6N} + \frac{2\pi v}{N}\left(\frac{(S^z)^2 + D^2}{2}\right),$$

(345)

with $S^z = (h^+ + h^-)/2$ being the magnetization, and $D = (h^+ - h^-)/2$ measuring the anisotropy of the rapidity distribution. Observing, that in (234)

$$\sum_{\alpha}^{r} 2\tan^{-1} x_\alpha = \sum_{\alpha}^{r} \frac{2\pi}{N} J_\alpha \,, \tag{346}$$

the momentum of the state is easyly obtained:

$$P - P_0 = \pi(D - S^z) - \frac{2\pi}{N} D S^z \,. \tag{347}$$

The $\pi(D - S^z)$ can be eliminated in the continuum limit by redefining the lattice so that one elementary cell contains two sites (i.e. one point of the continuum consists of two lattice sites). Finally we can see, that in (306)

$$c = 1, \quad x = \frac{(S^z)^2 + D^2}{2}, \text{ and } s = -D S^z \,, \tag{348}$$

i.e.:

$$\Delta^\pm = \frac{1}{2} (S^z \mp D)^2 \,. \tag{349}$$

We have to note, that by this method we can describe those towers, for which $S^z \geq |D|$. There are, however, towers, for which $S^z < |D|$. In these towers, even for the state corresponding to the apex, there are complex rapidities, and their description needs more subtle considerations ([Woynarovich 1987]).

Finally we have to note, that (315) ((316)) applied to the present problem takes into account the leading terms of a nonconvergent (asymptotic) expansion: as $\rho(B^+)$ is of the order of $1/N$, all the next terms not taken into account are of the order of $1/N$. Nevertheless the result is reliable, as it is well indicated by the numerical calculations for very long chains ([Woynarovich and Eckle 1987]), but also by results obtained by other methods ([Klümper and Pearce 1991, Pearce and Klümper 1991]).

References

Alkaraz F C, Barber M N, Batchelor M T, Baxter R J, Quispel G R W 1987; J.Phys.A **20** (1987) 6397

Andrei N 1992; The BAE are derived in this way in 'Integrabele Models in Condensed Matter Physics' by N. Andrei in ICTP Summer course on low-dimensional quantum field theories for condensed matter physicists (Trieste, 1992)

Andrei N, Furuya K, Lowenstein J H 1983; Rev.Mod.Phys. **55** (1983) 331

Andrei N, Lowenstein J H 1979; Phys.Rev.Lett. **43** (1979) 1698,

Andrei A, Lowenstein J H 1980a; Phys.Lett. **90B** (1980) 106

Andrei N, Lowenstein J H 1980b; Phys.Lett. **91B** (1980) 401

Babelon O, de Vega H J, Viallet C M 1993; Nucl.Phys. B **220** [FS8] (1983) 13

Bethe H 1931; Z.Physik **71** (1931) 205

Cardy J 1987; in Phase Transitions and Critical Phenomena, ed. C.Domb, J.L.Lebowitz; Academic, New York, 1987.

Choy T C, Haldane F D M 1982; Phys.Lett. **90A** (1982) 84
des Cloisoux J, Gaudin M 1966; J.Math.Phys. **7** (1966) 1384
Destri C, Lowenstein J H 1982; Nucl.Phys.B205[FS5] (1982) 369
Faddeev L D, Takhtajan L A 1981; Phys.Lett. **85A** (1981) 375
Fung M K 1981; J.Math.Phys. **22** (1981) 2017
Gaudin M 1971; Phys.Rev.Lett. **26** (1971) 1301
Hamer C J, Quispel G R W, Batchelor M T 1987; J.Phys. **A20** (1987) 5677;
Johnson J D, Krinsky S, McCoy B 1973; Phys.Rev. **A8** (1973) 2526
Klümper A, Pearce P A 1991; Phys.Rev.Lett. **66** (1991) 974,
Klümper A, Zittartz J 1988; Z. Phys. B **71** (1988) 495
Korepin 1980 V E; Theor. Math. Phys. 76 (1980) 165
Lieb E, Liniger W 1963; Phys.Rev. **130** (1963) 1605
Lieb E, Wu F Y 1968; Phys.Rev.Lett. **20** (1968) 1445
Orbach R 1958; Phys.Rev. **112** (1958) 309
Pearce P A, Klümper A; J.Stat.Phys. **64** (1991) 13
Shastry B S, Sutherland B 1990; Phys.Rev.Lett. **65** (1990) 243
Sutherland B 1995; Phys.Rev.Lett. **74** (1995) 816
Takahashi M 1971; Prog.Theor.Phys. **46** (1971) 401
Takahashi M, Suzuki M 1972; Prog.Theor.Phys. **48** (1972) 2187
Tsvelick A M, Wiegmann P B 1983; Advances in Physics **32** (1983) 453
Virosztek A, Woynarovich F 1984; J.Phys.A **17** (1984) 3029
Woynarovich F 1982; J.Phys.A **15** (1982) 2985
Woynarovich F 1987; Phys.Rev.Lett. **59** (1987) 259, (Erratum: Phys.Rev.Lett. **59** (1987) 1264.)
Woynarovich F, Eckle H-P 1987; J.Phys.A20 (1987) L97
Yang C N 1967; Phys.Rev.Lett. **19** (1967) 1312
Yang C N, Yang C P 1966; Phys.Rev. **150** (1966) 323; Phys.Rev. **150** (1966) 327; Phys.Rev. **151** (1966) 259
Zamolodchikov Al B 1990; Nucl.Phys.B **342** (1990) 695

Thermodynamical Bethe Ansatz and Condensed Matter

Minoru Takahashi

Institute for Solid State Physics, University of Tokyo,
Roppongi, Minato-ku, Tokyo 106, Japan

Abstract. The basics of the thermodynamic Bethe ansatz equation are given. The simplest case is repulsive delta function bosons, the thermodynamic equation contains only one unknown function. We also treat the XXX model with spin 1/2 and the XXZ model and the XYZ model. This method is very useful for the investigation of the low temperature thermodynamics of solvable systems.

1 Introduction

Thermodynamic Bethe ansatz equations were first introduced by Yang and Yang for the model of repulsive delta-function bosons. Later this method was extended to the other Bethe ansatz soluble models such as the XXX model, the XXZ model, the XYZ model, delta-function fermions, the Hubbard model, the t-J model and so on. In these theories the 0 of strings and holes play an essential role. The number of kinds of strings is generally infinite except in some special cases. We can construct a set of non-linear integral equations for these models. These are very useful for the investigation of the low-temperature thermodynamics of soluble models. Nowadays thermodynamic Bethe ansatz equations have been derived for almost all solvable one-dimensional quantum models.

In chapter 2, Yang and Yang's theory for bosons is introduced. Here the number of unknown functions is 1. In chapter 3 we treat the XXX model. Here the number of unknown functions is infinite. But we can treat this set of integral equations analytically in some special cases such as the infinite temperature limit, the zero temperature limit, and the high magnetic field limit. In chapter 3 we treat the XXZ model. The case $|\Delta| > 1$ is almost identical to the XXX model ($\Delta = 1$). On the other hand the case $|\Delta| < 1$ is more complicated, the number of kinds of strings is dependent on a parameter. If $(\cos^{-1}\Delta)/\pi$ is a rational number, the number of kinds of strings is finite and thermodynamic Bethe ansatz equations contain a finite number of unknown functions. In chapter 5 we treat the XYZ model in zero magnetic field. Here, the number of unknown functions is also dependent on parameters. The equations are obtained by a slight modification of XXZ model at $|\Delta| < 1$. In chapter 6 we review numerical calculations of these Bethe ansatz equations.

2   Repulsive Delta-Function Bosons

2.1   Bethe Ansatz Equations and Uniqueness of the Solution

Here we consider the system

$$\mathcal{H} = -\sum_{i=1}^{N} \frac{\partial^2}{\partial x_i^2} + 2c \sum_{i<j} \delta(x_i - x_j), \tag{1}$$

the problem of bosons interacting via a repulsive delta function potential with periodic boundary conditions. Lieb and Liniger showed that this model is solvable by Bethe ansatz and calculated the ground state energy and elementary excitations ([Lieb and Liniger (1963)], [Lieb (1963)]). The wave function is assumed to be a linear combination of $N!$ plane waves. One must determine $N!$ coefficients $A(P)$ with eigenvalue E at $x_1 \leq x_2 \leq ... \leq x_N$

$$f = \sum_P A(P) \exp[i(k_{P1}x_1 + k_{P2}x_2 + ... + k_{PN}x_N)],$$

$$A(P) = C\epsilon(P) \prod_{j<k}(k_{Pj} - k_{Pk} + ic),$$

$$E = \sum_{j=1}^{N} k_j^2. \tag{2}$$

Here P's are permutations and $\epsilon(P)$ is the parity of permutation P. The total momentum is given by

$$K = \sum_{j=1}^{N} k_j.$$

k_j's are called the quasi momenta. If we apply periodic boundary conditions, k_j's should be solutions of the following coupled equations

$$e^{ik_jL} = -\prod_{l=1}^{N}\left(\frac{k_j - k_l + ic}{k_j - k_l - ic}\right). \tag{3}$$

The logarithm of these equations is

$$k_jL = 2\pi I_j - \sum_l 2\hat{T}_n^{a-1}\left(\frac{k_j - k_l}{c}\right). \tag{4}$$

Here the I_j's form a set of distinct but otherwise arbitrary integers (half-odd integers) for odd(even) N. The quasi-momenta k_j are determined by eq. (4) if a set of quantum numbers I_j is given. The ground state energy is obtained from a Fredholm type integral equation ([Lieb and Liniger (1963), Lieb (1963)]). For this model Yang and Yang found a non-linear integral equation which gives the free energy at a given temperature ([Yang and Yang (1969)]). This result is important because it forms the basic theory for thermodynamics of other solvable models.

At first we should show that the solution of equation (4) is unique for a given set of $\bar{I}_j$'s. The equations (4) are equivalent to the following equations

$$\frac{\partial B(k_1, k_2, ..., k_N)}{\partial k_j} = 0,$$

$$B(k_1, k_2, ..., k_N) \equiv \frac{L}{2} \sum k_j^2 - 2\pi I_j k_j + \sum_{j<l} \theta_1(k_j - k_l),$$

$$\theta_1(x) = x\hat{T}_n^{a-1}(x/c) - \frac{c}{2}\ln(1 + (x/c)^2). \tag{5}$$

The extremal point of the function B with N variables is determined by the solution of (4). It gives an eigenstate of the Hamiltonian. Let us consider the following $N \times N$ matrix

$$B_{jl} = \frac{\partial^2 B}{\partial k_j \partial k_l} = \delta_{jl}(L + \sum_m \frac{2c}{c^2 + (k_j - k_m)^2}) - \frac{2c}{c^2 + (k_j - k_l)^2}. \tag{6}$$

This matrix is always positive definite because

$$\sum_{lj} u_l B_{lj} u_j = L \sum u_l^2 + \sum_{l<j} \frac{2c}{c^2 + (k_j - k_l)^2}(u_j - u_l)^2 \geq 0, \tag{7}$$

for an arbitrary real vector $\{u_j\}$. The function B is a concave function in N dimensional space. Thus a solution of equation (4) for a given set of $\{I_j\}$ is unique.

2.2   Holes of Quasi Momenta and Their Distribution Function

In the limit of $c \to \infty$ this set of eigenstates is complete. Let us consider the following function

$$h(k) = k + \frac{2}{L} \sum_j \hat{T}_n^{a-1} \frac{k - k_j}{c}. \tag{8}$$

The position of a hole is defined by

$$h(k_h) = \frac{2\pi}{L} \times \text{unoccupied (half - odd) integer.}$$

We define the distribution function of holes $\rho^h(k)$ such that the number of holes between k and $k + dk$ is $\rho^h(k)Ldk$ and the density of particles $\rho(k)$ such that the number of particles between k and $k + dk$ is $\rho(k)Ldk$. In the thermodynamic limit we have

$$2\pi \int^k \rho(t) + \rho^h(t)dt = h(k) = k + 2\int \hat{T}_n^{a-1} \frac{k - k'}{c} \rho(k')dk'. \tag{9}$$

Differentiating this yields

$$2\pi(\rho(k) + \rho^h(k)) = 1 + 2\int_{-\infty}^{\infty} \frac{c\rho(k')dk'}{c^2 + (k - k')^2}. \tag{10}$$

From equation (2) one obtains the energy and the particle number per unit length

$$e = \int_{-\infty}^{\infty} k^2 \rho(k)dk, \quad n = \int_{-\infty}^{\infty} \rho(k)dk. \tag{11}$$

The entropy of the distribution between k and $k + dk$ is the logarithm of number of orderings of $L\rho(k)dk$ particles and $L\rho^h(k)dk$ holes

$$\ln \frac{[L(\rho(k) + \rho^h(k))dk]!}{[L\rho(k)]![L\rho^h(k)]!} = Ldk[(\rho(k) + \rho^h(k))\ln(\rho(k) + \rho^h(k))$$
$$-\rho(k)\ln\rho(k) - \rho^h(k)\ln\rho^h(k)].$$

Here we use the Stirling formula $\ln(n!) \simeq n(\ln n - 1)$. Then the entropy per unit length is

$$s = \int_{-\infty}^{\infty} (\rho(k) + \rho^h(k))\ln(\rho(k) + \rho^h(k)) - \rho(k)\ln\rho(k) - \rho^h(k)\ln\rho^h(k)dk. \tag{12}$$

2.3   Thermodynamic Equilibrium

The free energy per unit length $f = e - Ts$ must be minimized under the condition that n is constant. f and n are functionals of $\rho(k)$ and $\rho^h(k)$. Thus we should determine $\rho(k)$ and $\rho^h(k)$ to minimize $f - An$. Next we look for parameter A such that n takes on its required value. In variational calculus A is called a Lagrange multiplier. At the minimum point, the variation of $f - An$ must be zero for any infinitesimal variation of functions

$$0 = \delta \int_{-\infty}^{\infty} (k^2 - A)\rho(k) - T\{(\rho(k) + \rho^h(k))\ln(\rho(k) + \rho^h(k))$$
$$-\rho(k)\ln\rho(k) - \rho^h(k)\ln\rho^h(k)\}dk$$
$$= \int \left\{k^2 - A - T\ln\left(\frac{\rho(k) + \rho^h(k)}{\rho(k)}\right)\right\}\delta\rho(k) - T\ln\left(\frac{\rho(k) + \rho^h(k)}{\rho^h(k)}\right)\delta\rho^h(k)dk. \tag{13}$$

One should note that $\delta\rho(k)$ and $\delta\rho^h(k)$ are not independent. From (10) we have

$$\delta\rho^h(k) = -\delta\rho(k) + \frac{1}{\pi}\int_{-\infty}^{\infty} \frac{c\delta\rho(k')dk'}{c^2 + (k - k')^2}.$$

Substituting this into (13), one obtains

$$0 = \int_{-\infty}^{\infty} \delta\rho(k)$$

$$\times \{k^2 - A - T\ln(\frac{\rho^h(k)}{\rho(k)}) - \frac{T}{\pi}\int_{-\infty}^{\infty} dq \frac{c}{c^2 + (k-q)^2}\ln(1 + \frac{\rho(q)}{\rho^h(q)})dq\}. \quad (14)$$

This equation must hold for any arbitrary infinitesimal change of the function $\rho(k)$. Thus

$$T\ln(\frac{\rho^h(k)}{\rho(k)}) = k^2 - A - \frac{T}{\pi}\int_{-\infty}^{\infty} dq \frac{c}{c^2 + (k-q)^2}\ln(1 + \frac{\rho(q)}{\rho^h(q)})dq, \quad (15)$$

must be satisfied at thermodynamic equilibrium. If we put
$\epsilon(k) = T\ln(\rho^h(k)/\rho(k))$, this becomes

$$\epsilon(k) = k^2 - A - \frac{T}{\pi}\int_{-\infty}^{\infty} dq \frac{c}{c^2 + (k-q)^2}\ln(1 + \exp(-\frac{\epsilon(q)}{T}))dq. \quad (16)$$

$\epsilon(k)$ can be determined by iteration. This function has the physical meaning as the excitation energy for an elementary excitation, as will be shown in next section. From equation (10) we have

$$2\pi\rho(k)(1 + \exp(\frac{\epsilon(k)}{T})) = 1 + 2c\int_{-\infty}^{\infty} \frac{\rho(q)dq}{c^2 + (k-q)^2}. \quad (17)$$

From this equation we can determine $\rho(k)$.

If equation (16) is differentiated with respect to the chemical potential A, one obtains a linear integral equation

$$\frac{\partial\epsilon(k, A)}{\partial A} = -1 + \int dq \frac{c}{c^2 + (k-q)^2} \frac{1}{1 + \exp(\frac{\epsilon(k)}{T})} \frac{\partial\epsilon(q, A)}{\partial A}. \quad (18)$$

Comparing this with (17) one finds that

$$\frac{\partial\epsilon(k, A)}{\partial A} = -2\pi\rho(k)(1 + \exp(\frac{\epsilon(k)}{T})) = -2\pi(\rho(k) + \rho^h(k)). \quad (19)$$

The entropy density (12) can be written

$$s = \int [(\rho + \rho^h)\ln(1 + \exp(-\epsilon(k)/T)) + \rho(k)\epsilon(k)/T]dk.$$

The free energy density is

$$f = \int (k^2 - \epsilon(k))\rho(k) - T(\rho(k) + \rho^h(k))\ln(1 + \exp(-\frac{\epsilon(k)}{T}))dk. \quad (20)$$

Substituting (16) into (20) one obtains

$$
f = \int [A + \frac{T}{\pi} \int \frac{dqc}{c^2 + (k-q)^2} \ln(1 + \exp(-\frac{\epsilon(q)}{T}))]\rho(k)
$$

$$
- T(\rho(k) + \rho^h(k)) \ln(1 + \exp(-\frac{\epsilon(k)}{T}))dk
$$

$$
= \int A\rho(k)dk - T \int \ln(1 + \exp(-\frac{\epsilon(k)}{T}))
$$

$$
\times [\rho(k) + \rho^h(k) - \frac{1}{\pi} \int \frac{c}{c^2 + (k-q)^2} \rho(q)dq]dk. \tag{21}
$$

Substituting (10) we have a very simple expression for the free energy density

$$
f = An - T \int \frac{dk}{2\pi} \ln(1 + \exp(-\frac{\epsilon(k)}{T})). \tag{22}
$$

The thermodynamic potential density $g = f - An$ is

$$
g(T, A) = -T \int \ln(1 + \exp(-\frac{\epsilon(k)}{T})) \frac{dk}{2\pi}. \tag{23}
$$

For a given temperature T and chemical potential A, $\epsilon(k)$ is determined through the non-linear integral equation (16) and obtain thermodynamic potential g from (23). All thermodynamic quantities are derived from $g(T, A)$ through thermodynamic relations

$$
n = -\frac{\partial g}{\partial A}, \quad s = -\frac{\partial g}{\partial T}, \quad e = g + An + Ts, ... \tag{24}
$$

The pressure p is $-g$.

2.4   Elementary Excitations

We consider the change of energy and momentum when one particle is moved from the thermodynamic equilibrium. Assume that the l-th particle is removed and one particle is added between the m-th and $m + 1$-th particle

$$
\{I_1 > I_2 > I_3 > ... > I_N\} \rightarrow
$$
$$
\{I_1 > I_2 > ... > I_m > I_l' > I_{m+1}...I_{l-1} > I_{l+1} > ... > I_N\}
$$

The Bethe ansatz equations for the original state are

$$
Lk_j = 2\pi I_j - \sum_{n=1}^{N} 2\hat{T}_n^{-1} \frac{k_j - k_n}{c} \quad j = 1, ..., N. \tag{25}
$$

For the excited state they are

$$Lk'_j = 2\pi I_j - 2\hat{T}_n^{a-1}\frac{k'_j - k'_l}{c} - \sum_{n \neq m} 2\hat{T}_n^{a-1}\frac{k'_j - k'_n}{c}, \quad j \neq l$$

(26)

$$Lk'_l = 2\pi I'_l - \sum_{n \neq m} 2\hat{T}_n^{a-1}\frac{k'_l - k'_n}{c}.$$

(27)

The change of momentum and energy are

$$\Delta K = k'_l - k_l + \sum_{n \neq l}(k'_n - k_n),$$

$$\Delta E = k'^2_l - k^2_l + \sum_{n \neq l}(k'^2_n - k^2_n)$$

(28)

Subtracting (25) from (26) we have

$$\Delta k_j L + \sum \frac{2c}{c^2 + (k_j - k_l)^2}(\Delta k_j - \Delta k_l)$$

$$= 2\hat{T}_n^{a-1}\frac{k_j - k_l}{c} - 2\hat{T}_n^{a-1}\frac{k_j - k'_l}{c}.$$

(29)

Then equation for the back-flow $J(k) = \rho(k)\Delta kL$ is

$$\left(1 + \frac{\rho^h(k)}{\rho(k)}\right)J(k; k'_l, k_l) - \int \frac{c}{\pi(c^2 + (k - k')^2)}J(k'; k'_l, k_l)dk'$$

$$= \frac{1}{\pi}\left(\hat{T}_n^{a-1}\frac{k - k_l}{c} - \hat{T}_n^a\frac{k - k'_l}{c}\right).$$

(30)

Energy and momentum changes are

$$\Delta K(k'_l, k_l) = k'_l - k_l + \int J(k; k'_l, k_l)dk,$$

$$\Delta E(k'_l, k_l) = k'^2_l - k^2_l + \int 2kJ(k; k'_l, k_l)dk.$$

(31)

It is clear that $\Delta K(k_l, k_l) = \Delta E(k_l, k_l) = 0$. The differentiation of these with respect to k'_l is

$$\frac{\partial \Delta K}{\partial k'_l} = 1 + \int u(k; k'_l)dk$$

$$\frac{\partial \Delta E}{\partial k'_l} = 2k'_l + \int 2ku(k; k'_l)dk,$$

(32)

where

$$u(k; k_l') \equiv \frac{\partial}{\partial k_l'} J(k; k_l', k_l),$$

$$(1 + \eta(k))u(k; k_l') - \int a(k - k')u(k'; k_l')dk' = a(k - k_l'). \tag{33}$$

Using this integral equation we find that $u(k; k_l')$ is given by the infinite series

$$u(k; k_l') = \frac{1}{1 + \eta(k)} \left[a(k - k_l') + \int dk_1 \frac{a(k - k_1)}{1 + \eta(k_1)} a(k_1 - k_l') \right.$$
$$\left. + \int \int dk_1 dk_2 \frac{a(k - k_1)}{1 + \eta(k_1)} \frac{a(k_1 - k_2)}{1 + \eta(k_2)} a(k_2 - k_l') + ... \right]. \tag{34}$$

Substituting this into (32) we get an expression of $\frac{\partial \Delta K}{\partial k_l'}$ and $\frac{\partial \Delta E}{\partial k_l'}$ as an infinite series

$$\frac{\partial \Delta K}{\partial k_l'} = 1 + \int dk_1 a(k_l' - k_1) \frac{1}{1 + \eta(k_1)}$$
$$+ \int \int dk_1 dk_2 a(k_l' - k_1) \frac{1}{1 + \eta(k_1)} a(k_1 - k_2) \frac{1}{1 + \eta(k_2)} + ...$$
$$\frac{\partial \Delta E}{\partial k_l'} = 2k_l' + \int dk_1 a(k_l' - k_1) \frac{1}{1 + \eta(k_1)} 2k_1$$
$$+ \int \int dk_1 dk_2 a(k_l' - k_1) \frac{1}{1 + \eta(k_1)} a(k_1 - k_2) \frac{1}{1 + \eta(k_2)} 2k_2 + ...$$

From these infinite series linear integral equations for $\frac{\partial \Delta K}{\partial k_l'}$ and $\frac{\partial \Delta E}{\partial k_l'}$ are obtained,

$$\frac{\partial \Delta K(k_l', k_l)}{\partial k_l'} = 1 + \int a(k_l' - k_1) \frac{1}{1 + \eta(k_1)} \frac{\partial \Delta K(k_1, k_l)}{\partial k_1},$$
$$\frac{\partial \Delta E(k_l', k_l)}{\partial k_l'} = 2k_l' + \int a(k_l' - k_1) \frac{1}{1 + \eta(k_1)} \frac{\partial \Delta E(k_1, k_l)}{\partial k_1}. \tag{35}$$

On the other hand we have equations for $2\pi(\rho(k) + \rho^h(k))$ and $\partial \epsilon(k)/\partial k$ from (10) and (16)

$$2\pi(\rho(k) + \rho^h(k)) = 1 + \int a(k - k') \frac{2\pi(\rho(k') + \rho^h(k'))}{1 + \eta(k')} dk',$$
$$\frac{\partial \epsilon(k)}{\partial k} = 2k + \int a(k - k') \frac{1}{1 + \eta(k')} \frac{\partial \epsilon(k')}{\partial k'} dk'. \tag{36}$$

The solution of non-singular linear integral equation is unique. Thus one obtains

$$\frac{\partial \Delta K(k_l', k_l)}{\partial k_l'} = 2\pi(\rho(k_l') + \rho^h(k_l')),$$
$$\frac{\partial \Delta E(k_l', k_l)}{\partial k_l'} = \frac{\partial \epsilon(k_l')}{\partial k_l'}. \tag{37}$$

Integrating these we have

$$\Delta K(k'_l, k_l) = 2\pi \int_{k_l}^{k'_l} \rho(k) + \rho^h(k)dk,$$

$$\Delta E(k'_l, k_l) = \epsilon(k'_l) - \epsilon(k_l). \tag{38}$$

Thus $\epsilon(k) = T\ln(\rho^h(k)/\rho(k))$ has the physical meaning as the energy of the elementary excitation.

2.5   Some Special Limits

$c = \infty$ **Limit** In the thermodynamic limit the system is equivalent to ideal spinless fermions

$$p = \frac{1}{2\pi} \int \frac{k^2 dk}{\exp(k^2 - A)/T + 1} =$$

$$-\frac{T}{2\pi} \int \ln(1 + \exp[(A - k^2)/T])dk = \lim_{L \to \infty} -G/L, \tag{39}$$

by the partial differentiation with respect to k. In this limit $c/(c^2 + (k-q)^2)$ in the integrand of (16) is zero. Then we have have very simple solution $\epsilon(k) = k^2 - A$. From equation (17) we have

$$\rho(k) = \frac{1}{2\pi} \frac{1}{\exp[(k^2 - A)/T] + 1}. \tag{40}$$

From (23) the thermodynamic potential per unit length is

$$g = -\frac{1}{2\pi} \int dk \ln(1 + \exp(-\frac{k^2 - A}{T})). \tag{41}$$

This is equivalent to (39).

$c = 0+$ **Limit** In this limit integration kernel $c/(c^2 + (k - q)^2)$ can be replaced by $\pi\delta(k - q)$. Then (16) becomes $\epsilon(k) = k^2 - A - T\ln(1 + \exp(-\epsilon(k)/T))$. Then we obtain

$$\epsilon(k) = T\ln(\exp((k^2 - A)/T) - 1). \tag{42}$$

The Gibbs free energy per site g, pressure p, $\rho(k)$ and $\rho^h(k)$ are

$$g = -p = T \int \ln(1 - \exp(-\frac{k^2 - A}{T}))\frac{dk}{2\pi}, \quad \rho^h(k) = \frac{1}{2\pi},$$

$$\rho(k) = \frac{1}{2\pi} \frac{1}{\exp((k^2 - A)/T) - 1}. \tag{43}$$

This result coincide with that for ideal bosons.

$T = 0+$ Limit Generally speaking $\epsilon(k)$ is a monotonically increasing function of k^2. At $T = 0+$ we assume that $\epsilon(\pm q_0) = 0$. Thus we have

$$\rho(k) = 0 \quad \text{for} \quad k^2 > q_0^2, \quad \rho^h(k) = 0 \quad \text{for} \quad k^2 < q_0^2.$$

Equations (10) and (16) for $k < q_0$ are

$$2\pi\rho(k) = 1 + 2c \int_{-q_0}^{q_0} \frac{\rho(q)dq}{c^2 + (k-q)^2},$$

$$\epsilon(k) = k^2 - A + \frac{c}{\pi} \int_{-q_0}^{q_0} \frac{\epsilon(q)dq}{c^2 + (k-q)^2}. \tag{44}$$

The first equation is equivalent with Lieb-Liniger equation([Lieb and Liniger (1963)]),

$$\rho(k) = \frac{1}{2\pi} + \int_{-B}^{B} \frac{c/\pi}{c^2 + (k-q)^2} \rho(q)dq. \tag{45}$$

The above theory was introduced by Yang-Yang ([Yang and Yang (1969)]). C.P. Yang solved numerically this equation([C.P. Yang 1970]). Very surprisingly it seems that this simple non-linear equation gives the exact free energy in the thermodynamic limit of the 1D repulsive Bosons. The next problem is to find thermodynamic Bethe ansatz equation for other soluble models. As the next simplest case we treat the $S = 1/2$ XXX chain.

3 Thermodynamics of the XXX Chain

3.1 Wave Functions of the XXX Chain

The Heisenberg model was the first model to be treated by the method of Bethe ansatz ([Bethe (1931)]). In the beginning of the 1930's only the ferromagnetic case was considered

$$\mathcal{H} = -J \sum_{l=1}^{N} S_l^x S_{l+1}^x + S_l^y S_{l+1}^y + S_l^z S_{l+1}^z - 2h \sum_{l=1}^{N} S_l^z,$$

$$h \geq 0, \quad \mathbf{S}_{N+1} \equiv \mathbf{S}_1. \tag{46}$$

This Hamiltonian is defined on a 2^N dimensional vector space. The space is classified by the total $S^z = \sum S_l^z$. The ground state is the state where all spins are up and $S^z = N/2$

$$\mathcal{H}|0> = E_0|0>, \quad E_0 = -JN/4 - Nh. \tag{47}$$

Write a general state $|\Psi>$ in terms of a wave function f,

$$|\Psi> = \sum f(n_1, n_2, ..., n_M) S_{n_1}^- S_{n_2}^- ... S_{n_M}^- |0>, \tag{48}$$

where $1 \leq n_1 < n_2 < ... < n_M \leq N$ and $2M \leq N$. The eigenvalue equation is

$$
-\frac{J}{2}\sum_j (1 - \delta_{n_j+1,n_{j+1}})\Big\{f(n_1, ..., n_j + 1, n_{j+1}, ..., n_M)
$$

$$
+ f(n_1, ..., n_j, n_{j+1} - 1, ..., n_m)\Big\}
$$

$$
+ \Big\{E_0 - E + (J + 2h)M - J\sum_j \delta_{n_j+1,n_{j+1}}\Big\}f(n_1, n_2, ..., n_M) = 0.
$$

$$(49)$$

Next we assume that the wave function is of the following form

$$
f(n_1, n_2, ..., n_M) = \sum_P^{M!} A(P)\exp(i\sum_{j=1}^{M} k_{Pj} n_j). \tag{50}
$$

Choosing

$$
A(P) = \epsilon(P)\prod_{j<l}(e^{i(k_{Pj}+k_{Pl})} + 1 - 2e^{ik_{Pl}}), \tag{51}
$$

$$
E - E_0 = 2hM + J\sum_{j=1}^{M}(1 - \cos k_j). \tag{52}
$$

insures that f satisfies (49). If we put $e^{ik_j} = (x_j + i)/(x_j - i)$, the wave function and energy are written as follows

$$
f(n_1, n_2, ..., n_M) = \sum_P^{M!} A(P)\prod_{j=1}^{M}(\frac{x_{Pj} + i}{x_{Pj} - i})^{n_j}, \tag{53}
$$

$$
A(P) = D\epsilon(P)\prod_{j<l}(x_{Pj} - x_{Pl} - 2i), \tag{54}
$$

$$
E - E_0 = 2hM + 2J\sum_{j=1}^{M}\frac{1}{x_j^2 + 1}. \tag{55}
$$

The x_j's are called rapidities. The periodic boundary condition

$$
f(x_1, x_2, ..., x_M) = f(x_2, x_3, ..., x_M, x_1 + N),
$$

implies the x_j satisfy

$$
\Big(\frac{x_j + i}{x_j - i}\Big)^N = \prod_{l\neq j}\Big(\frac{x_j - x_l + 2i}{x_j - x_l - 2i}\Big), \quad j = 1, ..., M. \tag{56}
$$

In terms of the rapidity variable, the periodic boundary conditions take on a very simple form.

3.2   Hulthen's Solution for the Antiferromagnet

In actual magnetic substances, the ferromagnetic case is rare. Usual one- dimensional magnetic substances are antiferromagnetic (the $J < 0$ case of equation (46)). The logarithm of equation (56) is

$$2N\hat{T}_n^{a-1} x_j = 2\pi I_j + 2 \sum_{l=1}^{M} \hat{T}_n^{a-1} \frac{x_j - x_l}{2}, \tag{57}$$

where I_j is an integer(half-odd integer) for odd(even) $N - M$. The total momentum is given by

$$K = \pi(1 - (-1)^M)/2 - \frac{2\pi}{N} \sum_j I_j.$$

For simplicity we set N to be even. One can show that the lowest energy state in the subspace of total $S_z = N/2 - M$ is given by

$$I_j = (M + 1 - 2j)/2, \quad j = 1, 2, ..., M. \tag{58}$$

In the thermodynamic limit x_j's distribute from $-B$ to B. From equation (57) we have

$$\hat{T}_n^{a-1} x = \pi \int^x \rho(t)dt + \int_{-B}^{B} \hat{T}_n^{a-1} \frac{x - y}{2} \rho(y)dy. \tag{59}$$

Differentiating with respect to x yields

$$\rho(x) = \frac{1}{\pi} \frac{1}{x^2 + 1} - \int_{-B}^{B} \frac{1}{\pi} \frac{2}{(x - y)^2 + 4} \rho(y)dy. \tag{60}$$

The energy and magnetization per site are

$$\frac{E}{N} = \frac{|J|}{4} - h + \int_{-B}^{B} [2h - \frac{2|J|}{x^2 + 1}]\rho(x)dx, \tag{61}$$

$$\frac{S_z}{N} = \frac{1}{2} - \int_{-B}^{B} \rho(x)dx. \tag{62}$$

This integral equation can be solved in the case of infinite B. We define the Fourier transform of $\rho(x)$ as follows

$$\tilde{\rho}(\omega) = \int_{-\infty}^{\infty} e^{-i\omega x} \rho(x)dx. \tag{63}$$

Using the formula $\int \pi^{-1}n/(x^2+n^2) \exp(-ix\omega)dx = \exp(-n|\omega|)$, one can rewrite (60) as follows

$$\rho(\omega)(1 + e^{-2|\omega|}) = e^{-|\omega|}.$$

Thus $\tilde{\rho}(\omega) = 1/(2\cosh\omega)$ and

$$\rho(x) = \frac{1}{2\pi} \int_{-\infty}^{\infty} e^{i\omega x} \tilde{\rho}(\omega)d\omega = \frac{1}{4}\text{sech}(\frac{\pi x}{2}). \tag{64}$$

Substituting this into (62) we have

$$e = -|J|(\ln 2 - \frac{1}{4}) = -0.443147|J|, \quad s_z = 0. \tag{65}$$

It can be shown that the case $B = \infty$ is the true ground state at $h = 0$ and that the magnetization is zero. Thus an analytical result for a one-dimensional antiferromagnet is obtained ([Hulthen (1938)]). The first neighbor correlation is derived from this result

$$< S_i^z S_{i+1}^z > = \frac{1}{12}(1 - 4\ln 2) = -0.14771573. \tag{66}$$

An analytic expression for the second neighbor correlation function is also known for this model ([Takahashi (1977)]),

$$< S_i^z S_{i+2}^z > = \frac{1}{12}(1 - 16\ln 2 + 9\zeta(3)) = 0.06067977. \tag{67}$$

This was calculated from the ground state energy of the Hubbard model. Hulthen's solution is the ground state of **XXX** antiferromagnet at zero magnetic field. Griffiths calculated the magnetization curve of this model ([Griffiths (1964)]). des Cloizeaux and Pearson calculated the elementary excitation away from this ground state ([des Cloizeaux and Pearson (1962)]).

3.3   String Solution of Infinite System

The elementary excitation for the $M = 1$ case is

$$E - E_0 = J(1 - \cos K) + 2h. \tag{68}$$

This is the spin wave excitation of the ferromagnetic Heisenberg model. Bethe found that the bound state of spin waves exists (Bethe, 1931),

$$x_j = \alpha + i(n + 1 - 2j), j = 1, 2, ..., n. \tag{69}$$

The energy and momentum of this excitation are

$$E = E_0 + \frac{2nJ}{\alpha^2 + n^2} + 2nh, \quad K = \frac{1}{i}\ln\left(\frac{\alpha + in}{\alpha - in}\right). \tag{70}$$

Thus the dispersion relation is

$$E = E_0 + \frac{2J}{n}(1 - \cos K) + 2nh. \tag{71}$$

These string solutions play an essential role in the thermodynamics of soluble models. The author obtained the thermodynamic Bethe ansatz equation for the XXX model ([Takahashi (1971a)]). The wave function is

$$f(n_1, n_2, ..., n_M) = \sum_P A(P) \prod_{j=1}^{M} \left(\frac{x_{Pj} + i}{x_{Pj} - i} \right)^{n_j},$$

$$A(P) = D\epsilon(P) \prod_{j<l} (x_{Pj} - x_{Pl} - 2i). \tag{72}$$

We consider the wave function in the equation (72) assuming that N is infinity. Particle coordinates n_j move from $-\infty$ to ∞. Assume that $\Im k_1 \geq \Im k_2 \geq ... \geq \Im k_n$. The wave function is written as follows

$$f(n_1, n_2, ..., n_M) = (z_1 z_2 .. z_M)^{n_1} \sum_P A(P) \prod_{j=2}^{M} (\prod_{l=j}^{M} z_{Pl})^{n_{j+1} - n_j},$$

$$A(P) = \epsilon(P) \prod_{j<l} (x_{Pj} - x_{Pl} - 2i), \quad z_j = e^{ik_j} = \left(\frac{x_j + i}{x_j - i} \right). \tag{73}$$

In the infinite system f satisfies the following boundary conditions

$$| \lim_{n_1, ..., n_r \to -\infty} f(n_1, n_2, ..., n_M)| < \infty,$$

$$| \lim_{n_{M-r+1}, ..., n_M \to \infty} f(n_1, n_2, ..., n_M)| < \infty, \tag{74}$$

From this condition we find that $|z_1 z_2 .. z_M| = 1$ and that $A(P) = 0$ if one of $|\prod_{l=j}^{M} z_{Pl}|$ is greater than 1. From the normalizability condition of the wave function we have

$$A(I) \neq 0, \quad A(P \neq I) = 0, \quad |z_1 z_2 ... z_n| = 1. \tag{75}$$

These conditions are satisfied only if

$$x_j = \alpha + (n + 1 - 2j)i, \quad j = 1, 2, ..., n. \tag{76}$$

From $|z_1 z_2 .. z_M| = |\frac{\alpha + ni}{\alpha - ni}| = 1$, we find that α must be real. The condition

$$(\prod_{l=j+1}^{M} z_l) = |\frac{\alpha + (n - 2j)i}{\alpha - ni}| \leq 1, \quad j = 1, 2, ..., n - 1$$

is automatically satisfied for general n. Thus strings with arbitrary length are possible for the XXX Heisenberg chain. Moreover, the following type of string is impossible

$$x_1 = \alpha + (2 + \beta)i, x_2 = \alpha + \beta i, x_3 = \alpha - \beta i, x_4 = \alpha - (2 + \beta)i, \tag{77}$$

where α, β are real and $\beta \neq 1$.

3.4   String Hypothesis for a Long XXX Ring

We assume that all rapidities x_j belong to bound states with $n = 1, 2, \dots$. For a bound states of n-x's the real parts of all x's are the same and imaginary parts are $(n-1)i, (n-3)i, \dots, -(n-1)i$ within the accuracy of $O(\exp(-\delta N))$. This assumption seems to be too strong and there are some counter examples in special cases. This is a very controversial point of thermodynamic Bethe-ansatz equations for soluble models, except in the repulsive Boson case, which has no the string solution. But equations obtained using the string hypothesis seem to give the correct free energy and other thermodynamic quantities.

Consider the case where M_n bound states of $n - x$'s exist. We designate x's as

$$x_\alpha^{n,j}, \quad \alpha = 1, 2, \dots, M_n,$$

$$x_\alpha^{n,j} = x_\alpha^n + i(n+1-2j) + \text{deviation}. \tag{78}$$

From (56),

$$e^N(x_\alpha^{n,j}) = \prod_{(m,\beta) \neq (n,\alpha)} e\left(\frac{x_\alpha^{n,j} - x_\beta^m}{m-1}\right) e\left(\frac{x_\alpha^{n,j} - x_\beta^m}{m+1}\right) \prod_{j' \neq j} e\left(\frac{x_\alpha^{n,j} - x_\alpha^{n,j'}}{2}\right),$$

$$j = 1, 2, \dots, n. \tag{79}$$

Here $e(x) \equiv (x+i)/(x-i)$. The last product is delicate because the numerator or denominator may become very small. If we take the product of these n equations, these delicate terms are canceled and we have

$$e^N(x_\alpha^n/n) = \prod_{j=1}^n e^N(x_\alpha^{n,j}) = \prod_{(m,\beta) \neq (n,\alpha)} E_{nm}(x_\alpha^n - x_\beta^m), \tag{80}$$

where

$$E_{nm}(x) \equiv \begin{cases} e\left(\frac{x}{|n-m|}\right) e^2\left(\frac{x}{|n-m|+2}\right) e^2\left(\frac{x}{|n-m|+4}\right) \dots e^2\left(\frac{x}{n+m-2}\right) e\left(\frac{x}{n+m}\right) \\ \qquad\qquad\qquad\qquad \text{for} \quad n \neq m, \\ e^2\left(\frac{x}{2}\right) e^2\left(\frac{x}{4}\right) \dots e^2\left(\frac{x}{2n-2}\right) e\left(\frac{x}{2n}\right) \quad \text{for} \quad n = m. \end{cases} \tag{81}$$

The logarithm of these equations gives

$$N\theta(x_\alpha^n/n) = 2\pi I_\alpha^n + \sum_{(m,\beta) \neq (n,\alpha)} \Theta_{nm}(x_\alpha^n - x_\beta^m), \tag{82}$$

where

$$\theta(x) \equiv 2\hat{T}_n^{a^{-1}}(x), \tag{83}$$

and

$$\Theta_{nm}(x) \equiv \begin{cases} \theta\left(\frac{x}{|n-m|}\right) + 2\theta\left(\frac{x}{|n-m|+2}\right) + \dots + 2\theta\left(\frac{x}{n+m-2}\right) + \theta\left(\frac{x}{n+m}\right) \\ \qquad\qquad\qquad\qquad \text{for} \quad n \neq m, \\ 2\theta\left(\frac{x}{2}\right) + 2\theta\left(\frac{x}{4}\right) + \dots + 2\theta\left(\frac{x}{2n-2}\right) + \theta\left(\frac{x}{2n}\right) \quad \text{for} \quad n = m. \end{cases} \tag{84}$$

I_α^n is an integer (half-odd integer) if $N - M_n$ is odd (even) and should satisfy

$$|I_\alpha^n| \leq \frac{1}{2}(N - 1 - \sum_{m=1}^{\infty} t_{nm}M_m), \quad t_{nm} \equiv 2Min(n, m) - \delta_{nm}. \qquad (85)$$

We can prove the number of sets $\{I_\alpha^n\}$ is $C_M^N - C_{M-1}^N$, under the condition $M = \sum_{n=1}^{\infty} nM_n$. Here C_M^N is the binomial coefficient defined by $N!/(M!(N - M)!)$. For details, see Appendix A of ([Takahashi (1971a)]). The energy of this state is given by

$$E(\{I_\alpha^n\}) = N(-h - \frac{J}{4}) + \sum_{n,\alpha}(\frac{2Jn}{(x_\alpha^n)^2 + n^2} + 2hn). \qquad (86)$$

We can construct wave functions through (48), (50) and (54) at $S = S_z = N/2 - M$. The wave functions for $S_z = S - 1, S - 2, ..., -S$ are obtained by applying the operator S_{total}^-. The energy for these cases are $E - 2h, E - 4h, ..., E - 2(N - 2M)h$. Then the total number of states which are generated by the string assumption and the descending operator at $S_z = N/2 - M$ is C_M^N. Therefore the total number of states is 2^N. This coincides with the true total number of states. It is expected that all eigen functions constructed in the above way should be a complete set. The partition function of this system is written as follows

$$\mathcal{Z} = \sum_{M=0}^{[N/2]} \frac{1 - \exp(-2(N + 1 - 2M)h/T)}{1 - \exp(-2h/T)} \sum_{\{I_\alpha^n\}} \exp[-T^{-1}E(\{I_\alpha^n\})]. \qquad (87)$$

The free energy is given by $G = -T \ln \mathcal{Z}$. We define functions $h_n(x)$ by

$$h_n(x) \equiv \theta_n(x) - N^{-1} \sum_{(m,\alpha)} \Theta_{nm}(x - x_\alpha^m). \qquad (88)$$

We can define holes in n-string sea by the solution of

$$2\pi J_\beta^n/N = h_n(x_\beta^n), \qquad (89)$$

where J_β^n are omitted integers or half-odd integers in the region.

3.5   Thermodynamic Bethe-Ansatz Equations for XXX Chain

In the thermodynamic limit, define distribution functions of n-strings and holes of n-string as $\rho_n(x)$ and $\rho_n^h(x)$. The number of strings and holes between x and $x + dx$ is $\rho_n(x)Ndx$ and $\rho_n^h(x)Ndx$, respectively. Thus we have

$$2\pi \int^x \rho_n(t) + \rho_n^h(t)dt = \theta_n(x) - \sum_{m=1}^{\infty} \int_{-\infty}^{\infty} \Theta_{nm}(x - y)\rho_m(y)dy. \qquad (90)$$

Differentiating by x, we obtain the integral equation

$$a_n(x) = \rho_n(x) + \rho_n^h(x) + \sum_m \int_{-\infty}^{\infty} T_{nm}(x-y)\rho_m(y)dy, \qquad (91)$$

where $T_{nm}(x)$ is a function defined by

$$T_{nm}(x) \equiv \begin{cases} a_{|n-m|}(x) + 2a_{|n-m|+2}(x) + 2a_{|n-m|+4}(x) + \cdots \\ \qquad +2a_{n+m-2}(x) + a_{n+m}(x) \quad \text{for} \quad n \neq m, \\ 2a_2(x) + 2a_4(x) + \cdots + 2a_{2n-2}(x) + a_{2n}(x) \quad \text{for} \quad n = m. \end{cases} \qquad (92)$$

$a_n(x)$ is a function defined by

$$a_n(x) \equiv \frac{1}{\pi} \frac{n}{x^2 + n^2}, \quad a_0(x) \equiv \delta(x). \qquad (93)$$

The energy per site is

$$e = -(\frac{J}{4} + h) + \sum_{n=1}^{\infty} \int_{-\infty}^{\infty} g_n(x)\rho_n(x)dx,$$
$$g_n(x) \equiv 2\pi J a_n(x) + 2nh. \qquad (94)$$

The total entropy per site is

$$s = \sum_{n=1}^{\infty} \int_{-\infty}^{\infty} \rho_n(x)\ln(1 + \frac{\rho_n^h(x)}{\rho_n(x)}) + \rho_n^h(x)\ln(1 + \frac{\rho_n(x)}{\rho_n^h(x)})dx. \qquad (95)$$

$e - Ts$ should be minimized at the thermodynamic equilibrium. Consider the functional variation of the free energy with respect to $\rho_n(k)$ and $\rho_n^h(k)$,

$$0 = \delta e - T\delta s = \sum_{n=1}^{\infty} \int dx$$
$$[g_n(x) - T\ln(1 + \frac{\rho_n^h(x)}{\rho_n(x)})]\delta\rho_n(x) - T\ln(1 + \frac{\rho_n(x)}{\rho_n^h(x)})\delta\rho_n^h(x).$$
$$(96)$$

From equation (88) we have

$$\delta\rho_n^h(x) = -\delta\rho_n(x) - \sum_m \int T_{nm}(x-y)\delta\rho_m(y)dy. \qquad (97)$$

Substituting these into (96) yields

$$0 = T\sum_{n=1}^{\infty} \int \left\{ \frac{g_n(x)}{T} - \ln\eta_n(x) + \sum_{m=1}^{\infty} \int T_{nm}(x-y)\ln(1 + \eta_m^{-1}(y))dy \right\}\delta\rho_n(x)dx,$$
$$(98)$$

where $\eta_n(x) \equiv \rho_n^h(x)/\rho_n(x)$. Thus we have integral equations for an infinite number of unknown $\eta_n(x)$

$$\ln \eta_n(x) = \frac{g_n(x)}{T} + \sum_{m=1}^{\infty} T_{nm} * \ln(1 + \eta_m^{-1}(x)). \tag{99}$$

In the theory of Bethe ansatz equations we encounter very frequently the integration of the following type

$$\int_{-\infty}^{\infty} a(x - y)b(y)dy.$$

This is a convolution of two functions $a(x)$ and $b(x)$. In the above and hereafter we write the convolution as $a * b(x)$,

$$a * b(x) \equiv \int_{-\infty}^{\infty} a(x - y)b(y)dy. \tag{100}$$

The free energy per site becomes as follows

$$f = e - Ts = -(\frac{J}{4} + h) + \sum_{n=1}^{\infty} \int g_n \rho_n - T[\rho_n \ln(1 + \eta_n) + \rho_n^h \ln(1 + \eta_n^{-1})]dx.$$

We eliminate ρ_n^h using (91)

$$f = -(\frac{J}{4} + h) - T\sum_{n=1}^{\infty} \int \ln(1 + \eta_n^{-1})a_n(x)$$

$$+\rho_n[\ln \eta_n - \frac{g_n}{T} - T_{nm} * \ln(1 + \eta_m^{-1})]dx. \tag{101}$$

Using (99), one sees the inside of the bracket on r.h.s. is zero. So we have

$$f = -(\frac{J}{4} + h) - T\sum_{n=1}^{\infty} \int a_n(x) \ln(1 + \eta_n^{-1}(x))dx. \tag{102}$$

From the $n = 1$ case of (99) we have

$$\ln(1 + \eta_1) = \frac{2\pi J a_1(x) + 2h}{T} + \sum_{l=1}^{\infty} (a_{l-1} + a_{l+1}) * \ln(1 + \eta_l^{-1}). \tag{103}$$

Operating $\int dx s(x)$ on this equation yields

$$\int dx s(x) \ln(1 + \eta_1) = \frac{2\pi J}{T} \int s(x)a_1(x)dx + \frac{h}{T}$$

$$+ \sum_{l=1}^{\infty} \int a_l(x) \ln(1 + \eta_l^{-1}(x))dx.$$

Then equation (102) is transformed as follows

$$f = J(\ln 2 - \frac{1}{4}) - T \int s(x) \ln(1 + \eta_1(x))dx. \qquad (104)$$

Solutions η_n of (99) are functions of x, J, T and h. Differentiating (99) with respect to J yields

$$\frac{2\pi J a_n(x)}{T} = \frac{1}{\eta_n} \frac{\partial \eta_n}{\partial J} + \sum_m T_{nm} * \frac{1}{(1 + \eta_m)\eta_m} \frac{\partial \eta_m}{\partial J}. \qquad (105)$$

Comparing this with (91) we have

$$\rho_n = \frac{T}{2\pi} \frac{1}{(1 + \eta_n)\eta_n} \frac{\partial \eta_n}{\partial J}, \quad \rho_n + \rho_n^h = \frac{T}{2\pi} \frac{\partial \ln \eta_n}{\partial J}. \qquad (106)$$

By the definition (92),

$$a_1 * (T_{n-1,m} + T_{n+1,m}) - (a_0 + a_2) * T_{n,m} = (\delta_{n-1,m} + \delta_{n+1,m})a_1. \qquad (107)$$

Using equations (99), (107) yields

$$(a_0 + a_2) * \ln \eta_1(x) = \frac{2\pi J a_1(x)}{T} + a_1 * \ln(1 + \eta_2(x)), \qquad (108)$$

$$(a_0 + a_2) * \ln \eta_n(x) = a_1 * \ln(1 + \eta_{n-1}(x))(1 + \eta_{n+1}(x)), \quad n = 2, 3, ... \qquad (109)$$

Equations (108) and (109) are not complete to determine all of the $\eta_n(x)$, as they do not contain h. Take the $n = 1$ case of (99)

$$\ln \eta_1 = \frac{2\pi J a_1(x) + 2h}{T} + a_2 * \ln(1 + \eta_1^{-1}) + \sum_{j=2}^{\infty}(a_{j-1} + a_{j+1}) * \ln(1 + \eta_j^{-1}). \qquad (110)$$

Substituting (108), (109) we can eliminate $\eta_j, j < n$ for a given integer n,

$$\frac{2h}{T} = a_n * \ln \eta_{n+1} - a_{n+1} * \ln(1 + \eta_n) - a_{n+2} * \ln(1 + \eta_{n+1}^{-1})$$
$$- \sum_{l=n+2}^{\infty} (a_{l-1} + a_{l+1}) * \ln(1 + \eta_l^{-1}). \qquad (111)$$

Thus we have

$$\ln \eta_{n+1} = \frac{2h}{T} + a_1 * \ln \eta_n + a_2 * \ln(1 + \eta_{n+1}^{-1})$$
$$+ \sum_{l=n+2}^{\infty} (a_{l-n-1} + a_{l-n+1}) * \ln(1 + \eta_l^{-1}). \qquad (112)$$

For large n, $\ln(1 + \eta_n^{-1}) \simeq o(n^{-2})$ and therefore:

$$\lim_{n \to \infty} \ln \eta_{n+1} - a_1 * \ln \eta_n = \frac{2h}{T}, \qquad (113)$$

or

$$\lim_{n\to\infty} \frac{\ln \eta_n}{n} = \frac{2h}{T}. \tag{114}$$

Thus the following equations determine η_n,

$$\ln \eta_1(x) = \frac{2\pi J}{T} s(x) + s * \ln(1 + \eta_2(x)), \tag{115}$$

$$\ln \eta_n(x) = s * \ln(1 + \eta_{n-1}(x))(1 + \eta_{n+1}(x)), \tag{116}$$

$$\lim_{n\to\infty} \frac{\ln \eta_n}{n} = \frac{2h}{T}, \tag{117}$$

where

$$s(x) = \frac{1}{4} \mathrm{sech}(\frac{\pi x}{2}). \tag{118}$$

3.6   Some Special Cases and Expansions

$J/T \to 0$ **Case** In the limit $J/T \to 0$ and $h/T \geq 0$ we can expect that $\eta_n(x)$ is independent of x, because there are no x dependent terms in equations (115), (116 and (117). As $\int dx s(x) = 1/2$, equations (115,116,117) become

$$\eta_n^2 = (1 + \eta_{n-1})(1 + \eta_{n+1}), \tag{119}$$

$$\eta_1^2 = 1 + \eta_2, \quad \lim_{n\to\infty} \ln \eta_n/n = 2h/T. \tag{120}$$

Equation (119) is a difference equation of second order. It is similar to a differential equation of the second order and contains two arbitrary parameters. The general solution of this equation is

$$\eta_n = (\frac{az^n - a^{-1}z^{-n}}{z - z^{-1}})^2 - 1. \tag{121}$$

Parameters a and z are determined by (120) and we have $a = z, z = \exp(h/T)$ and

$$\eta_n = (\frac{\sinh[(n+1)h/T]}{\sinh[h/T]})^2 - 1 \quad \text{for} \quad h > 0$$

$$\eta_n = (n+1)^2 - 1 \quad \text{for} \quad h = 0. \tag{122}$$

Substituting this into (104) we obtain the free energy, magnetization and entropy

$$f = -T\ln[2\cosh h/T], \quad m = 2s_z = -\partial f/\partial h = \tanh h/T,$$

$$s = -\partial f/\partial T = \ln[2\cosh(h/T)] - (h/T)\tanh(h/T). \tag{123}$$

At $h = 0$ the entropy per site is $\ln 2$. This corresponds to the fact that the number of states per site is two.

High Temperature Expansion or Small J Expansion For the XXX chain, we can perform the high temperature expansion of the free energy density from the definition

$$f/T = -N^{-1} \ln \operatorname{Tr} \exp(-\mathcal{H}/T). \tag{124}$$

This is expanded as a power series of $1/T$. Assume that $\mathcal{H} = \mathcal{H}_0 + \mathcal{H}_1$ where $\mathcal{H}_0$ and $\mathcal{H}_1$ commute each other. Then the exponential operator of $\mathcal{H}$ can be expanded as follows

$$\exp(-\mathcal{H}/T) = \exp(-\mathcal{H}_0/T)\left(1 - T^{-1}\frac{\mathcal{H}_1}{1!} + T^{-2}\frac{\mathcal{H}_2}{2!} - +...\right). \tag{125}$$

Thus,

$$f/T = -N^{-1} \ln \operatorname{Tr} \exp(-\mathcal{H}_0/T) + \frac{<\mathcal{H}_1>}{NT} - \frac{<\mathcal{H}_1^2> - <\mathcal{H}_1>^2}{2!NT^2}$$

$$+ \frac{<\mathcal{H}_1^3> -3 <\mathcal{H}_1>^2<\mathcal{H}_1> +2 <\mathcal{H}_1>^3}{3!NT^3} - +..., \tag{126}$$

where

$$< X > \equiv \frac{\operatorname{Tr} \exp(-\mathcal{H}_0/T)X}{\operatorname{Tr} \exp(-\mathcal{H}_0/T)}. \tag{127}$$

In the beginning of 1930's only the ferromagnetic model was considered. If we set

$$\mathcal{H}_0 = -2h \sum_{l=1} S_l^z, \quad \mathcal{H}_1 = -J \sum_{l=1}^{N} S_l^x S_{l+1}^x + S_l^y S_{l+1}^y + S_l^z S_{l+1}^z, \tag{128}$$

for the Hamiltonian (46), we obtain the J/T expansion of free energy at fixed h/T

$$f/T = -\ln(2 \cosh h/T) - \frac{J}{4T} \tanh^2(h/T)$$

$$-\frac{J^2}{32T^2}(3 + 2\tanh^2(h/T) - 3\tanh^4(h/T)) + O((J/T)^3). \tag{129}$$

The calculation to higher orders can be done by the use of linked cluster expansion. Higher order terms are polynomials of $\tanh(h/T)$.

Apparently the expression of the free energy in (123) coincides with the first term of the above expansion. Writing $\ln(\eta_n + 1)$ as the expansion

$$\ln(\eta_n(x) + 1) = \ln[\frac{1}{\alpha_n - 1}] + \sum_{l=1}^{\infty} f_n^{(l)}[\frac{J}{T}]^l,$$

$$\alpha_n \equiv \frac{\sinh^2(h(n+1)/T)}{\sinh(hn/T)\sinh(h(n+2)/T)}, \tag{130}$$

we obtain an expansion of $\ln \eta_n(x)$

$$\ln \eta_n(x) = \ln \frac{\alpha_n}{\alpha_n - 1} + [\frac{J}{T}]\alpha_n f_n^{(1)}$$
$$+ [\frac{J}{T}]^2 (\alpha_n f_n^{(2)} + (\alpha_n - \alpha_n^2)\frac{(f_n^{(1)})^2}{2}) + O([\frac{J}{T}]^3). \tag{131}$$

Substituting these expansions into (115,116,117,118) and taking first order terms in J/T, linear integral equations for $f_n^{(1)}(x)$ are obtained

$$\alpha_1 f_1^{(1)}(x) - s * f_2^{(1)}(x) = 2\pi s(x), \tag{132}$$
$$\alpha_n f_n^{(1)}(x) - s * (f_{n-1}^{(1)}(x) + f_{n+1}^{(1)}(x)) = 0, \tag{133}$$
$$\lim_{n \to \infty} \frac{\alpha_n f_n(x)}{n} = 0. \tag{134}$$

The r.h.s. of these equations are inhomogeneous terms of the integral equations. The Fourier transform of these equations are

$$(e^{|\omega|} + e^{-|\omega|})\alpha_n \tilde{f}_n^{(1)}(\omega) = \tilde{f}_{n-1}^{(1)}(\omega) + \tilde{f}_{n+1}^{(1)}(\omega). \tag{135}$$

The general solution of this difference equation is

$$\tilde{f}_n^{(1)}(\omega) =$$
$$A(\omega)[\frac{\sinh((n+2)h/T)}{\sinh((n+1)h/T)}e^{-n|\omega|} - \frac{\sinh(nh/T)}{\sinh((n+1)h/T)}e^{-(n+2)|\omega|}]$$
$$+ B(\omega)[\frac{\sinh((n+2)h/T)}{\sinh((n+1)h/T)}e^{n|\omega|} - \frac{\sinh(nh/T)}{\sinh((n+1)h/T)}e^{(n+2)|\omega|}]. \tag{136}$$

From the boundary conditions we have

$$A(\omega) = \frac{\pi}{\cosh h/T}, \quad B(\omega) = 0. \tag{137}$$

Thus

$$\tilde{f}_1^{(1)}(\omega) = \frac{\pi}{\cosh(h/T)}[\frac{\sinh 3h/T}{\sinh 2h/T}e^{-|\omega|} - \frac{\sinh h/T}{\sinh 2h/T}e^{-3|\omega|}],$$
$$f_1^{(1)}(x) = \frac{\pi}{\cosh(h/T)}[\frac{\sinh 3h/T}{\sinh 2h/T}a_1(x) - \frac{\sinh h/T}{\sinh 2h/T}a_3(x)]. \tag{138}$$

Substituting this into (104) we obtain the second term of the J/T expansion (129). The higher order terms can be calculated by solving the linear integral equations for $f_n^{(2)}$, $f_n^{(3)}$, ... The equations are similar to (132,133,134) except for the inhomogeneous terms, which are given by lower order $f_n^{(l)}$.

Low Temperature Limit At low temperature, $\ln \eta_n$ diverges as $1/T$. So we should define the following functions

$$\epsilon_n(x) = T \ln \eta_n(x). \tag{139}$$

The integral equations become

$$\epsilon_1(x) = 2\pi J s(x) + s * T \ln(1 + \exp(\frac{\epsilon_2(x)}{T})),$$

$$\epsilon_n(x) = s * T \ln(1 + \exp(\frac{\epsilon_{n-1}(x)}{T}))(1 + \exp(\frac{\epsilon_{n+1}(x)}{T})),$$

$$\lim_{n \to \infty} \frac{\epsilon_n(x)}{n} = 2h. \tag{140}$$

The free energy expression becomes

$$f = -(\frac{J}{4} + h) - T \sum_{n=1}^{\infty} \int a_n(x) \ln(1 + \exp(-\epsilon_n(x)/T)) dx$$

$$= J(\ln 2 - \frac{1}{4}) - T \int s(x) \ln(1 + \exp(\epsilon_1(x)/T)) dx. \tag{141}$$

The $T = 0$ limit of these equations is

$$\epsilon_1(x) = 2\pi J s(x) + s * \epsilon_2^+(x),$$
$$\epsilon_n(x) = s * (\epsilon_{n-1}^+(x) + \epsilon_{n+1}^+(x)),$$

$$\lim_{n \to \infty} \frac{\epsilon_n(x)}{n} = 2h, \tag{142}$$

$$f = -(\frac{J}{4} + h) + \sum_{n=1}^{\infty} \int a_n(x) \epsilon_n^-(x) dx = J(\ln 2 - \frac{1}{4}) - \int s(x) \epsilon_1^+(x) dx,$$

$$\epsilon_n^+(x) \equiv \begin{cases} \epsilon_n(x), & \text{for } \epsilon_n(x) \geq 0, \\ 0, & \text{for } \epsilon_n(x) < 0, \end{cases}$$

$$\epsilon_n^-(x) \equiv \begin{cases} 0, & \text{for } \epsilon_n(x) \geq 0, \\ \epsilon_n(x), & \text{for } \epsilon_n(x) < 0. \end{cases} \tag{143}$$

In the ferromagnetic case $J > 0$ we have

$$\epsilon_n(x) = \epsilon_n^+(x) = 2\pi J a_n(x) + 2hn, \quad n = 1, 2, ..., \tag{144}$$

and therefore $f = -(\frac{J}{4} + h)$. This is the ground state energy of the ferromagnetic case.

In the antiferromagnetic case $J < 0$ we have

$$\epsilon_n(x) = \epsilon_n^+(x) = a_{n-1} * \epsilon_1^+(x) + 2(n-1)h, \quad n = 2, 3, ... \tag{145}$$

The equation which determines ϵ_1 is

$$\epsilon_1(x) = -2\pi|J|s(x) + h + \int_{|y|>B} R(x - y)\epsilon_1(y)dy, \quad \epsilon_1(\pm B) = 0. \tag{146}$$

In the limit of $h \to 0$ B becomes infinite. We have $\epsilon_1(x) = -2\pi|J|s(x)$ and $f = J(\ln 2 - \frac{1}{4})$.

Fugacity Expansion In the case of very large h the free energy can be expanded as a power series of $z = \exp(-h/T)$, z is the called the fugacity. From equation (94) and (99) we have expansions of η_n^{-1} as follows

$$\eta_n^{-1} = z^{2n} \exp(-\frac{2\pi J}{T} a_n(x)) \exp[-T_{nm} * (\eta_m^{-1} - \frac{1}{2}\eta_m^{-2} + -...)]. \qquad (147)$$

The expansion of η_1^{-1} and η_2^{-1} up to z^4 is

$$\eta_1^{-1} = z^2 \exp(-\frac{2\pi J}{T} a_1(x))(1 - z^2 \int a_2(x-y)\exp(-\frac{2\pi J}{T} a_1(y))dy) + O(z^6),$$

$$\eta_2^{-1} = z^4 \exp(-\frac{2\pi J}{T} a_2(x)) + O(z^6). \qquad (148)$$

As η_n^{-1} becomes small, (102) is more convenient than (104)

$$f = -\frac{J}{4} - h - T\sum_{n=1}^{\infty} \int dx a_n(x)(\eta_n^{-1} - \frac{1}{2}\eta_n^{-2} + -...).$$

Substituting (148), we obtain

$$f = -\frac{J}{4} - h - z^2 T \int a_1(x)\exp(-\frac{2\pi J}{T} a_1(x))dx$$

$$-z^4 T\left\{ \int a_2(x)\exp(-\frac{2\pi J}{T} a_2(x)) - \frac{1}{2}a_1(x)\exp(-\frac{4\pi J}{T} a_1(x))dx \right.$$

$$\left. - \int dx \int dy a_1(x)\exp\left[-\frac{2\pi J}{T}(a_1(x) + a_1(y))\right]a_2(x-y)\right\} + O(z^6). \qquad (149)$$

Putting $x = \hat{T}_n^a u, y = \hat{T}_n^a v$,

$$(f + \frac{J}{4} + h)/T$$

$$= z^2 e^{-K} I_0(K) + z^4 \left\{ -\frac{1}{2}e^{-2K} I_0(2K) + e^{-K/2} I_0(K/2) \right.$$

$$\left. -\frac{1}{\pi^2} \int_0^\pi \int_0^\pi \frac{e^{2K(1-\cos\omega_1 \cos\omega_2)}(1 - \cos\omega_1 \cos\omega_2)}{1 - 2\cos\omega_1 \cos\omega_2 + \cos^2 \omega_1} d\omega_1 d\omega_2 \right\}$$

$$+O(z^6), \qquad (150)$$

where $K \equiv J/T$, $\omega_1 = \pi + u + v$, $\omega_2 = u - v$ and $I_0(x)$ is modified Bessel function. This result is the same as that of Katsura ([Katsura (1965)]).

The strings are stable in a chain of infinite length in the case of very few down spins. Some counter examples of string assumption are found in some special limit. Nevertheless the thermodynamic Bethe ansatz equation seems to give the exact free energy in the case where the density of down-spins is comparable to that of up-spins. This non-linear integral equation contains an infinite number of unknown functions. To solve this equation one needs to do numerical calculations by computer.

4   Thermodynamics of the XXZ Model

4.1   Symmetry of the Hamiltonian

In this section we consider the following Hamiltonian

$$\mathcal{H}(J, \Delta, h) = -J \sum_{l=1}^{N} S_l^x S_{l+1}^x + S_l^y S_{l+1}^y + \Delta S_l^z S_{l+1}^z - 2h \sum_{l=1} S_l^z,$$

$$h \geq 0, \quad \mathbf{S}_{N+1} \equiv \mathbf{S}_1. \tag{151}$$

This Hamiltonian contains an additional parameter Δ. The case $\Delta = 0$ is called the XY model, which can be mapped to non-interacting fermions making it possible to calculate many physical quantities ([Lieb Schultz and Mattis (1961)], [Katsura (1962)]). The case $\Delta = 1$ is the XXX model and was treated in the previous chapter. The limit of very large Δ is the Ising model. The generalization of Bethe's method to $\Delta \neq 1$ was done by Orbach and Walker ([Orbach 1958] [Walker 1959]). Yang and Yang investigated the ground state of this model in detail ([Yang and Yang (1966)]). Bonner and Fisher investigated this model using the diagonalization method up to $N = 12$ ([Bonner and Fisher (1964)]). In this Hamiltonian the magnetic field is applied in the z-direction. For a magnetic field in a different direction, the exact solution is not known.

Let us consider the following unitary transformation:

$$\mathcal{H}(J, \Delta, h) = U_1 \mathcal{H}(J, \Delta, -h) U_1^{-1}, \quad U_1 \equiv \prod_{l=1}^{\infty} 2S_l^x = U_1^{-1}. \tag{152}$$

By this unitary transformation S_{total}^z changes its sign and we can treat the $N \geq M > N/2$ case. In the case of even N we can show that

$$\mathcal{H}(-J, -\Delta, h) = U_2 \mathcal{H}(J, \Delta, h) U_2^{-1}, \quad U_2 \equiv \prod_{l=even} 2S_l^z = U_2^{-1}. \tag{153}$$

By this unitary transformation S_l^x, S_l^y, S_l^z change to $-S_l^x, -S_l^y, S_l^z$ at $l = even$.

4.2   Bethe Ansatz Wave Function

Consider the state where all spins are up and the total S^z is $N/2$

$$\mathcal{H}|0> = E_0|0>, \quad E_0 = -J\Delta N/4 - Nh. \tag{154}$$

Writing a general state $|\Psi>$ in terms of a wave function f, as in equation (48), the eigenvalue condition can be expressed as

$$0 = -\frac{J}{2} \sum_j (1 - \delta_{n_j+1, n_{j+1}}) \Big\{ f(n_1, ., n_j + 1, n_{j+1}, ., n_M)$$

$$+ f(n_1, ., n_j, n_{j+1} - 1, ., n_m) \Big\}$$

$$+ \Big\{ E_0 - E + (J\Delta + 2h)M - J\Delta \sum_j \delta_{n_j+1, n_{j+1}} \Big\} f(n_1, n_2, ., n_M).$$

$$\tag{155}$$

For a wave function of the type eq. (50) to be the eigenstate, set E and $A(P)$ to be

$$E = E_0 + \sum_{j=1}^{M} [J(\Delta - \cos k_j) + h], \tag{156}$$

$$0 = A(P)(e^{ik_{Pj}} + e^{-k_{P(j+1)}} - 2\Delta)e^{k_{P(j+1)}}$$
$$+A(P(j,j+1))(e^{ik_{P(j+1)}} + e^{-k_{Pj}} - 2\Delta)e^{k_{Pj}}, \tag{157}$$

$$A(P) = \epsilon(P) \prod_{l<j}(e^{i(k_{Pl}+k_{Pj})} + 1 - 2\Delta e^{ik_{Pl}}). \tag{158}$$

The periodic boundary condition is as follows,

$$\exp(ik_j N) = (-1)^{M-1} \prod_{l \neq j} \frac{\exp[i(k_j + k_l)] + 1 - 2\Delta \exp(ik_j)}{\exp[i(k_j + k_l)] + 1 - 2\Delta \exp(ik_l)},$$
$$j = 1, 2, ..., M. \tag{159}$$

This is a set of complicated coupled equations for M unknowns. If we have a solution of this set, we have one eigenstate and its energy eigenvalue and total momentum. If we set rapidity parameters x_j as $\cot(k_j/2)$

$$\exp(ik_j) = \left(\frac{x_j + i}{x_j - i}\right), \tag{160}$$

the phase factor

$$\frac{\exp[i(k_j + k_l)] + 1 - 2\Delta \exp(ik_j)}{\exp[i(k_j + k_l)] + 1 - 2\Delta \exp(ik_l)}$$

cannot be written as a function of $x_j - x_l$ except in the case $\Delta = 1$. This is not convenient. If we set

$$\exp(ik_j) = \frac{\sin \frac{\varphi}{2}(x_j + i)}{\sin \frac{\varphi}{2}(x_j - i)}, \tag{161}$$

in place of (160), the phase factor becomes

$$\frac{\cos \frac{\varphi}{2}(x_j + x_l)(\cosh \varphi - \Delta) + (\Delta \cos \frac{\varphi}{2}(x_j - x_l + 2i) - \cos \frac{\varphi}{2}(x_j - x_l))}{\cos \frac{\varphi}{2}(x_j + x_l)(\cosh \varphi - \Delta) + (\Delta \cos \frac{\varphi}{2}(x_l - x_j + 2i) - \cos \frac{\varphi}{2}(x_l - x_j))}.$$

At $\cosh \varphi - \Delta = 0$, this phase factor becomes a function of $x_j - x_l$ and independent of $x_j + x_l$. It is written as

$$\sin \frac{\varphi}{2}(x_j - x_l + 2i) / \sin \frac{\varphi}{2}(x_j - x_l - 2i).$$

Then for $\Delta > 1$ (159) becomes

$$\left(\frac{\sin \frac{\varphi}{2}(x_j + i)}{\sin \frac{\varphi}{2}(x_j - i)}\right)^N = \prod_{l \neq j} \frac{\sin \frac{\varphi}{2}(x_j - x_l + 2i)}{\sin \frac{\varphi}{2}(x_j - x_l - 2i)},$$
$$\varphi = \cosh^{-1} \Delta, \quad \varphi > 0. \tag{162}$$

For $\varDelta < -1$ we set

$$\exp(ik_j) = -\frac{\sin\frac{\varphi}{2}(x_j + i)}{\sin\frac{\varphi}{2}(x_j - i)}. \tag{163}$$

Equation (159) becomes

$$\left(\frac{\sin\frac{\varphi}{2}(x_j + i)}{\sin\frac{\varphi}{2}(x_j - i)}\right)^N = \prod_{l \neq j} \frac{\sin\frac{\varphi}{2}(x_j - x_l + 2i)}{\sin\frac{\varphi}{2}(x_j - x_l - 2i)},$$

$$\varphi = \cosh^{-1}(-\varDelta), \quad \varphi > 0. \tag{164}$$

In the case $1 > \varDelta > -1$ we set

$$\exp(ik_j) = -\frac{\sinh\frac{\gamma}{2}(x_j + i)}{\sinh\frac{\gamma}{2}(x_j - i)}, \tag{165}$$

yielding the Bethe ansatz equations

$$\left(\frac{\sinh\frac{\gamma}{2}(x_j + i)}{\sinh\frac{\gamma}{2}(x_j - i)}\right)^N = \prod_{l \neq j} \frac{\sinh\frac{\gamma}{2}(x_j - x_l + 2i)}{\sinh\frac{\gamma}{2}(x_j - x_l - 2i)},$$

$$\gamma = \cos^{-1}(-\varDelta), \quad \pi > \gamma > 0. \tag{166}$$

4.3   String Solutions at $\varDelta > 1$

By the transformation (161) the wave function and eigenvalue are written as follows,

$$f(n_1, n_2, ..., n_M)$$

$$= \sum_P \epsilon(P) \prod_{j<l} \sin\frac{\varphi}{2}(x_{Pj} - x_{Pl} + 2i) \prod_{j=1}^M \left(\frac{\sin\frac{\varphi}{2}(x_{Pj} + i)}{\sin\frac{\varphi}{2}(x_{Pj} - i)}\right)^{n_j}, \tag{167}$$

$$E = E_0 + \sum_{j=1}^M \left(2h + \frac{J\sinh^2\varphi}{\cosh\varphi - \cos\varphi x_j}\right), \quad K = \sum_j 2\cot^{-1}\frac{\hat{T}_n^a(\varphi x_j/2)}{\tanh(\varphi/2)}. \tag{168}$$

The following string solutions are possible for complex x_j's for the $N = \infty$ case from the normalizability of the wave function,

$$x_j = \alpha + (M + 1 - 2j)i, \tag{169}$$

where α is a real number at $-Q < \alpha \leq Q$, $Q \equiv \pi/\varphi$. The total momentum and energy is given by

$$K = 2\cot^{-1}\frac{\hat{T}_n^a(\varphi\alpha/2)}{\tanh(M\varphi/2)}, \quad E = E_0 + \frac{J\sinh\varphi\sinh M\varphi}{\cosh M\varphi - \cos\varphi\alpha} + 2Mh. \tag{170}$$

Thus the dispersion is

$$E = E_0 + 2Mh + J\sinh\varphi\left[\frac{\cosh M\varphi - \cos K}{\sinh M\varphi}\right]. \tag{171}$$

This excitation energy gives (71) in the limit $\Delta \to 1$. In the limit of large Δ the energy is $J\Delta + 2Mh$. This is M successive down spins in a sea of up spins. The lowest energy state at $\Delta > 1$ with M down spins is the M string state given by (171) with zero total momentum. So the energy of the ground state for $\Delta > 1$ is

$$E = -\frac{JN\Delta}{4} - (N - 2M)h + J\sinh\varphi\tanh\frac{M\varphi}{2}. \tag{172}$$

4.4   Thermodynamic Equations for the XXZ Model for $\Delta > 1$

Gaudin derived a set of thermodynamic Bethe ansatz equations at $\Delta > 1$ ([Gaudin (1971)]). The wave function for M down spins in the infinite lattice is

$$f(n_1, n_2, ..., n_M) = (z_1 z_2..z_M)^{n_1} \sum_P A(P) \prod_{j=2}^{M} (\prod_{l=j}^{M} z_{Pl})^{n_{j+1}-n_j},$$

$$z_j = e^{ik_j} = (\frac{\sin\frac{\varphi}{2}(x_j + i)}{\sin\frac{\varphi}{2}(x_j - i)}). \tag{173}$$

This corresponds to (73). From the normalizability condition of the wave function we have

$$A(I) \neq 0, A(P \neq I) = 0, \quad |z_1 z_2 ... z_M| = 1$$

$$|\prod_{l=j+1}^{M} z_l| \leq 1, \quad j = 1, 2, 3, ..., M - 1. \tag{174}$$

These conditions are satisfied only if

$$x_j = \alpha + (M + 1 - 2j)i, \quad j = 1, 2, ..., M, \quad Q \geq \alpha > -Q. \tag{175}$$

We can show that

$$|\prod_{l=j}^{M} z_l| = |\frac{\sin\frac{\varphi}{2}(\alpha + i(M - 2j))}{\sin\frac{\varphi}{2}(\alpha - iM)}| = \sqrt{\frac{\cosh\varphi(M - 2j) - \cos\varphi\alpha}{\cosh\varphi n - \cos\varphi\alpha}} \leq 1,$$

$$1 \leq j \leq M - 1,$$

for arbitrary M. Thus a string with arbitrary length is possible for the XXZ chain at $|\Delta| > 1$. In the case $|\Delta| < 1$ the string condition is more complicated than the case $|\Delta| \geq 1$. From (162) we have the following equation corresponding to (80),

$$e_n^N(x_\alpha^n) = \prod_{j=1}^{n} e^N(x_\alpha^{n,j}) = \prod_{(m,\beta)\neq(n,\alpha)} E_{nm}(x_\alpha^n - x_\beta^m). \tag{176}$$

Here

$$e_n(x) = \frac{\sin\frac{\varphi}{2}(x + in)}{\sin\frac{\varphi}{2}(x - in)}, \tag{177}$$

$$E_{nm}(x) \equiv \begin{cases} e_{|n-m|}(x)e^2_{|n-m|+2}(x)e^2_{|n-m|+4}(x)...e^2_{n+m-2}(x)e_{n+m}(x) \\ \qquad\qquad\qquad \text{for} \quad n \neq m, \\ e^2_2(x)e^2_4(x)...e^2_{2n-2}(x)e_{2n}(x) \quad \text{for} \quad n = m. \end{cases} \qquad (178)$$

x^n_α is the real part of α-th string in the strings of length n. The logarithm of (176) is

$$N\theta_n(x^n_\alpha) = 2\pi I^n_\alpha + \sum_{(m,\beta)\neq(n,\alpha)} \Theta_{nm}(x^n_\alpha - x^m_\beta), \qquad (179)$$

where

$$\theta_n(x) = 2\hat{T}^a_n{}^{-1}(\frac{\hat{T}^a_n \frac{x\varphi}{2}}{\tanh \frac{n\varphi}{2}}) + 2\pi[\frac{x+Q}{2Q}],$$

and

$$\Theta_{nm}(x) \equiv \begin{cases} \theta_{|n-m|}(x) + 2\theta_{|n-m|+2}(x) + ... + 2\theta_{n+m-2}(x) + \theta_{n+m}(x) \\ \qquad\qquad\qquad \text{for} \quad n \neq m, \\ 2\theta_2(x) + 2\theta_4(x) + ... + 2\theta_{2n-2}(x) + \theta_{2n}(x) \quad \text{for} \quad n = m. \end{cases} \qquad (180)$$

The function $\theta_n(x)$ is a quasi periodic function which satisfies

$$\theta_n(x + 2jQ) = \theta_n(x) + 2\pi j, \quad j = \text{integer}.$$

We consider the energy of general eigenstates which is given by the set of quantum numbers $\{I^n_\alpha\}$,

$$E(\{I^n_\alpha\}) = N(-h - \frac{J\Delta}{4}) + \sum_{n,\alpha}(\frac{2\pi J \sinh \varphi}{\varphi}a_n(x^n_\alpha) + 2hn), \qquad (181)$$

where

$$a_n(x) = \frac{1}{2\pi}\frac{\varphi \sinh n\varphi}{\cosh n\varphi - \cos \varphi x}. \qquad (182)$$

The partition function of the XXZ model is as follows,

$$Z = \sum_{M=0}^{[N/2]}(1 + (1 - \delta_{N,2M})\exp -\frac{(N - 2M)h}{T}) \sum_{\{I^n_\alpha\}} \exp[-T^{-1}E(\{I^n_\alpha\})]. \qquad (183)$$

Corresponding to (88) we define the following functions,

$$h_n(x) \equiv \theta_n(x) - N^{-1} \sum_{(m,\alpha)} \Theta_{nm}(x - x^m_\alpha). \qquad (184)$$

Using this function we can determine the position of holes for n-strings. We define the distribution functions of particles and holes of n-strings as $\rho_n(x)$ and $\rho^h_n(x)$. By the equation (179) we have the conditions for these two kinds of functions

$$a_n(x) = \rho_n(x) + \rho^h_n(x) + \sum_m T_{nm} * \rho_m(x). \qquad (185)$$

Here $\mathbf{a}_n$ was defined in (182) and

$$\mathbf{T}_{nm}(x) \equiv \begin{cases} \mathbf{a}_{|n-m|}(x) + 2\mathbf{a}_{|n-m|+2}(x) + 2\mathbf{a}_{|n-m|+4}(x) + \cdots \\ \qquad + 2\mathbf{a}_{n+m-2}(x) + \mathbf{a}_{n+m}(x) \quad \text{for} \quad n \neq m, \\ 2\mathbf{a}_2(x) + 2\mathbf{a}_4(x) + \cdots + 2\mathbf{a}_{2n-2}(x) + \mathbf{a}_{2n}(x) \quad \text{for} \quad n = m. \end{cases} \tag{186}$$

Here the meaning of convolution of two periodic functions $\mathbf{a}$ and $\mathbf{b}$ with periodicity $2Q$ is redefined

$$\mathbf{a} * \mathbf{b}(x) \equiv \int_{-Q}^{Q} \mathbf{a}(x-y)\mathbf{b}(y)dy, \tag{187}$$

The energy per site is

$$e = -(\frac{J\Delta}{4} + h) + \sum_{n=1}^{\infty} \int_{-Q}^{Q} g_n(x)\rho_n(x)dx,$$

$$g_n(x) \equiv \frac{2\pi J \sinh\varphi}{\varphi} \mathbf{a}_n(x) + 2nh. \tag{188}$$

The entropy per site s is

$$s = \sum_{n=1}^{\infty} \int_{-Q}^{Q} \rho_n(x)\ln(1 + \frac{\rho_n^h(x)}{\rho_n(x)}) + \rho_n^h(x)\ln(1 + \frac{\rho_n(x)}{\rho_n^h(x)})dx. \tag{189}$$

The condition of minimizing the free energy $e - Ts$ yields equations for $\eta_n(x) \equiv \rho_n^h(x)/\rho_n(x)$,

$$\ln \eta_n(x) = \frac{g_n(x)}{T} + \sum_{m=1}^{\infty} \mathbf{T}_{nm} * \ln(1 + \eta_m^{-1}(x)). \tag{190}$$

This set of equations is equivalent to the following one

$$\ln \eta_1(x) = \frac{2\pi J \sinh\varphi}{T\varphi} s(x) + s * \ln(1 + \eta_2(x)), \tag{191}$$

$$\ln \eta_n(x) = s * \ln(1 + \eta_{n-1}(x))(1 + \eta_{n+1}(x)), \tag{192}$$

$$\lim_{n \to \infty} \frac{\ln \eta_n}{n} = \frac{2h}{T}, \tag{193}$$

where

$$s(x) = \frac{1}{4} \sum_{n=-\infty}^{\infty} \text{sech}(\frac{\pi(x - 2nQ)}{2}). \tag{194}$$

The free energy per site is

$$f = -(\frac{J\Delta}{4} + h) - T\sum_{n=1}^{\infty} \int_{-Q}^{Q} \mathbf{a}_n(x)\ln(1 + \eta_n^{-1}(x))dx. \tag{195}$$

Corresponding to (104) we have another expression for the free energy,

$$f = J\left[\frac{2\pi \sinh\varphi}{\varphi} \int_{-Q}^{Q} \mathbf{a}_1(x)s(x)dx - \frac{\Delta}{4}\right] - T\int_{-Q}^{Q} s(x)\ln(1 + \eta_1(x))dx. \tag{196}$$

(115-117) and (191-193) have the almost same structure. These equations are called Gaudin-Takahashi equation ([Gaudin (1971)], [Takahashi (1971a)]).

4.5   Theory for $|\Delta| < 1$ XXZ Model

String Solution of an Infinite XXZ Model with $|\Delta| < 1$ The shapes
of strings for $|\Delta| < 1$ are quite different from those at $|\Delta| \geq 1$. Takahashi
and Suzuki proposed a condition of the strings and constructed thermodynamic
integral equations ([Takahashi and Suzuki (1972)]). Later Hida, Fowler and
Zotos derived these conditions from the normalizability condition of the string
wave function for an infinite chain ([Hida (1981)], [Fowler and Zotos (1981)]).
For $|\Delta| < 1$ there are two kinds of strings, one of which has the center on the
real axis and the other is centered on the $p_0 i$ axis,

$$x_j = \alpha + (n + 1 - 2j)i, \quad j = 1, 2, ..., n, \tag{197}$$

$$x_j = \alpha + (n + 1 - 2j)i + p_0 i, \quad j = 1, 2, ..., n. \tag{198}$$

We designate that the string of the former type has parity $v = 1$ and that the
latter has parity $v = -1$. Applying the normalizability condition of the form
(197) yields

$$1 > |\prod_{l=j+1}^{n} z_l| = |\frac{\sinh \frac{\gamma}{2}(\alpha + i(n - 2j))}{\sinh \frac{\gamma}{2}(\alpha - in)}| = \sqrt{\frac{\cosh \gamma\alpha - \cos \gamma(n - 2j)}{\cosh \gamma\alpha - \cos \gamma n}}.$$

Thus $\cos \gamma n < \cos \gamma(n - 2j)$ for $j = 1, 2, 3, ..., n - 1$. For (198) we have

$$1 > |\prod_{l=j+1}^{n} z_l| = |\frac{\cosh \frac{\gamma}{2}(\alpha + i(n - 2j))}{\cosh \frac{\gamma}{2}(\alpha - in)}| = \sqrt{\frac{\cosh \gamma\alpha + \cos \gamma(n - 2j)}{\cosh \gamma\alpha + \cos \gamma n}},$$

and therefore $\cos \gamma n > \cos \gamma(n - 2j)$ for $j = 1, 2, 3, ..., n - 1$. Then from the
normalizability condition we get

$$0 < v(\cos((n - 2j)\gamma) - \cos(n\gamma))$$
$$= 2v \sin((n - j)\gamma) \sin(j\gamma), \quad \text{for } j = 1, 2, ..., n - 1. \tag{199}$$

This is equivalent to

$$(-1)^{[\frac{(n-j)\gamma}{\pi}]+[\frac{j\gamma}{\pi}]} = v, \quad \text{for } j = 1, 2, ..., n - 1, \tag{200}$$

$$\frac{j\gamma}{\pi} \neq [\frac{j\gamma}{\pi}], \quad \text{for } j = 1, 2, ..., n - 1, \tag{201}$$

where $[x]$ denotes the maximum integer less than or equal to x (Gauss' symbol).
For rational $p_0 = \pi/\gamma$, (201) is a strong condition. If $p_0 = n_1/n_2$, and n_1 and
n_2 are coprime, the string with length greater than n_1 cannot satisfy at least
one of (201). Thus $n \geq n_1 + 1$ strings are forbidden. Moreover for a $n = n_1$
string, the momentum is always 0 or π. So this string has also no meaning for
the thermodynamics. Next we seek the number n and parity v which satisfies
(200) within $n < n_1$. Equation (200) is equivalent to

$$[\frac{(n-j)\gamma}{\pi}] + [\frac{j\gamma}{\pi}] \equiv [\frac{(n-j-1)\gamma}{\pi}] + [\frac{(j+1)\gamma}{\pi}] (\text{Mod}2), \quad j = 1, 2, ..., n - 2,$$

$$[\frac{(n-1)\gamma}{\pi}] \equiv \frac{1-v}{2} \quad (\text{Mod} 2).$$

As $[\frac{(n-j)\gamma}{\pi}] - [\frac{(n-j-1)\gamma}{\pi}]$ is 0 or 1 and $[\frac{j\gamma}{\pi}] - [\frac{(j+1)\gamma}{\pi}]$ is 0 or -1, we obtain

$$[\frac{(n-j)\gamma}{\pi}] + [\frac{j\gamma}{\pi}] = [\frac{(n-j-1)\gamma}{\pi}] + [\frac{(j+1)\gamma}{\pi}], \quad j = 1, 2, ..., n-2.$$

These are strong restrictions on the parity v and the length of the string n. The above conditions are equivalent to the following conditions which were given in ([Takahashi and Suzuki (1972)]). The length n of a string should satisfy

$$2\sum_{j=1}^{n-1}[j\gamma/\pi] = (n-1)[(n-1)\gamma/\pi], \qquad (202)$$

$$v\sin\{(n-1)\gamma\} \geq 0. \qquad (203)$$

This condition was first introduced under the assumption that these strings form a complete half-filled state ([Takahashi and Suzuki (1972)]). Later Hida, Fowler and Zotos showed that conditions (202,203) can be rederived from the normalizability conditions of the wave function for $N \to \infty$ and finite M ([Hida (1981)], [Fowler and Zotos (1981)]). For a given value of Δ (or γ) we can determine the series of n which satisfies the conditions (202) and (203). If γ/π is a rational number, this series becomes finite and the number of unknown functions is also finite. We consider the $\gamma = \pi/v$, $v =$ integer case. In this case $n = 1, 2, ..., v$ satisfy (202). For $n = 1$ both $v = 1$ and -1 are possible. For $n = 2, 3, ..., v-1$, only $v = +1$ states are possible. These excitations have the following energy and momentum

$$E = -2J\frac{\sin\gamma\sin(n\gamma)}{v\cosh(\alpha\gamma) - \cos n\gamma} + 2nh, \qquad (204)$$

$$K = -i\ln\left(-\frac{\sinh\frac{1}{2}(\alpha\gamma + i(1-v)\pi/2 + in\gamma)}{\sinh\frac{1}{2}(\alpha\gamma + i(1-v)\pi/2 - in\gamma)}\right). \qquad (205)$$

The energy and momentum have the following relation,

$$E = -J\sin\gamma\frac{\cos n\gamma - \cos K}{\sin n\gamma} + 2nh.$$

The momentum is restricted to the region

$$|K| < \pi - (n\gamma - \pi[\frac{n\gamma}{\pi}]) \quad \text{for} \quad v = 1,$$

$$\pi \geq |K| > \pi - (n\gamma - \pi[\frac{n\gamma}{\pi}]) \quad \text{for} \quad v = -1.$$

Then for $n = v$, the energy and momentum are always zero. Only one state is obtained from this string solution. So we exclude this $n = v$ state from the thermodynamics of this case. So v string states $(1,+), (2,+), ..., (v-1,+), (1,-)$ play important roles. Especially at $\Delta = 0$, $\gamma = \pi/2$, $v = 2$ we have only string states $(1, +), (1, -)$. These are single states at momentum $|K| < \pi/2$ and $|K| > \pi/2$.

Next we consider the $\gamma = \pi/(\nu_1 + 1/\nu_2)$ case. $(1, +), (2, +), ..., (\nu_1 - 1, +)$, $(1, -), (1 + \nu_1, +), (1 + 2\nu_1, -), ...(1 + (\nu_2 - 1)\nu_1, (-1)^{\nu_2 - 1}), (\nu_1, (-1)^{\nu_2})$ satisfy conditions (202, 203). Thus $\nu_1 + \nu_2$ strings are necessary to describe the thermodynamics of this case.

For a general rational number between 0 and 1, we can express it by a continued fraction with length l,

$$\frac{\gamma}{\pi} = \frac{1|}{|\nu_1} + \frac{1|}{|\nu_2} + ... + \frac{1|}{|\nu_l}, \quad \nu_1, \nu_2, ..., \nu_{l-1} \geq 1, \quad \nu_l \geq 2. \tag{206}$$

We define the following series of numbers $y_{-1}, y_0, y_1, ..., y_l$ and $m_0, m_1, ..., m_l$ as

$$y_{-1} = 0, \quad y_0 = 1, \quad y_1 = \nu_1 \text{ and } y_i = y_{i-2} + \nu_i y_{i-1},$$

$$m_0 = 0, \quad m_i = \sum_{k=1}^{i} \nu_k. \tag{207}$$

The general rule to determine the parity v and length n is as follows

$$n_j = y_{i-1} + (j - m_i)y_i, \quad v_j = (-1)^{[(n_j - 1)/p_0]} \text{ for } m_i < j < m_{i+1},$$

$$n_{m_l} = y_{l-1}, \quad v_{m_l} = (-1)^l. \tag{208}$$

The number of strings is m_l. We give examples for some rational numbers in Tables (1), (2). We put x_j^α as the real part of strings with parity v_j and length n_j. α takes values from 1 to M_j. We find the following relations for these series of numbers

$$n_j = \frac{1}{2}[(1 - 2\delta_{m_i, j})n_{j-1} + n_{j+1}], \quad \text{for } m_i \leq j \leq m_{i+1} - 2,$$

$$n_j = (1 - 2\delta_{m_{i-1}, j})n_{j-1} + n_{j+1}, \quad \text{for } j = m_i - 1, \quad i < l,$$

$$n_0 = 0, \quad n_{m_l} + n_{m_l - 1} = y_l. \tag{209}$$

Table 1. Length n_j, parity v_j and q_j of strings for some rational values of γ/π

j	$\frac{1}{5}$			$\frac{3}{16} =$		$\frac{1}{5+\frac{1}{3}}$	$\frac{13}{69} =$		$\frac{1}{5+\frac{1}{3+\frac{1}{4}}}$
1	1	+	4	1	+	13/3	1	+	56/13
2	2	+	3	2	+	10/3	2	+	43/13
3	3	+	2	3	+	7/3	3	+	30/13
4	4	+	1	4	+	4/3	4	+	17/13
5	1	−	−1	1	−	−3/3	1	−	−13/13
6				6	+	−2/3	6	+	−9/13
7				11	−	−1/3	11	−	−5/13
8				5	+	1/3	5	+	4/13
9							21	−	3/13
10							37	+	2/13
11							53	−	1/13
12							16	−	−1/13

Table 2. Length n_j, parity v_j and q_j of strings for conjugate values of γ/π in previous table

j	$\frac{4}{5} =$		$\frac{1}{1+\frac{1}{4}}$	$\frac{13}{16} =$		$\frac{1}{1+\frac{1}{4+\frac{1}{3}}}$	$\frac{56}{69} =$		$\frac{1}{1+\frac{1}{4+\frac{1}{3+\frac{1}{4}}}}$
1	1	−	−4/4	1	−	−13/13	1	−	−56/56
2	2	+	−3/4	2	+	−10/13	2	+	−43/56
3	3	−	−2/4	3	−	−7/13	3	−	−30/56
4	4	+	−1/4	4	+	−4/13	4	+	−17/56
5	1	+	1/4	1	+	3/13	1	+	13/56
6				6	+	2/13	6	+	9/56
7				11	+	1/13	11	+	5/56
8				5	−	−1/13	5	−	−4/56
9							21	+	−3/56
10							37	−	−2/56
11							53	+	−1/56
12							16	+	1/56

Scattering Phase Shift Among Strings Corresponding to (80) and (176) we have Bethe ansatz equations for strings as follows

$$\{e_j(x_\alpha^j)\}^N = -\prod_{k=1}^{m_l}\prod_{\beta=1}^{M_k} E_{j,k}(x_\alpha^j - x_\beta^k), \tag{210}$$

where

$$e_j(x) = g(x; n_j, v_j), \tag{211}$$

$$E_{j,k}(x) = \begin{cases} g(x; 2n_j, v_j v_k)\prod_{l=1}^{n_j-1} g^2(x; 2l, v_j v_k) & \text{for } n_j = n_k, \\ g(x; (n_j + n_k), v_j v_k)g(x; |n_j - n_k|, v_j v_k) \\ \times \prod_{l=1}^{Min(n_j,n_k)-1} g^2(x; |n_j - n_k| + 2l, v_j v_k) & \text{for } n_j \neq n_k, \end{cases} \tag{212}$$

a) $\gamma = \pi/5$

$p_0 = 5$

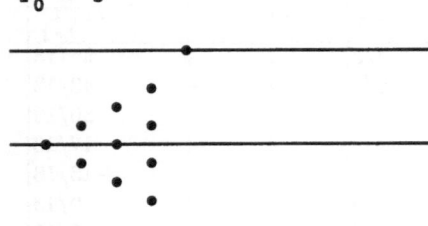

b) $\gamma = 4\pi/5$

$p_0 = \frac{5}{4}$

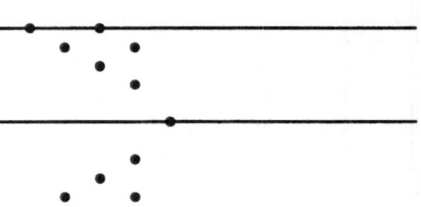

c) $\gamma = 3\pi/16$

$p_0 = \frac{16}{3}$

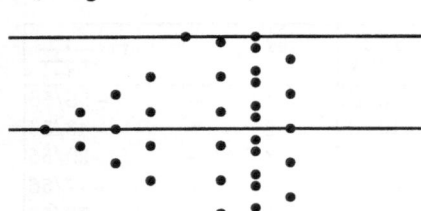

d) $\gamma = 13\pi/16$

$p_0 = \frac{16}{13}$

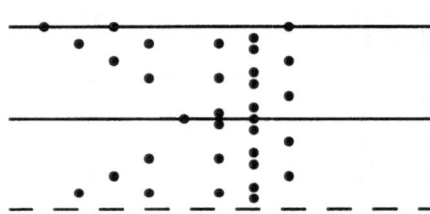

e) $\gamma = 13\pi/69$

$p_0 = \frac{69}{13}$

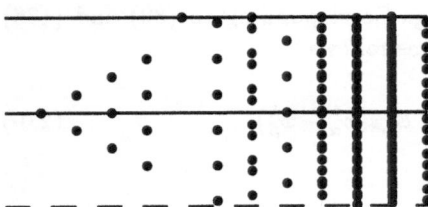

f) $\gamma = 56\pi/69$

$p_0 = \frac{69}{56}$

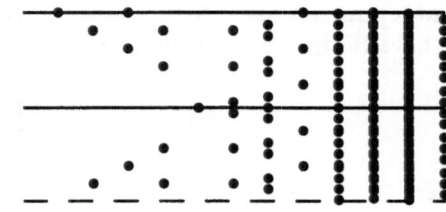

Fig. 1. Strings on the complex plane for given values of $p_0 = \pi/\gamma$. If we change $\gamma \rightarrow \pi - \gamma$, the strings are almost the same. But imaginary parts of γx_j shift by π.

$$g(x; n, +) = \frac{\sinh \frac{\gamma}{2}(x + in)}{\sinh \frac{\gamma}{2}(x - in)}, \quad g(x; n, -) = -\frac{\cosh \frac{\gamma}{2}(x + in)}{\cosh \frac{\gamma}{2}(x - in)}. \tag{213}$$

The logarithm of (210) yields

$$N\theta_j(x_\alpha^j) = 2\pi I_\alpha^j + \sum_{k=1}^{m_l} \sum_{\beta=1}^{M_k} \Theta_{j,k}(x_\alpha^j - x_\beta^k), \quad \alpha = 1, 2, ..., M_j, \tag{214}$$

where

$$\theta_j(x) = f(x; n_j, v_j), \quad \Theta_{jk}(x) = f(x; |n_j - n_k|, v_j v_k) +$$

$$f(x; n_j + n_k, v_j v_k) + 2 \sum_{i=1}^{Min(n_j, n_k)-1} f(x; |n_j - n_k| + 2i, v_j v_k), \tag{215}$$

and

$$f(x; n, v) \equiv \begin{cases} 0 & \text{for } n/p_0 = \text{integer}, \\ 2v\hat{T}_n^{a-1}\{(\cot(n\pi/2p_0))^v \tanh(\pi x/2p_0)\} & \text{otherwise.} \end{cases} \tag{216}$$

The quantity I_α^j is an integer (half-odd integer) for M_j odd (even), which is located in the region

$$|I_\alpha^j| < \frac{1}{2\pi} |N\theta_j(\infty) - \sum_{k=1}^{m_l} M_k \Theta_{j,k}(\infty)|. \tag{217}$$

The function $f(x; n_j, v_j)$ is a monotonically increasing function for $m_{2i} \leq j < m_{2i+1}$ and a monotonically decreasing function for $m_{2i-1} \leq j < m_{2i}$.

Bethe-Ansatz Equation for XXZ Model with $|\Delta| < 1$ Following Yang and Yang we define particles and holes of strings. We obtain an integral equation for distribution functions ρ_j and ρ_j^h of particles and holes of strings in the thermodynamic limit,

$$a_j(x) = \text{sign}(q_j)(\rho_j(x) + \rho_j^h(x)) + \sum_{k=1}^{m_l} T_{j,k} * \rho_k(x). \tag{218}$$

Here

$$a_j(x) \equiv (2\pi)^{-1} \frac{d}{dx} \theta_j(x), \quad T_{j,k}(x) \equiv (2\pi)^{-1} \frac{d}{dx} \Theta_{j,k}(x). \tag{219}$$

The symbol $a * b$ denotes the convolution of $a(x)$ and $b(x)$ as follows

$$a * b(x) = \int_{-\infty}^{\infty} a(x - y)b(y)dy. \tag{220}$$

The functions $a_j(x)$ and their Fourier transforms $\tilde{a}_j(\omega)$ are written as

$$a_j(x) = \frac{1}{2\pi} \frac{\gamma \sin \gamma q_j}{\cosh \gamma x + \cos \gamma q_j}, \quad \tilde{a}_j(\omega) = \frac{\sinh q_j \omega}{\sinh p_0 \omega}, \tag{221}$$

$$q_j \equiv (-1)^i (p_i - (j - m_i) p_{i+1}), \quad \text{for } m_i \leq j < m_{i+1}, \tag{222}$$

where the p_i, $i = 0, 1, ..., l$ are defined by

$$p_0 = \pi/\gamma, \quad p_1 = 1, \quad \nu_i = [p_{i-1}/p_i], \quad p_i = p_{i-2} - p_{i-1}\nu_{i-1}. \tag{223}$$

For the series q_j we find the following relations

$$q_j = \frac{1}{2}[(1 - 2\delta_{m_i,j})q_{j-1} + q_{j+1}], \quad \text{for } m_i \leq j \leq m_{i+1} - 2,$$

$$q_j = (1 - 2\delta_{m_{i-1},j})q_{j-1} + q_{j+1}, \quad \text{for } j = m_i - 1, \quad i < l,$$

$$q_0 = p_0, \quad n_{m_l} + n_{m_l-1} = 0. \tag{224}$$

The Fourier transform of $T_{j,k}(x)$ is given by

$$\tilde{T}_{j,k}(\omega) = \tilde{T}_{k,j}(\omega)$$

$$= 2\mathrm{sign}(q_j) \coth(p_{i+1}\omega) \frac{\sinh((p_0 - |q_j|)\omega)\sinh(q_k\omega)}{\sinh(p_0\omega)}$$

$$+\delta_{j,m_l-1}\delta_{k,m_l} - \delta_{j,k}, \quad \text{for } j \leq k, \quad m_i < j \leq m_{i+1}. \tag{225}$$

At $j = 1$ we have

$$\tilde{T}_{1,k}(\omega) = \mathrm{sign}(q_1)2\cosh\omega\tilde{a}_k(\omega) - \delta_{1,k}. \tag{226}$$

At $j = m_l$ we have

$$\tilde{T}_{m_l,k}(\omega) = -\tilde{T}_{m_l-1,k}(\omega),$$

$$\tilde{T}_{m_l,m_l}(\omega) = -\tilde{T}_{m_l-1,m_l}(\omega) = \frac{\sinh((p_0 - 2p_l)\omega)}{\sinh(p_0\omega)}. \tag{227}$$

The energy and entropy per site are given by

$$e = -(\frac{J\Delta}{4} + h) + \sum_{j=1}^{m_l} \int_{-\infty}^{\infty} g_j(x)\rho_j(x)dx,$$

$$g_j(x) \equiv -\frac{2\pi J \sin\gamma}{\gamma}a_j(x) + 2n_j h, \tag{228}$$

and

$$s = \int_{-\infty}^{\infty} \rho_j \ln(1 + \frac{\rho_j^h}{\rho_j}) + \rho_j^h \ln(1 + \frac{\rho_j}{\rho_j^h})dx. \tag{229}$$

To minimize the free energy density $e - Ts$ with respect to ρ_j, we have

$$\delta(e - Ts) = \sum_j \int_{-\infty}^{\infty} g_j(x)\delta\rho_j(x) - T\left\{\delta\rho_j \ln(1 + \frac{\rho_j^h}{\rho_j}) + \delta\rho_j^h \ln(1 + \frac{\rho_j}{\rho_j^h})\right\}dx.$$

The variation of (218) gives

$$\delta\rho_j^h = -\delta\rho_j - \text{sign}(q_j)\sum T_{jk} * \delta\rho_k.$$

Thus we obtain

$$\delta(e - Ts) = T\sum_j \int_{-\infty}^{\infty} dx \delta\rho_j(x)$$

$$\left\{\frac{g_j(x)}{T} - \ln(\frac{\rho_j^h}{\rho_j}) + \sum_k \text{sign}(q_k)T_{j,k} * \ln(1 + \frac{\rho_k}{\rho_k^h})\right\}.$$

At the thermodynamic equilibrium one obtains the following non-linear equations determining $\eta_j(x) \equiv \rho_j^h(x)/\rho_j(x)$,

$$\ln \eta_j(x) = g_j(x)/T + \sum_{k=1}^{m_l} \text{sign}(q_k)T_{k,j} * \ln(1 + \eta_k^{-1}(x)), \quad j = 1, ..., m_l. \quad (230)$$

The free energy is given as follows,

$$f = e - Ts = -(\frac{J\Delta}{4} + h)$$

$$+ \sum_{j=1}^{m_l} \int_{-\infty}^{\infty} \rho_j(x)[g_j(x) - T\ln\eta_j(x)] - T[\rho_j + \rho_j^h]\ln(1 + \eta_j^{-1}).$$

If we substitute (230) into the first bracket and (218) into the second, the $T_{j,k}$ terms are cancelled and we get

$$f = -(\frac{J\Delta}{4} + h) - T\sum_{j=1}^{m_l} \text{sign}(q_j)\int_{-\infty}^{\infty} a_j(x)\ln(1 + \eta_j^{-1}(x))dx. \quad (231)$$

If one uses the $j = 1$ case of equation (230) and (226), one obtains

$$f = -\frac{J\Delta}{4} - \text{sign}(q_1)\frac{2\pi J \sin\gamma}{\gamma}\int_{-\infty}^{\infty} a_1(x)s_1(x)dx$$

$$-T\int_{-\infty}^{\infty} s_1(x)\ln(1 + \eta_1(x))dx. \quad (232)$$

From equation (221-223) we get the following relations,

$$a_j - s_i * ((1 - 2\delta_{m_{i-1},j})a_{j-1} + a_{j+1}) = 0$$
for $m_{i-1} \le j \le m_i - 2$,
$$a_{m_i-1} - (1 - 2\delta_{m_{i-1},m_i-1})s_i * a_{m_i-2} - d_i * a_{m_i-1} - s_{i+1} * a_{m_i} = 0$$
for $i < l$,
$$a_{m_l-1}(x) = -a_{m_l}(x) = s_l * a_{m_l-2}, \quad (233)$$

where

$$a_0(x) = \delta(x),$$

$$s_i(x) \equiv \int_{-\infty}^{\infty} \frac{d\omega}{4\pi} \frac{e^{i\omega x}}{\cosh(p_i\omega)} = \frac{1}{4p_i}\operatorname{sech}\frac{\pi x}{2p_i},$$

$$d_i(x) \equiv \int_{-\infty}^{\infty} \frac{d\omega}{4\pi} \frac{e^{i\omega x}\cosh((p_i - p_{i+1})\omega)}{\cosh(p_i\omega)\cosh(p_{i+1}\omega)}. \tag{234}$$

Using (225) one can show the following relations,

$$T_{j,k} - s_i * ((1 - 2\delta_{m_{i-1},j})T_{j-1,k} + T_{j+1,k})$$
$$= (-1)^{i+1}(\delta_{j-1,k} + \delta_{j+1,k})s_i,$$
$$\text{for } m_{i-1} \leq j \leq m_i - 2,$$
$$T_{m_i-1,k} - (1 - 2\delta_{m_{i-1},m_i-1})s_i * T_{m_i-2,k} - d_i * T_{m_i-1,k}$$
$$-s_{i+1} * T_{m_i,k} = (-1)^{i+1}(\delta_{m_i-2,k}s_i + \delta_{m_i-1,k}d_i - \delta_{m_i,k}s_{i+1}),$$
$$\text{for } i = 1, 2, ., l - 1,$$
$$T_{m_l-1,k} = -T_{m_l,k} = s_l * T_{m_l-2,k} + \operatorname{sign}(q_k)\delta_{m_l-2,k}s_l, \tag{235}$$

with $T_{0,k} = 0$. Using (233) and (235) one can rewrite (218) as follows

$$\rho_j + \rho_j^h = s_i * (\rho_{j-1}^h + \rho_{j+1}^h) \text{ for } m_{i-1} \leq j \leq m_i - 2,$$
$$\rho_{m_i-1} + \rho_{m_i-1}^h = s_i * \rho_{m_i-2}^h + d_i * \rho_{m_i-1}^h - s_{i+1} * \rho_{m_i}^h,$$
$$\rho_{m_l-1} + \rho_{m_l-1}^h = \rho_{m_l} + \rho_{m_l}^h = s_l * \rho_{m_l-1}^h, \tag{236}$$

with $\rho_0^h = \delta(x)$. Equations (230) are rewritten as

$$\ln(1 + \eta_0) = -\frac{2\pi J \sin\gamma}{\gamma T}\delta(x),$$
$$\ln\eta_j = (1 - 2\delta_{m_{i-1},j})s_i * \ln(1 + \eta_{j-1}) + s_i * \ln(1 + \eta_{j+1}),$$
$$\text{for } m_{i-1} \leq j \leq m_i - 2, j \neq m_l - 2$$
$$\ln\eta_{m_i-1} = (1 - 2\delta_{m_{i-1},m_i-1})s_i * \ln(1 + \eta_{m_i-2})$$
$$+d_i * \ln(1 + \eta_{m_i-1}) + s_{i+1} * \ln(1 + \eta_{m_i}), \text{ for } i < l$$
$$\ln\eta_{m_l-2} = (1 - 2\delta_{m_{l-1},m_l-2})s_l * \ln(1 + \eta_{m_l-3})$$
$$+s_l * \ln((1 + \eta_{m_l-1})(1 + \eta_{m_l}^{-1})),$$
$$\ln\eta_{m_l-1} - y_l h/T = y_l h/T - \ln\eta_{m_l}$$
$$= s_l * \ln(1 + \eta_{m_l-2}). \tag{237}$$

Then if we write $\ln\kappa(x) = \ln\eta_{m_l-1} - y_l h/T$ we have integral equations with $m_l - 1$ unknown functions

$$\ln(1 + \eta_0) = -\frac{2\pi J \sin\gamma}{\gamma T}\delta(x),$$
$$\ln\eta_j = (1 - 2\delta_{m_{i-1},j})s_i * \ln(1 + \eta_{j-1}) + s_i * \ln(1 + \eta_{j+1}),$$

for $m_{i-1} \leq j \leq m_i - 2, j \neq m_l - 2$

$\ln \eta_{m_i-1} = (1 - 2\delta_{m_{i-1},m_i-1})s_i * \ln(1 + \eta_{m_i-2})$

$+d_i * \ln(1 + \eta_{m_i-1}) + s_{i+1} * \ln(1 + \eta_{m_i})$, for $i < l$

$\ln \eta_{m_l-2} = (1 - 2\delta_{m_{l-1},m_l-2})s_l * \ln(1 + \eta_{m_l-3})$

$+s_l * \ln(1 + 2\cosh(y_l h/T)\kappa + \kappa^2)$,

$\ln \kappa(x) = s_l * \ln(1 + \eta_{m_l-2})$.   (238)

4.6   Some Special Limits

$T \rightarrow \infty$ or $J \rightarrow 0$ **Limit** In equations (238), $\ln(1 + \eta_0)$ becomes zero and $\eta_j(x)$ are all independent of x. This yields the following difference equation

$$\eta_j^2 = (1 + \eta_{j-1})^{1-2\delta_{m_{i-1},j}}(1 + \eta_{j+1})$$

for $m_{i-1} \leq j \leq m_i - 2, j \neq m_l - 2$

$$\eta_{m_i-1}^2 = (1 + \eta_{m_i-2})^{1-2\delta_{m_{i-1},m_i-1}}(1 + \eta_{m_i-1})\ln(1 + \eta_{m_i}),$$

for $i < l$

$$\eta_{m_l-2}^2 = (1 + \eta_{m_l-3})^{1-2\delta_{m_{l-1},m_l-2}}(1 + 2\cosh(y_l h/T)\kappa + \kappa^2),$$

$$\kappa^2 = (1 + \eta_{m_l-2}).$$ (239)

The solution of this set of equations is

$$\eta_j = \left(\frac{\sinh(n_j + y_{i-1})h/T}{\sinh(y_{i-1}h/T)}\right)^2 - 1$$

for $m_{i-1} < j \leq m_i$, $j \leq m_l - 2$,

$$\kappa = \frac{\sinh(n_{m_l-2} + y_l)h/T}{\sinh(y_l h/T)}.$$ (240)

For $j = 1$ we have $\eta_1 = (2\cosh h/T)^2 - 1$. Substituting this into (232) we find the free energy

$$f/T = -\ln(2\cosh h/T).$$ (241)

At $h = 0$ this gives that entropy per site is $\ln 2$, as it should be.

$J > 0$, $T \rightarrow 0$ **Limit** We define $\epsilon_j(x) = T\ln \eta_j(x)$ and $\epsilon_j^+(x) = T\ln(1 + \eta_j(x))$. One can show that ϵ_j, $j \geq 2$ is always positive. The equation (230) gives

$$\epsilon_1(x) = -\frac{2\pi J \sin \gamma}{\gamma}a(x, 1) + 2h - \int_{-\infty}^{\infty} a(x - y, 2)\epsilon_1^-(y)dy.$$ (242)

If $\epsilon(x) < 0$ at $|x| < B$ and $\epsilon(x) > 0$ at $|x| > B$, then one obtains a linear integral equation for $\rho_1(x)$,

$$\rho_1(x) + \int_{-B}^{B} a(x - y, 2)\rho_1(y)dy = a(x, 1),$$ (243)

where

$$a(x, n) \equiv \frac{1}{2\pi} \frac{\gamma \sin n\gamma}{\cosh \gamma x - \cos n\gamma}. \tag{244}$$

5 Thermodynamics of the XYZ Model

5.1 Bethe-Ansatz Equations for the XYZ Model

Here we consider the symmetry of the following Hamiltonian

$$\mathcal{H} = -\sum_{l=1}^{N} J_x S_l^x S_{l+1}^x + J_y S_l^y S_{l+1}^y + J_z S_l^z S_{l+1}^z. \tag{245}$$

We assume N is even. By the transformation

$$U_2 \mathcal{H} U_2^{-1}, \quad U_2 = \prod_{l=even} 2S_l^z,$$

$\mathcal{H}(J_x, J_y, J_z) \to \mathcal{H}(-J_x, -J_y, J_z)$. In the same way
$\mathcal{H}(J_x, J_y, J_z) \to \mathcal{H}(J_x, -J_y, -J_z)$ and $\mathcal{H}(J_x, J_y, J_z) \to \mathcal{H}(J_x, -J_y, -J_z)$. Namely
the energy spectrum of this Hamiltonian is unchanged for reversing signs of two
J_α's. It is evident that the spectrum is unchanged for exchanging J_α's. Thus it
is sufficient to treat only the case $1 \geq J_y/J_z \geq |J_x|/J_z \geq 0$. Baxter solved the
eight-vertex model and also the XYZ model ([Baxter (1972a)], [Baxter (1972b)]).
The Bethe ansatz equation for this model is

$$\left(\frac{H_l(i\zeta(x_l + i))}{H_l(i\zeta(x_l - i))}\right)^N = -e^{-2\pi i\nu'/p_0} \prod_{j=1}^{N/2} \frac{H_l(i\zeta(x_l - x_j + 2i))}{H_l(i\zeta(x_l - x_j - 2i))},$$

$$\sum_{l}^{N/2} x_l = Q\nu' + ip_0\nu, \quad Q = K(l')/\zeta, \quad p_0 = K(l)/\zeta. \tag{246}$$

Here the modulus l and the parameter ζ are determined by

$$l = \sqrt{\frac{J_z^2 - J_y^2}{J_z^2 - J_x^2}}, \quad cn(2\zeta, l) = -J_x/J_z.$$

There are $N/2$ rapidities. $H_l(x)$ is the Jacobian elliptic function defined by

$$H_l(x) \equiv 2 \sum_{n=1}^{\infty} (-1)^{n+1} q^{n(n-1)+1/4} \sin(2n - 1)\frac{\pi x}{2K}, \quad q \equiv \exp(-\frac{\pi K(l')}{K(l)}).$$

This function has the following properties,

$$H_l(x) = -H_l(x + 2K(l)) = -qe^{i\pi x/K(l)}H_l(x + 2iK(l')). \tag{247}$$

In the Bethe ansatz equation (166), the function $\sinh \frac{\gamma}{2} x$ is merely replaced by the elliptic theta function. The energy is given by

$$E = -\frac{NJ_z}{4}[1 - \frac{\pi \mathrm{sn}2\zeta}{\zeta}(\mathbf{a}(0,1) + \mathbf{a}(Q,1))] - \frac{J_z \pi \mathrm{sn}2\zeta}{\zeta} \sum_{l=1}^{N/2} \mathbf{a}(x_l, 1), \qquad (248)$$

where

$$\mathbf{a}(x, l) \equiv \frac{1}{2\pi i} \frac{d}{dx} \ln\left(\frac{H_l(i\zeta(x + il))}{H_l(i\zeta(x - il))}\right). \qquad (249)$$

5.2   Strings and the Thermodynamic Bethe Ansatz Equation for the XYZ Model

In the limit $l \to 0$ equation (246) becomes

$$\left(\frac{\sin(i\zeta(x_l + i))}{\sin(i\zeta(x_l - i))}\right)^N = -e^{-2\pi i \nu'/p_0} \prod_{j=1}^{N/2} \frac{\sin(i\zeta(x_l - x_j + 2i))}{\sin(i\zeta(x_l - x_j - 2i))}. \qquad (250)$$

This equation is equivalent to (166), if we assume that the x_l's are finite and $\nu' = 0$. ζ becomes $\gamma/2$ and K_l becomes $\pi/2$. So it is natural to assume the same types of strings can be determined using $p_0 = K_l/\zeta$,

$$\begin{aligned} x_j &= \alpha + (n + 1 - 2j)i, \\ x_j &= \alpha + (n + 1 - 2j)i + p_0 i, \quad Q \geq \alpha > -Q. \end{aligned} \qquad (251)$$

The solution becomes doubly periodic. So we consider the distribution of solutions at $-Q < \Re x \leq Q$ and $-p_0 < \Im x \leq p_0$. It is expected that the same kind of strings appear in the case of XXZ model at $\pi/\gamma = p_0$. We can determine n_j's and q_j's via (209) and (222),

$$\{e_j(x_\alpha^j)\}^N = -\exp(-2\pi i \nu'/p_0) \prod_{k=1}^{m_l} \prod_{\beta=1}^{M_k} E_{jk}(x_\alpha^j - x_\beta^k),$$

$$\nu' = \frac{1}{Q} \sum_{j=1}^{m_l} \sum_{\alpha=1}^{M_j} n_j x_\alpha^j, \qquad (252)$$

where

$$e_j(x) = g(x; n_j, v_j), \qquad (253)$$

$$g(x; n, +) = \frac{H_l(i\zeta(x + in))}{H_l(i\zeta(x - in))},$$

$$g(x; n, -) = -\frac{H_l(K_l + i\zeta(x + in))}{H_l(K_l + i\zeta(x - in))}, \qquad (254)$$

$$E_{jk}(x) = \begin{cases} g(x; 2n_j, v_j v_k) \prod_{l=1}^{n_j - 1} g^2(x; 2l, v_j v_k) & \text{for } n_j = n_k, \\ g(x; (n_j + n_k), v_j v_k) g(x; |n_j - n_k|, v_j v_k) \\ \quad \times \prod_{l=1}^{Min(n_j, n_k) - 1} g^2(x; |n_j - n_k| + 2l, v_j v_k) \\ \qquad \text{for } n_j \neq n_k, \end{cases} \qquad (255)$$

Taking the logarithm of (252) we have

$$N\theta_j(x_\alpha^j) = 2\pi I_\alpha^j - 2\pi\nu'/p_0 + \sum_{k=1}^{m_l}\sum_{\beta=1}^{M_k}\Theta_{jk}(x_\alpha^j - x_\beta^k), \quad \alpha = 1, 2, ..., M_j. \quad (256)$$

Here

$$\theta_j(x) = \mathbf{f}(x; n_j, v_j), \quad \Theta_{jk}(x) = \mathbf{f}(x; |n_j - n_k|, v_jv_k) +$$
$$\mathbf{f}(x; n_j + n_k, v_jv_k) + 2\sum_{i=1}^{Min(n_j,n_k)-1}\mathbf{f}(x; |n_j - n_k| + 2i, v_jv_k), \quad (257)$$

and $\mathbf{f}(x, n, v)$ is defined by

$$\mathbf{f}(x, n, v) = f(x, n, v) + \sum_{l=1}^{\infty}f(x - 2lQ, n, v) + f(x + 2lQ, n, v).$$

$f(x, n, v)$ was defined in (216). An eigenstate should be identified by the set of quantum numbers I_j^α. From (248) the energy must be

$$E = -NJ_zR - \frac{J_z\pi\mathrm{sn}2\zeta}{\zeta}\sum_{l=1}^{N/2}\mathbf{a}(x_l, 1) = -NJ_zR - \frac{J_z\pi\mathrm{sn}2\zeta}{\zeta}\sum_{j=1}^{m_l}\sum_{\alpha=1}^{M_j}\mathbf{a}_j(x_\alpha),$$

$$R \equiv \frac{1}{4}[1 - \frac{\pi\mathrm{sn}2\zeta}{\zeta}(\mathbf{a}(0, 1) + \mathbf{a}(Q, 1))],$$

$$\mathbf{a}_j(x) \equiv \frac{1}{2Q}\left[\frac{q_j}{p_0} + 2\sum_{l=1}^{\infty}\frac{\sinh(q_j\pi l/Q)}{\sinh(p_0\pi l/Q)}\cos(\pi jx/Q)\right], \quad (258)$$

where q_j was defined in (222) The number of zeros must be $N/2$, so

$$N/2 = \sum_{j=1}^{m_l}n_jM_j. \quad (259)$$

Thus the energy per site is given by

$$e = -J_zR - \frac{J_z\pi\mathrm{sn}2\zeta}{\zeta}\sum_{j=1}^{m_l}\int_{-Q}^{Q}\mathbf{a}_j(x)\rho_j(x)dx. \quad (260)$$

The entropy per site is

$$s = \sum_{j=1}^{m_l}\int_{-Q}^{Q}\rho_j\ln(1 + \frac{\rho_j^h}{\rho_j}) + \rho_j^h\ln(1 + \frac{\rho_j}{\rho_j^h})dx. \quad (261)$$

From (257) we have the relation between $\rho_j(x)$ and $\rho_j^h(x)$,

$$\mathbf{a}_j(x) = \mathrm{sign}(q_j)(\rho_j(x) + \rho_j^h(x)) + \sum_{k=1}^{m_l}T_{j,k} * \rho_k(x). \quad (262)$$

Moreover, from (259),

$$\frac{1}{2} = m \equiv \sum_{j=1}^{m_l} n_j \int_{-Q}^{Q} \rho_j(x) dx. \tag{263}$$

Next we need a Lagrange multiplier to guarantee the condition (263). One should minimize $e - Ts + 2hm$ under conditions (262), and after that the multiplier h should be chosen so that (263) is satisfied. Just in the same way as before we get the integral equations for $\eta_j(x) = \rho_j^h(x)/\rho_j(x)$,

$$\ln \eta_j(x) = g_j(x)/T + \sum_{k=1}^{m_l} \text{sign}(q_k) \mathbf{T}_{k,j} * \ln(1 + \eta_k^{-1}(x)), \quad j = 1, ..., m_l. \tag{264}$$

Here $*$, $g_j(x)$ and $\mathbf{T}_{j,k}(x)$ are

$$f * g(x) = \int_{-Q}^{Q} f(x-y)g(y) dy,$$

$$g_j(x) \equiv -\frac{J_z \pi \text{sn} 2\zeta}{\zeta} a_j(x) + 2n_j h,$$

$$\mathbf{T}_{j,k}(x) = \frac{1}{2Q} \sum_{l=-\infty}^{\infty} e^{i\pi n x/Q} \tilde{T}_{j,k}\left(\frac{\pi l}{Q}\right) = \sum_{l=-\infty}^{\infty} T_{j,k}(x - 2lQ).$$

The quantity $g \equiv e - Ts + 2hm$ is given as follows

$$g(J_z, T, h) = -J_z R$$
$$+ \sum_{j=1}^{m_l} \int_{-Q}^{Q} \rho_j(x) \left[g_j(x) - T \ln \eta_j(x) \right] - T \left[\rho_j + \rho_j^h \right] \ln(1 + \eta_j^{-1}) dx$$
$$= -J_z R - T \sum_{j=1}^{m_l} \text{sign}(q_j) \int_{-Q}^{Q} a_j(x) \ln(1 + \eta_j^{-1}(x)) dx. \tag{265}$$

Corresponding to (232) this is

$$g = -J_z R + h - \text{sign}(q_1) \frac{\pi J_z \text{sn} 2\zeta}{\zeta} \int_{-Q}^{Q} a_1(x) s_1(x) dx$$
$$- T \int_{-Q}^{Q} s_1(x) \ln(1 + \eta_1(x)) dx. \tag{266}$$

Then m should be determinded by

$$m = \frac{1}{2} \frac{\partial g}{\partial h} = \frac{1}{2} - \frac{1}{2} \int_{-Q}^{Q} s_1(x)(1 + \eta_1(x))^{-1} \frac{\partial \eta_1(x)}{\partial h} dx. \tag{267}$$

The equation (264) is also equivalent to the following block tridiagonal equations,

$$\ln(1 + \eta_0) = -\frac{\pi J_z \mathrm{sn} 2\zeta}{\zeta T} \delta(x),$$

$$\ln \eta_j = (1 - 2\delta_{m_{i-1},j}) \mathrm{s}_i * \ln(1 + \eta_{j-1}) + \mathrm{s}_i * \ln(1 + \eta_{j+1})$$

$$\text{for } m_{i-1} \leq j \leq m_i - 2, j \neq m_l - 2,$$

$$\ln \eta_{m_i - 1} = (1 - 2\delta_{m_{i-1}, m_i - 1}) \mathrm{s}_i * \ln(1 + \eta_{m_i - 2})$$

$$+ \mathbf{d}_i * \ln(1 + \eta_{m_i - 1}) + \mathrm{s}_{i+1} * \ln(1 + \eta_{m_i}) \text{ for } i < l,$$

$$\ln \eta_{m_l - 2} = (1 - 2\delta_{m_{l-1}, m_l - 2}) \mathrm{s}_l * \ln(1 + \eta_{m_l - 3})$$

$$+ \mathrm{s}_l * \ln(1 + 2\cosh(y_l h/T)\kappa + \kappa^2),$$

$$\ln \kappa = \mathrm{s}_l * \ln(1 + \eta_{m_l - 2}). \tag{268}$$

In this equation parameter h appears only in $\cosh(y_l h/T)$ term. So $\eta_1(x, h)$ is even function of h and $\frac{\partial \eta_1(x)}{\partial h}|_{h=0} = 0$. Using (267) we find $m = 1/2$ and the condition (263) is satisfied at $h = 0$. Thus equation (268) becomes

$$\ln(1 + \eta_0) = -\frac{\pi J_z \mathrm{sn} 2\zeta}{\zeta T} \delta(x),$$

$$\ln \eta_j = (1 - 2\delta_{m_{i-1},j}) \mathrm{s}_i * \ln(1 + \eta_{j-1}) + \mathrm{s}_i * \ln(1 + \eta_{j+1})$$

$$\text{for } m_{i-1} \leq j \leq m_i - 2, j \neq m_l - 2,$$

$$\ln \eta_{m_i - 1} = (1 - 2\delta_{m_{i-1}, m_i - 1}) \mathrm{s}_i * \ln(1 + \eta_{m_i - 2})$$

$$+ \mathbf{d}_i * \ln(1 + \eta_{m_i - 1}) + \mathrm{s}_{i+1} * \ln(1 + \eta_{m_i}) \text{ for } i < l,$$

$$\ln \eta_{m_l - 2} = (1 - 2\delta_{m_{l-1}, m_l - 2}) \mathrm{s}_l * \ln(1 + \eta_{m_l - 3})$$

$$+ 2\mathrm{s}_l * \ln(1 + \kappa),$$

$$\ln \kappa = \mathrm{s}_l * \ln(1 + \eta_{m_l - 2}). \tag{269}$$

Corresponding to (232) the free energy is

$$f = -J_z R - \mathrm{sign}(q_1)\frac{\pi J_z \mathrm{sn} 2\zeta}{\zeta} \int_{-Q}^{Q} \mathrm{a}_1(x)\mathrm{s}_1(x) dx$$

$$-T \int_{-Q}^{Q} \mathrm{s}_1(x) \ln(1 + \eta_1(x)) dx. \tag{270}$$

We can calculate free energy of XYZ model in zero external field.

6   Numerical Calculation and Recent Developments

In this lecture we restricted ourselves to the unnested Bethe ansatz. Fermions with δ-function interactions and the Hubbard model belong to the nested Bethe ansatz which has several kinds of rapidities. For details see ([Takahashi (1971b)],

[Lai (1971)], [Takahashi (1972)]). But the method of derivation is essentially the same as the unnested cases.

The equations introduced in this lecture note have been solved numerically. One can calculate the specific heat, magnetic susceptibility, magnetization curve for the XXZ model, and the specific heat for the XYZ model at $p_0 = 2, 3$ ([Takahashi (1974a)]). If we increase p_0 the problem approaches to the $\Delta = -1$ case. For the analysis of thermodynamic quantities not at low temperature, we can use the high-temperature expansion method or exact diagonalization. For the investigation of low-temperature thermodynamics of solvable models the Bethe ansatz method is the only way. For the spin 1/2 ferromagnetic XXX chain, the susceptibility diverges as $T^{-\gamma}$ and the specific heat behaves as $T^{-\alpha}$. The estimations of exponent γ had been done by many authors. Baker et. al. estimated $\gamma = 1.66$ using high temperature series expansions ([Baker et al (1964)]). Lyklema obtained $\gamma = 1.75$ using quantum Monte Carlo calculations ([Lyklema (1983)]). By the numerical calculation of thermodynamic Bethe ansatz equations it was established that $\gamma = 2$ and $\alpha = 1/2$ ([Takahashi and Yamada (1985)], [Yamada and Takahashi (1986)], [Schlottmann (1985)]). This investigation continued to the spin-wave theory for low-dimensional magnets ([Takahashi (1986)], [Takahashi (1987)], [Takahashi (1989)]).

In many cases of one-dimensional quantum systems, one can define the quantum transfer matrix. The largest eigenvalue of this matrix gives the free energy. The ratio of the largest and the second largest eigenvalues gives the correlation length. This method was developed mainly by Japanese theorists within the last ten years. The XYZ model and the Hubbard model were investigated by this method, for other solvable models this method is expected to be applicable. Logarithmic anomalies in the low-temperature susceptibility of the XXX antiferromagnet were found by the numerical calculation of Bethe ansatz equations for the largest eigenvalue of the quantum transfer matrix ([Eggert, Affleck and Takahashi (1994)]).

References

Baker Jr., G.A., Rushbrooke G.S. and Gilbert, H.E. (1964) Phys. Rev. **135** A 1272.
Baxter, R.(1972a). Ann. Phys.(N.Y.) **70**, 193.
Baxter, R.(1972b). Ann. Phys.(N.Y.) **70**, 323.
Bethe, H.A.(1931). Z. Physik **71**, 205.
Bonner J.C. and Fisher,M.E. (1964). Phys. Rev.**135**, A640
des Cloizeaux, J. and Pearson, J.(1962). Phys. Rev. **128**, 2131.
Eggert, S. Affleck, I. and Takahashi, M. (1994) Phys. Rev. Lett. **73** 332.
Fowler, M. and Zotos, X.(1981) Phys. Rev. B **24**,2634.
Gaudin, M.(1971). Phys. Rev. Lett. **26**, 1301.
Griffiths, R.B.(1964). Phys. Rev. **133A**, 768.
Hida, K. (1981) Phys. Lett. **84A**,338.
Hulthen, L.(1938). Arkiv Math. Astron. Fys. **26A**, No11.
Katsura, S. (1962). Phys. Rev. **127**, 1508.

Katsura, S. (1965). Ann. of Phys. **31**, 325.

Lai, C.K.(1971). Phys. Rev. Lett. **26**, 1472.

Lieb, E.H.(1963). Phys. Rev. **130**, 1616.

Lieb, E.H. and Liniger, W.(1963). Phys. Rev. **130**, 1605.

Lieb, E.H., Schultz, T. and Mattis, D. (1961). Ann. Phys. (N.Y.) **16**, 417.

Lyklema, J.W. (1983) Phys. Rev. **27**, 3108.

Orbach, R.(1958). Phys. Rev. **112**, 309.

Schlottmann, P.(1985). Phys. Rev. Lett. **54**, 2131.

Schlottmann, P.(1986). Phys. Rev. B **33** 4880 .

Takahashi, M.(1971a). Prog. Theor. Phys. **46**, 401.

Takahashi, M.(1971b). Prog. Theor. Phys. **46**, 1388.

Takahashi, M.(1972). Prog. Theor. Phys. **47**, 69.

Takahashi, M.(1973). Prog. Theor. Phys. **50**, 1519.

Takahashi, M.(1974a). Prog. Theor. Phys. **51**, 1348.

Takahashi, M.(1974b). Prog. Theor. Phys. **52**, 103.

Takahashi, M.(1977). J. of Phys. C **10**, 1289.

Takahashi, M.(1986). Prog. Theor. Phys. Suppl. **87**, 233.

Takahashi, M.(1987). Phys. Rev. Lett.,**58** 168,

Takahashi, M.(1989) Phys. Rev. B **40** 2494.

Takahashi, M. and Suzuki, M.(1972). Prog. Theor. Phys. **46**, 2187.

Takahashi, M. and Yamada, M.(1985). J. Phys. Soc. Jpn. **54**, 2808.

Walker, L.R. (1959). Phys. Rev. **116** 1089.

Yamada, M and Takahashi, M. (1986) J. Phys. Soc. Jpn., **55**, 2024.

Yang, C.N.(1967). Phys. Rev. Lett. **19**, 1312.

Yang, C.N. and Yang, C.P. (1966). Phys. Rev. **147** 303,**150** 321,327,**151** 258.

Yang, C.N. and Yang, C.P. (1969). J. Math. Phys. **10**, 1115.

Yang, C.P. (1970). Phys. Rev. A **2**, 154.

Subject Index

Lecture Notes in Physics

For information about Vols. 1–461
please contact your bookseller or Springer-Verlag

New Series m: Monographs